Undergraduate Lecture Notes in Physics

Undergraduate Lecture Notes in Physics (ULNP) publishes authoritative texts covering topics throughout pure and applied physics. Each title in the series is suitable as a basis for undergraduate instruction, typically containing practice problems, worked examples, chapter summaries, and suggestions for further reading.

ULNP titles must provide at least one of the following:

- An exceptionally clear and concise treatment of a standard undergraduate subject.
- A solid undergraduate-level introduction to a graduate, advanced, or non-standard subject.
- A novel perspective or an unusual approach to teaching a subject.

ULNP especially encourages new, original, and idiosyncratic approaches to physics teaching at the undergraduate level.

The purpose of ULNP is to provide intriguing, absorbing books that will continue to be the reader's preferred reference throughout their academic career.

Series editors

Neil Ashby
University of Colorado, Boulder, CO, USA

William Brantley
Department of Physics, Furman University, Greenville, SC, USA

Matthew Deady
Physics Program, Bard College, Annandale-on-Hudson, NY, USA

Michael Fowler
Department of Physics, University of Virginia, Charlottesville, VA, USA

Morten Hjorth-Jensen
Department of Physics, University of Oslo, Oslo, Norway

Michael Inglis
Department of Physical Sciences, SUNY Suffolk County Community College, Selden, NY, USA

More information about this series at http://www.springer.com/series/8917

Cosimo Bambi

Introduction to General Relativity

A Course for Undergraduate Students of Physics

 Springer

Cosimo Bambi
Department of Physics
Fudan University
Shanghai
China

and

Theoretical Astrophysics
Eberhard-Karls Universität Tübingen
Tübingen
Germany

ISSN 2192-4791 ISSN 2192-4805 (electronic)
Undergraduate Lecture Notes in Physics
ISBN 978-981-13-1089-8 ISBN 978-981-13-1090-4 (eBook)
https://doi.org/10.1007/978-981-13-1090-4

Library of Congress Control Number: 2018945069

This Springer imprint is published by the registered company Springer Nature Singapore Pte Ltd.
The registered company address is: 152 Beach Road, #21-01/04 Gateway East, Singapore 189721, Singapore

Fatti non foste a viver come bruti,
ma per seguir virtute e canoscenza.
(Dante Alighieri, Inferno, Canto XXVI)

Preface

The formulations of the theories of special and general relativity and of the theory of quantum mechanics in the first decades of the twentieth century are a fundamental milestone in science, not only for their profound implications in physics but also for the research methodology. In the same way, the courses of special and general relativity and of quantum mechanics represent an important milestone for every student of physics. These courses introduce a different approach to investigate physical phenomena, and students need some time to digest such a radical change.

In Newtonian mechanics and in Maxwell's theory of electrodynamics, the approach is quite empirical and natural. First, we infer a few fundamental laws from observations (e.g., Newton's Laws) and then we construct the whole theory (e.g., Newtonian mechanics). In modern physics, starting from special and general relativity and quantum mechanics, this approach may not be always possible. Observations and formulation of the theory may change order. This is because we may not have direct access to the basic laws governing a certain physical phenomenon. In such a case, we can formulate a number of theories, or we can introduce a number of ansatzes to explain a specific physical phenomenon within a certain theory if we already have the theory, and then we compare the predictions of the different solutions to check which one, if any, is consistent with observations.

For example, Newton's First, Second, and Third Laws can be directly inferred from experiments. Einstein's equations are instead obtained by imposing some "reasonable" requirements and they are then confirmed by comparing their predictions with the results of experiments. In modern physics, it is common that theorists develop theoretical models on the basis of "guesses" (motivated by theoretical arguments but without any experimental support), with the hope that it is possible to find predictions that can later be tested by experiments.

At the beginning, a student may be disappointed by this new approach and may not understand the introduction of *ad hoc* assumptions. In part, this is because we are condensing in a course the efforts of many physicists and many experiments, without discussing all the unsuccessful—but nevertheless necessary and important—attempts that eventually led to a theory in its final form. Moreover, different students may have different backgrounds, not only because they are students of different disciplines (e.g., theoretical physics, experimental physics, astrophysics, and mathematical physics) but also because undergraduate programs in different countries can be very different. Additionally, some textbooks may follow approaches appreciated by some students and not by others, who may instead prefer different textbooks. This point is quite important when we study for the first time the theories of special and general relativity and the theory of quantum mechanics, because there are some concepts that at the beginning are difficult to understand, and a different approach may make it easier or harder.

In the present textbook, the theories of special and general relativity are introduced with the help of the Lagrangian formalism. This is the approach employed in the famous textbook by Landau and Lifshitz. Here, we have tried to have a book more accessible to a larger number of students, starting from a short review of Newtonian mechanics, reducing the mathematics, presenting all the steps of most calculations, and considering some (hopefully illuminating) examples. The present textbook dedicates quite a lot of space to the astrophysical applications, discussing Solar System tests, black holes, cosmological models, and gravitational waves at a level adequate for an introductory course of general relativity. These lines of research have become very active in the past couple of decades and have attracted an increasing number of students. In the last chapter, students can get a quick overview of the problems of Einstein's gravity and current lines of research in theoretical physics.

The textbook has 13 chapters, and in a course of one semester (usually 13–15 weeks) every week may be devoted to the study of one chapter. Note, however, that Chaps. 1–9 are almost "mandatory" in any course of special and general relativity, while Chaps. 10–13 cover topics that are often omitted in an introductory course for undergraduate students. Exercises are proposed at the end of most chapters and are partially solved in Appendix I.

Acknowledgments. I am particularly grateful to Dimitry Ayzenberg for reading a preliminary version of the manuscript and providing useful feedback. I would like also to thank Ahmadjon Abdujabbarov and Leonardo Modesto for useful comments and suggestions. This work was supported by the National Natural Science Foundation of China (Grant No. U1531117), Fudan University (Grant No. IDH1512060), and the Alexander von Humboldt Foundation.

Shanghai, China   Cosimo Bambi
April 2018

The original version of the book was revised: Belated corrections have been incorporated. The correction to the book is available at https://doi.org/10.1007/978-981-13-1090-4_14

Contents

Conventions

There are several conventions in the literature and this, unfortunately, can sometimes generate confusion.

In this textbook, the spacetime metric has signature $(-+++)$ (convention of the gravity community). The Minkowski metric thus reads

$$||\eta_{\mu\nu}|| = \begin{pmatrix} -1 & 0 & 0 & 0 \\ 0 & 1 & 0 & 0 \\ 0 & 0 & 1 & 0 \\ 0 & 0 & 0 & 1 \end{pmatrix}, \tag{1}$$

where here and in the rest of the book the notation $||A_{\mu\nu}||$ is used to indicate the matrix of the tensor $A_{\mu\nu}$.

Greek letters (μ, ν, ρ,...) are used for spacetime indices and can assume the values $0, 1, 2,..., n$, where n is the number of spatial dimensions. Latin letters ($i, j, k,...$) are used for space indices and can assume the values $1, 2,..., n$. The time coordinate can be indicated either as t or as x^0. The index associated with the time coordinate can be indicated either as t or as 0, for example V^t or V^0.

The Riemann tensor is defined as

$$R^{\mu}{}_{\nu\rho\sigma} = \frac{\partial \Gamma^{\mu}_{\nu\sigma}}{\partial x^{\rho}} - \frac{\partial \Gamma^{\mu}_{\nu\rho}}{\partial x^{\sigma}} + \Gamma^{\mu}_{\lambda\rho}\Gamma^{\lambda}_{\nu\sigma} - \Gamma^{\mu}_{\lambda\sigma}\Gamma^{\lambda}_{\nu\rho},$$

where $\Gamma^{\mu}_{\nu\rho}$s are the Christoffel symbols

$$\Gamma^{\mu}_{\nu\rho} = \frac{1}{2}g^{\mu\lambda}\left(\frac{\partial g_{\lambda\rho}}{\partial x^{\nu}} + \frac{\partial g_{\nu\lambda}}{\partial x^{\rho}} - \frac{\partial g_{\nu\rho}}{\partial x^{\lambda}}\right).$$

The Ricci tensor is defined as $R_{\mu\nu} = R^{\lambda}_{\mu\lambda\nu}$. The Einstein equations read

$$G_{\mu\nu} = R_{\mu\nu} - \frac{1}{2}g_{\mu\nu}R = \frac{8\pi G_{N}}{c^4}T_{\mu\nu}.$$

Since the present textbook is intended to be an introductory course on special and general relativity, unless stated otherwise we will explicitly show the speed of light c, Newton's gravitational constant G_N, and Dirac's constant $\hbar$. In some parts (Chaps. 10 and 13 and Sects. 8.2 and 8.6), we will employ units in which $G_N = c = 1$ to simplify the formulas.

Note that ρ will be sometimes used to indicate the energy density, and sometimes it will indicate the mass density (so the associated energy density will be ρc^2).

Chapter 1
Introduction

This chapter briefly reviews the Euclidean geometry, Galilean transformations, the Lagrangian formalism, and Newton's gravity. While the reader is supposed to be already familiar with all these concepts, it is convenient to summarize them here because they will be used or generalized in the next chapters for the theories of special and general relativity. We end the chapter pointing out the inconsistency between Galilean transformations and Maxwell's equations and how this issue led to the theory of special relativity between the end of the 19th century and the beginning of the 20th century.

1.1 Special Principle of Relativity

Let us consider the motion of a point-like particle in an n-dimensional space. If we want to describe such a physical system, we intuitively need $n + 1$ variables, namely n space coordinates to describe the position of the particle in space at a certain value of the temporal coordinate (see Fig. 1.1). In order to assign the n space coordinates to the particle, we need to measure the distance and the direction of the particle from a certain reference point with a standard rod. The temporal coordinate is determined by measuring the time interval with respect to a certain reference time with a standard clock. The choices of the reference point and of the reference time, as well as those of the standard rod and of the standard clock, correspond to the choice of a certain observer, namely of a particular *reference frame*. Two natural questions are:

1. Should we choose some particular reference frame to describe the motion of the point-like particle? In other words, is there any preferred observer/reference frame, or any class of preferred observers/reference frames, or are the laws of physics independent of such a choice?

© Springer Nature Singapore Pte Ltd. 2018, corrected publication 2020
C. Bambi, *Introduction to General Relativity*, Undergraduate Lecture Notes
in Physics, https://doi.org/10.1007/978-981-13-1090-4_1

Fig. 1.1 Motion of a
point-like particle in a
2-dimensional space. x and y
are the space coordinates and
t is the time. The trajectory
of the particle is described by
the curve $(x(t), y(t))$

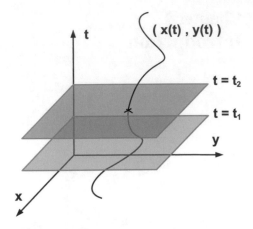

2. How are the physical quantities measured in a certain reference frame related to the same physical quantities measured in another reference frame?

Galileo Galilei was the first, in the 17th century, to discuss the issue of the choice of reference frame to describe physical phenomena. From simple observations, we can realize that there is a certain class of observers that turns out to be particularly suitable to describe physical phenomena. This is the class of inertial observers (or inertial reference frames).

> **Inertial Reference Frame**. An *inertial reference frame* is a reference frame in which the motion of a body not subject to forces either remains at rest or continues to move at a constant speed in a straight line.

While it is possible to describe physical phenomena even in non-inertial reference frames, namely in the reference frames not belonging to the class of inertial reference frames, the description is more complicated. In particular, it is usually necessary to introduce some (reference frame-dependent) corrections to the laws of physics.

Note that, strictly speaking, inertial reference frames do not exist in Nature, as in the Universe there are long-range forces that cannot be screened. Nevertheless, we can usually find reference frames that well approximate inertial ones.

With the concept of inertial reference frame we can introduce the Special Principle of Relativity.

> **Special Principle of Relativity**. The laws of physics are the same in all inertial reference frames.

As a principle, the Special Principle of Relativity cannot be proved by theoretical arguments, but only confirmed (or disproved) by experiments. Current experiments and observational data support this principle. However, there are still attempts today to test the Special Principle of Relativity with higher and higher precision or in different environments, as well as theoretical models in which this principle can be violated at some level.

1.2 Euclidean Space

Let us consider a 3-dimensional space with the system of Cartesian coordinates (x, y, z). Such a space can be "identified" with $\mathbb{R}^3$, because every point of the space can be characterized by three real numbers, which are the values of the coordinates (x, y, z).

The infinitesimal distance between the point $\mathbf{x}_A = (x_A, y_A, z_A)$ and the point $\mathbf{x}_B = (x_A + dx, y_A + dy, z_A + dz)$ is the square root of

$$dl^2 = dx^2 + dy^2 + dz^2 . \tag{1.1}$$

dl is called the *line element* and Eq. (1.1) simply follows from Pythagoras's theorem.

It is convenient to introduce the notation (x^1, x^2, x^3) to denote the coordinates of the space. In the case of Cartesian coordinates, we have $x^1 = x$, $x^2 = y$, and $x^3 = z$. Now Eq. (1.1) can be written in a more compact form as

$$dl^2 = \delta_{ij} dx^i dx^j , \tag{1.2}$$

where δ_{ij} is the Kronecker delta and we have adopted the Einstein convention of summation over repeated indices; that is,

$$\delta_{ij} dx^i dx^j \equiv \sum_{i,j=1}^{3} \delta_{ij} dx^i dx^j . \tag{1.3}$$

δ_{ij} is the *Euclidean metric* and can be written as the matrix

$$||\delta_{ij}|| = \begin{pmatrix} 1 & 0 & 0 \\ 0 & 1 & 0 \\ 0 & 0 & 1 \end{pmatrix} . \tag{1.4}$$

Our discussion can be easily extended to any n-dimensional space. The n-dimensional *Euclidean space* is $\mathbb{R}^n$ in which the square of the line element is[1]

---

[1]From the mathematical point of view, the n-dimensional Euclidean space is the differentiable manifold $\mathbb{R}^n$ equipped with the Euclidean metric δ_{ij} (See Appendix C).

$$dl^2 = \delta_{ij}dx^i dx^j \,, \tag{1.5}$$

where now i and j run from 1 to n.

Note that the infinitesimal distance between two points is independent of the coordinate system. The line element is thus an *invariant*; that is, it does not change with a change of coordinates. For an arbitrary coordinate system, we write the square of the line element as

$$dl^2 = g_{ij}dx^i dx^j \,, \tag{1.6}$$

where g_{ij} is called the *metric tensor* and, in general, is not δ_{ij}. g_{ij} must be symmetric in the indices, i.e. $g_{ij} = g_{ji}$. Generally speaking, if we move from the coordinate system (x^1, x^2, x^3) to the coordinate system (x'^1, x'^2, x'^3) we have

$$dx^i \rightarrow dx'^i = \frac{\partial x'^i}{\partial x^j}dx^j \,, \tag{1.7}$$

and therefore

$$g_{ij}dx^i dx^j = g'_{ij}dx'^i dx'^j = g'_{ij}\frac{\partial x'^i}{\partial x^m}dx^m \frac{\partial x'^j}{\partial x^n}dx^n \,. \tag{1.8}$$

We see that

$$g_{mn} = \frac{\partial x'^i}{\partial x^m}\frac{\partial x'^j}{\partial x^n}g'_{ij} \,. \tag{1.9}$$

We multiply both sides of this expression by $\partial x^m/\partial x'^p$ and $\partial x^n/\partial x'^q$, and we sum over repeated indices

$$\frac{\partial x^m}{\partial x'^p}\frac{\partial x^n}{\partial x'^q}g_{mn} = \frac{\partial x^m}{\partial x'^p}\frac{\partial x^n}{\partial x'^q}\frac{\partial x'^i}{\partial x^m}\frac{\partial x'^j}{\partial x^n}g'_{ij} = \delta^i_p \delta^j_q g'_{ij} = g'_{pq} \,. \tag{1.10}$$

The metric tensor thus transforms as

$$g_{ij} \rightarrow g'_{ij} = \frac{\partial x^m}{\partial x'^j}\frac{\partial x^n}{\partial x'^j}g_{mn} \,. \tag{1.11}$$

As an example, we can consider the spherical coordinates (r, θ, ϕ). The relation between Cartesian and spherical coordinates is

$$
\begin{aligned}
x &= r\sin\theta\cos\phi \,, \\
y &= r\sin\theta\sin\phi \,, \\
z &= r\cos\theta \,,
\end{aligned}
\tag{1.12}
$$

with inverse

$$r = \sqrt{x^2 + y^2 + z^2},$$

$$\theta = \arccos\left(\frac{z}{\sqrt{x^2 + y^2 + z^2}}\right),$$

$$\phi = \arctan\left(\frac{y}{x}\right). \tag{1.13}$$

It is straightforward to apply Eq. (1.11) and see that in spherical coordinates the square of the line element is

$$dl^2 = dr^2 + r^2 d\theta^2 + r^2 \sin^2\theta d\phi^2, \tag{1.14}$$

and therefore the corresponding metric tensor reads

$$\|g_{ij}\| = \begin{pmatrix} 1 & 0 & 0 \\ 0 & r^2 & 0 \\ 0 & 0 & r^2 \sin^2\theta \end{pmatrix}. \tag{1.15}$$

With the concept of a line element, we can measure the length of a curve. In the 3-dimensional space with Cartesian coordinates, a curve is a continuos function Γ: $t \in [t_1, t_2] \subset \mathbb{R} \to \mathbb{R}^3$. The points of the curve have the coordinates

$$\mathbf{x}(t) = \begin{pmatrix} x(t) \\ y(t) \\ z(t) \end{pmatrix}. \tag{1.16}$$

The length of the curve is

$$\ell = \int_\Gamma dl = \int_{t_1}^{t_2} \sqrt{\dot{x}^2 + \dot{y}^2 + \dot{z}^2} \, dt, \tag{1.17}$$

where here the dot indicates the derivative with respect to the parameter t. The length of a curve between two points of the space is an invariant as well.

As an example of the length of a curve, let us consider a circle in $\mathbb{R}^2$. The points of the circle have Cartesian coordinates

$$\mathbf{x}(t) = \begin{pmatrix} R\cos t \\ R\sin t \end{pmatrix}, \tag{1.18}$$

where R is the radius of the circle and $t \in [0, 2\pi)$. The length of the curve is

$$\ell = \int_0^{2\pi} \sqrt{R^2 \sin^2 t + R^2 \cos^2 t} \, dt = \int_0^{2\pi} R \, dt = 2\pi R. \tag{1.19}$$

1.3   Scalars, Vectors, and Tensors

A *scalar* ϕ is a quantity that does not change if we change coordinates: under the coordinate transformation $x^i \rightarrow x'^i$, we have

$$\phi \rightarrow \phi' = \phi. \tag{1.20}$$

For example, the line element dl is a scalar.

A *vector* is, strictly speaking, an element of a vector space, which is a set of objects in which we can define two operations (addition and multiplication) satisfying certain axioms. The reader is presumably already familiar with the concept of vectors, but more details can be found in Appendix A.2. For instance, the infinitesimal displacement between two nearby points of the space,

$$d\mathbf{x} = \left(dx^1, dx^2, \ldots, dx^n\right), \tag{1.21}$$

is a vector. Note that the use of upper and lower indices in the previous section was not accidental. Upper indices are employed for the components of vectors, which transform as

$$V^i \rightarrow V'^i = \frac{\partial x'^i}{\partial x^j} V^j, \tag{1.22}$$

for the change of coordinates[2] $x^i \rightarrow x'^i$.

Lower indices are used for the components of a *dual vector* (also called *cotangent vector* or *co-vector*), which transform as

$$V_i \rightarrow V_i' = \frac{\partial x^j}{\partial x'^i} V_j. \tag{1.23}$$

In this book, the dual vector of the vector $\mathbf{V} = (V^1, V^2, \ldots, V^n)$ is indicated as $\mathbf{V}^*$ and defined as the object with components

$$V_i \equiv g_{ij} V^j. \tag{1.24}$$

The dual vector can thus be seen as a function that requires as input a vector (let us write its components as W^i) and provides as output a real number

$$V_i(W^i) = g_{ij} W^i V^j. \tag{1.25}$$

---

[2]The coordinates of a space, $\{x^i\}$, are not the components of a vector even if they have upper indices. Indeed they do not transform with the rule (1.22) in general. For example, this is easy to check with the transformations between Cartesian and spherical coordinates in Eqs. (1.12) and (1.13). We write the space coordinates with upper indices because it is common to do so and we have to write the indices somewhere.

Note that a quantity like $V_i W^i$ is a scalar, namely it is invariant under a coordinate transformation

$$V_i W^i \rightarrow V_i' W'^i = \frac{\partial x^j}{\partial x'^i} V_j \frac{\partial x'^i}{\partial x^k} W^k = \delta_k^j V_j W^k = V_j W^j . \tag{1.26}$$

We say that upper indices are "lowered" by the metric tensor g_{ij}, as shown in Eq. (1.24), and lower indices are "raised" by the inverse of the metric tensor g^{ij}

$$V^i = g^{ij} V_j = g^{ij} g_{jm} V^m = \delta_m^i V^m = V^i , \tag{1.27}$$

where $g^{ij} g_{jm} = \delta_m^i$ by definition. When we use Cartesian coordinates, we have the Euclidean metric δ_{ij} and its action on a vector is trivial: if we have the vector $\mathbf{V} = (V^x, V^y, V^z)$, the dual vector is

$$\mathbf{V}^* = \begin{pmatrix} V_x \\ V_y \\ V_z \end{pmatrix} = \begin{pmatrix} V^x \\ V^y \\ V^z \end{pmatrix} . \tag{1.28}$$

However, this is not the general case. If we consider spherical coordinates, the dual vector of the vector $\mathbf{V} = (V^r, V^\theta, V^\phi)$ is

$$\mathbf{V}^* = \begin{pmatrix} V_r \\ V_\theta \\ V_\phi \end{pmatrix} = \begin{pmatrix} V^r \\ r^2 V^\theta \\ r^2 \sin^2 \theta V^\phi \end{pmatrix} . \tag{1.29}$$

Tensors are the generalization of vectors and dual vectors. They are multi-index objects. An example is the metric tensor g_{ij}. A tensor of type (r, s) and of order $r + s$ has r upper indices and s lower indices. The transformation rule for the components of a tensor is

$$T^{i_1 i_2 \ldots i_r}_{j_1 j_2 \ldots j_s} \rightarrow T'^{i_1 i_2 \ldots i_r}_{j_1 j_2 \ldots j_s} = \frac{\partial x'^{i_1}}{\partial x^{p_1}} \frac{\partial x'^{i_2}}{\partial x^{p_2}} \cdots \frac{\partial x'^{i_r}}{\partial x^{p_r}} \frac{\partial x^{q_1}}{\partial x'^{j_1}} \frac{\partial x^{q_2}}{\partial x'^{j_2}} \cdots \frac{\partial x^{q_s}}{\partial x'^{j_s}} T^{p_1 p_2 \ldots p_r}_{q_1 q_2 \ldots q_s} . \tag{1.30}$$

A scalar is a tensor of type $(0,0)$, a vector is a tensor of type $(1,0)$, and a dual vector is a tensor of type $(0,1)$. Upper indices can be lowered with g_{ij}, lower indices can be raised with g^{ij}. Some examples are[3]

$$T^{ijk} = g^{il} g^{jm} g^{kn} T_{lmn} , \quad T^i{}_{jk} = g^{il} g_{jm} T_l{}^m{}_k , \quad \ldots \tag{1.31}$$

---

[3]Note that, in general, the order of the indices is important. For instance, if we have the tensor A^{ab} and we lower the index a, we should write $A_a{}^b$. If we lower the index b, we should write $A^a{}_b$. The index a must remain the first from the left. If A^{ab} is a symmetric tensor, i.e. $A^{ab} = A^{ba}$, we have also $A_a{}^b = A^b{}_a$, the order does not matter, and we can simplify the notation writing A^a_b.

If we raise an index of the metric tensor, we get the Kronecker delta, $g^{ij} g_{jm} = g^i_{\ m} = \delta^i_{\ m}$.

When we sum over repeated upper and lower indices, we reduce the order of the tensor. For instance, if we have the tensor of type $(2,1)$ with components $T^{ij}_{\ \ k}$ and we "contract" the indices i and k, we get the vector

$$V^j = T^{ij}_{\ \ i}. \tag{1.32}$$

Indeed, if we consider the transformation of coordinates $x^i \to x'^i$, we have

$$V^j \to V'^j = T'^{ij}_{\ \ i} = \frac{\partial x'^i}{\partial x^l} \frac{\partial x'^j}{\partial x^m} \frac{\partial x^n}{\partial x'^i} T^{lm}_{\ \ n} = \delta^n_l \frac{\partial x'^j}{\partial x^m} T^{lm}_{\ \ n} = \frac{\partial x'^j}{\partial x^m} T^{lm}_{\ \ l}$$

$$= \frac{\partial x'^j}{\partial x^m} V^m, \tag{1.33}$$

and V^j transforms as a vector. If we have a tensor of type (r, s) and of order $r + s$ and we contract $2t$ indices, the new tensor is of type $(r - t, s - t)$ and of order $r + s - 2t$. We can also contract indices between two different tensors. For instance, if we have the vector with components V^i and the dual vector W_j and we contract over their indices, we get the scalar $V^i W_i$, which is a number that does not change under a change of coordinates.

If some physical quantity is described by a tensor of type (r, s) at every point of the space, we have a *tensor field* of type (r, s). A scalar field is thus a function of the form $\phi = \phi(x^1, x^2, \ldots, x^n)$. A vector field has the form

$$\mathbf{V} = \mathbf{V}(x^1, x^2, \ldots, x^n) = \begin{pmatrix} V^1(x^1, x^2, \ldots, x^n) \\ V^2(x^1, x^2, \ldots, x^n) \\ \vdots \\ V^n(x^1, x^2, \ldots, x^n) \end{pmatrix}, \tag{1.34}$$

that is, every component is a function of the space coordinates.

A more rigorous definition of vectors, dual vectors, and tensors can be found in Appendices A and C.

1.4   Galilean Transformations

Let us consider an inertial reference frame with the Cartesian coordinates $\mathbf{x} = (x, y, z)$ and the time t and another inertial reference frame with the Cartesian coordinates $\mathbf{x}' = (x', y', z')$ and the time t'. Let us also assume that the second reference frame is moving with constant velocity $\mathbf{v}$ with respect to the former. If the two Cartesian coordinates coincide at the time $t = t' = 0$, the transformation connecting the two reference frames is

$$\mathbf{x} \to \mathbf{x}' = \mathbf{x} - \mathbf{v}t \,,$$
$$t \to t' = t \,. \tag{1.35}$$

If, for instance, $\mathbf{v} = (v, 0, 0)$, we have

$$x \to x' = x - vt \,,$$
$$y \to y' = y \,,$$
$$z \to z' = z \,,$$
$$t \to t' = t \,. \tag{1.36}$$

The *Galilean transformations* are the transformations connecting the coordinates of two inertial reference frames which differ only by constant relative motion and have the form (1.35).

The inverse transformation of (1.35) is

$$\mathbf{x}' \to \mathbf{x} = \mathbf{x}' + \mathbf{v}t' \,,$$
$$t' \to t = t' \,, \tag{1.37}$$

which can be also obtained from Eq. (1.35) by replacing $\mathbf{v}$ with $-\mathbf{v}$ and exchanging primed and unprimed coordinates.

From Eq. (1.35) we can easily infer the relation between the velocities of a particle measured in the two reference frames. If $\mathbf{w} = \dot{\mathbf{x}}$ is the velocity of the particle in the reference frame with the Cartesian coordinates (x, y, z) and $\mathbf{w}' = \dot{\mathbf{x}}'$ is the velocity of the particle in the reference frame with the Cartesian coordinates (x', y', z'), where the dot $\dot{}$ indicates the derivative with respect to time, we have

$$\mathbf{w}' = \mathbf{w} - \mathbf{v} \,. \tag{1.38}$$

Two inertial reference frames may also differ by a translation or a rotation. The transformation connecting two inertial reference frames which differ only by a translation is

$$\mathbf{x} \to \mathbf{x}' = \mathbf{x} + \mathbf{T} \,, \tag{1.39}$$

where $\mathbf{T} = (T^1, T^2, T^3)$. In principle, a translation may also be applied to the temporal coordinate and we may have

$$t \to t' = t + t_0 \,. \tag{1.40}$$

Note that the time interval measured in different reference frames is the same, namely Δt is an invariant. This is one of the key assumptions in Newtonian mechanics, where there is an absolute time valid for any observer (inertial and non-inertial).

The transformation connecting two inertial reference frames which differ only by a rotation is

$$\mathbf{x} \to \mathbf{x}' = R\mathbf{x}, \tag{1.41}$$

where R is the rotation matrix of the transformation. The rotations about the x, y, and z axes (or, equivalently, the rotations in the yz, xz, and xy planes) of the angle θ have, respectively, the following form

$$R_{yz}(\theta) = \begin{pmatrix} 1 & 0 & 0 \\ 0 & \cos\theta & \sin\theta \\ 0 & -\sin\theta & \cos\theta \end{pmatrix},$$

$$R_{xz}(\theta) = \begin{pmatrix} \cos\theta & 0 & -\sin\theta \\ 0 & 1 & 0 \\ \sin\theta & 0 & \cos\theta \end{pmatrix},$$

$$R_{xy}(\theta) = \begin{pmatrix} \cos\theta & \sin\theta & 0 \\ -\sin\theta & \cos\theta & 0 \\ 0 & 0 & 1 \end{pmatrix}. \tag{1.42}$$

A general rotation can be written as a combination of elementary rotations about the x, y, and z axes, namely

$$R(\theta_{yz}, \theta_{xz}, \theta_{xy}) = R_{yz}(\theta_{yz}) \cdot R_{xz}(\theta_{xz}) \cdot R_{xy}(\theta_{xy}). \tag{1.43}$$

The Galilean transformations with the translations and the rotations form a group (see Appendix A.1), which is called the *Galilean group*. The combination of two or more transformations of coordinates is still a transformation of coordinates. The inverse element is the inverse transformation. The identity element is the trivial transformation

$$x \to x' = x, \quad y \to y' = y, \quad z \to z' = z, \quad t \to t' = t. \tag{1.44}$$

The Galilean group is the set of all possible transformations connecting different inertial frames in Newtonian mechanics. A generic transformation of the Galilean group has the form

$$\mathbf{x} \to \mathbf{x}' = R\mathbf{x} - \mathbf{v}t + \mathbf{T},$$
$$t \to t' = t + t_0. \tag{1.45}$$

Assuming that the transformations of the Galilean group are the correct transformations to move from an inertial reference frame to another inertial reference frame, the Special Principle of Relativity in Sect. 1.1 can be reformulated as follows:

Galileo's Principle of Relativity. The laws of physics are invariant under the Galilean group.

The physics of the 17 and 18th centuries was consistent with the Galileo Principle of Relativity. As shown at the end of this chapter, the situation changed in the 19th century with the study of electromagnetic phenomena. Maxwell's equations are not invariant under Galilean transformations and this was initially interpreted as an indication of the existence of a preferred reference frame. It was later realized that the problem was in the Galilean transformations, which can be used only when the relative velocity between two reference frames is much smaller than the speed of light.

1.5 Principle of Least Action

Let us consider a certain physical system. Its *action S* between the instants of time t_1 and t_2 is given by

$$S = \int_{t_1}^{t_2} L[\mathbf{q}(t), \dot{\mathbf{q}}(t), t] dt . \tag{1.46}$$

where $L[\mathbf{q}(t), \dot{\mathbf{q}}(t), t]$ is the *Lagrangian* of the system, $\mathbf{q} = (q^1, q^2, \ldots, q^n)$, q^i's are the Lagrangian coordinates and define the configuration of the system, and $\dot{\mathbf{q}}$ is the derivative of $\mathbf{q}$ with respect to t. For the moment we only say that the Lagrangian is a certain function capable of describing the dynamics of the system under consideration and we introduce the Principle of Least Action.

Principle of Least Action. The trajectory of a system between two times, say t_1 and t_2, is the one for which its action is stationary to first order.

The Principle of Least Action is an elegant way to infer the equations of motion of a system once its action is known. We consider small changes in the configuration of the system

$$\mathbf{q}(t) \rightarrow \tilde{\mathbf{q}}(t) = \mathbf{q}(t) + \delta\mathbf{q}(t) , \tag{1.47}$$

with the boundary conditions

$$\delta\mathbf{q}(t_1) = \delta\mathbf{q}(t_2) = \mathbf{0} . \tag{1.48}$$

The variation in the Lagrangian coordinates (1.47) produces a variation of the action

$$\delta S = \int_{t_1}^{t_2} \left(\frac{\partial L}{\partial q^i} \delta q^i + \frac{\partial L}{\partial \dot{q}^i} \delta \dot{q}^i \right) dt \,, \tag{1.49}$$

where we have used the convention of summation over repeated indices. Since

$$\delta \dot{q}^i = \dot{\tilde{q}}^i - \dot{q}^i = \frac{d}{dt} \delta q^i \,, \tag{1.50}$$

we can write

$$\frac{\partial L}{\partial \dot{q}^i} \delta \dot{q}^i = \frac{d}{dt} \left(\frac{\partial L}{\partial \dot{q}^i} \delta q^i \right) - \left(\frac{d}{dt} \frac{\partial L}{\partial \dot{q}^i} \right) \delta q^i \,. \tag{1.51}$$

From the boundary conditions in Eq. (1.48), the first term on the right hand side in Eq. (1.51) does not give any contribution when we integrate over t. Eq. (1.49) thus becomes

$$\delta S = \int_{t_1}^{t_2} \left(\frac{\partial L}{\partial q^i} - \frac{d}{dt} \frac{\partial L}{\partial \dot{q}^i} \right) \delta q^i \, dt \,. \tag{1.52}$$

Requiring that the action S is stationary for any small variation of the Lagrangian coordinates of the system, i.e. $\delta S = 0$ for any δq^i, we obtain the *Euler–Lagrange equations*

$$\frac{d}{dt} \frac{\partial L}{\partial \dot{q}^i} - \frac{\partial L}{\partial q^i} = 0 \,. \tag{1.53}$$

These are the equations of motion of the system.

Up to now we have not specified the Lagrangian L. However, in general there is no fundamental recipe to construct the Lagrangian of a specific system. The Lagrangian of a certain physical system is simply the one that provides the right equations of motion for that physical system. In other words, if we want to study a system, we can consider a number of Lagrangians, each of them representing a certain model for that system. We can then check (with experiments/observations) which Lagrangian can better describe the system, and thus find the best model.

The Principle of Least Action is a principle, so it cannot be proven. For the time being, all known physical systems can be treated with this formalism.

In particular classes of systems, it is straightforward to find their Lagrangians. The simplest example is a point-like particle moving in a potential V. In such a case, the Lagrangian of the system is simply given by the difference between the kinetic energy of the point-like particle T and its potential V. In three dimensions, we can write

$$L = T - V = \frac{1}{2}m\dot{\mathbf{x}}^2 - V, \tag{1.54}$$

where $\mathbf{x} = (x, y, z)$ are the Cartesian coordinates of the particle, $\dot{\mathbf{x}}$ is the particle velocity, and $\dot{\mathbf{x}}^2 = \dot{x}^2 + \dot{y}^2 + \dot{z}^2$. The Euler–Lagrange equations provide the equation of motion

$$m\ddot{\mathbf{x}} = -\nabla V. \tag{1.55}$$

Equation (1.55) is the well-known Newton's Second Law for a point-like particle in a potential V, and therefore the Lagrangian in Eq. (1.54) is the right one.

1.6 Constants of Motion

Let us now assume that the Lagrangian of a certain physical system does not explicitly depend on the time t, namely $L = L[\mathbf{q}(t), \dot{\mathbf{q}}(t)]$. In such a case

$$\frac{\partial L}{\partial t} = 0, \tag{1.56}$$

and therefore

$$\frac{dL}{dt} = \frac{\partial L}{\partial q^i}\dot{q}^i + \frac{\partial L}{\partial \dot{q}^i}\ddot{q}^i + \frac{\partial L}{\partial t} = \frac{\partial L}{\partial q^i}\dot{q}^i + \frac{\partial L}{\partial \dot{q}^i}\ddot{q}^i, \tag{1.57}$$

which we can rewrite as

$$\frac{\partial L}{\partial q^i}\dot{q}^i + \frac{\partial L}{\partial \dot{q}^i}\ddot{q}^i - \frac{dL}{dt} = 0. \tag{1.58}$$

From the Euler–Lagrange equations

$$\frac{\partial L}{\partial q^i} = \frac{d}{dt}\frac{\partial L}{\partial \dot{q}^i}, \tag{1.59}$$

and Eq. (1.58) can be rewritten as

$$\left(\frac{d}{dt}\frac{\partial L}{\partial \dot{q}^i}\right)\dot{q}^i + \frac{\partial L}{\partial \dot{q}^i}\ddot{q}^i - \frac{dL}{dt} = 0, \tag{1.60}$$

and also as

$$\frac{d}{dt}\left(\frac{\partial L}{\partial \dot{q}^i}\dot{q}^i - L\right) = 0. \tag{1.61}$$

The expression in brackets in Eq. (1.61) is a constant of motion

$$E = \frac{\partial L}{\partial \dot{q}^i} \dot{q}^i - L .$$

(1.62)

In the case of a point-like particle moving in the potential V, we have

$$\frac{\partial L}{\partial \dot{q}^i} \dot{q}^i = m\dot{\mathbf{x}}^2 = 2T ,$$

(1.63)

and

$$E = T + V .$$

(1.64)

E is the energy of the point-like particle.

Let us now consider the case in which the Lagrangian of a system does not depend on a certain Lagrangian coordinate, say q^i. Since $\partial L/\partial q^i = 0$, from the Euler–Lagrange equations it follows that

$$\frac{d}{dt} p_i = 0 ,$$

(1.65)

where p_i is the *conjugate momentum* defined as

$$p_i = \frac{\partial L}{\partial \dot{q}^i} .$$

(1.66)

p_i is a constant of motion of the system.

The simplest example is that of a free point-like particle. Since $V = 0$, the Lagrangian of the system is just the kinetic energy of the particle. In three dimensions and with Cartesian coordinates, we have

$$L = \frac{1}{2} m \left(\dot{x}^2 + \dot{y}^2 + \dot{z}^2 \right) .$$

(1.67)

The constants of motion are the three components of the momentum

$$p_x = m\dot{x} , \quad p_y = m\dot{y} , \quad p_z = m\dot{z} ,$$

(1.68)

as well as the energy $E = T$.

1.7 Geodesic Equations

In Newtonian mechanics, the Lagrangian of a free point-like particle is simply the particle kinetic energy and is given in Eq. (1.67). The equations of motion can be obtained by minimizing the action

$$S = \frac{1}{2}m \int_\Gamma \left(\dot{x}^2 + \dot{y}^2 + \dot{z}^2 \right) dt \,, \tag{1.69}$$

where Γ is the particle trajectory. The Euler–Lagrange equations are $\ddot{x} = \ddot{y} = \ddot{z} = 0$ and the solution is a constant velocity along a straight line.

If we consider a non-Cartesian coordinate system, for example spherical coordinates (r, θ, ϕ), the action can be written as

$$S = \frac{1}{2}m \int_\Gamma g_{ij}\dot{x}^i\dot{x}^j dt \,, \tag{1.70}$$

where g_{ij} is the metric tensor introduced in Sect. 1.2. Note that the line element $dl^2 = g_{ij}dx^idx^j$ is an invariant, but the square of the velocity $v^2 = g_{ij}\dot{x}^i\dot{x}^j$ is an invariant only if we do not consider reference frames with non-vanishing relative motion. This is simply because in dl^2 we are considering the infinitesimal distance between two specific points of the space, say $\mathbf{x}_A$ and $\mathbf{x}_B$, which exist independently of the coordinate system and have different coordinates in different reference frames. In v^2, $\dot{x}^i = dx^i/dt$ where dx^i is the change of the values of coordinates in the time dt in a certain reference frame, but the points of the space are different in the two reference frames with non-vanishing relative velocity.

Now the Euler–Lagrange equations are

$$\frac{d}{dt}\frac{\partial L}{\partial \dot{x}^k} - \frac{\partial L}{\partial x^k} = 0 \,,$$

$$\frac{d}{dt}\left(g_{ij}\delta_k^i\dot{x}^j + g_{ij}\delta_k^j\dot{x}^i \right) - \frac{\partial g_{ij}}{\partial x^k}\dot{x}^i\dot{x}^j = 0 \,,$$

$$\frac{d}{dt}\left(2g_{ik}\dot{x}^i \right) - \frac{\partial g_{ij}}{\partial x^k}\dot{x}^i\dot{x}^j = 0 \,,$$

$$2g_{ik}\ddot{x}^i + 2\frac{\partial g_{ik}}{\partial x^j}\dot{x}^i\dot{x}^j - \frac{\partial g_{ij}}{\partial x^k}\dot{x}^i\dot{x}^j = 0 \,,$$

$$2g_{ik}\ddot{x}^i + \frac{\partial g_{ik}}{\partial x^j}\dot{x}^i\dot{x}^j + \frac{\partial g_{jk}}{\partial x^i}\dot{x}^i\dot{x}^j - \frac{\partial g_{ij}}{\partial x^k}\dot{x}^i\dot{x}^j = 0 \,. \tag{1.71}$$

We multiply the last equation in (1.71) by g^{lk} (remember that $g^{ij}g_{jk} = \delta_k^i$)

$$\delta_i^l\ddot{x}^i + \frac{1}{2}g^{lk}\left(\frac{\partial g_{ik}}{\partial x^j}\dot{x}^i\dot{x}^j + \frac{\partial g_{jk}}{\partial x^i}\dot{x}^i\dot{x}^j - \frac{\partial g_{ij}}{\partial x^k}\dot{x}^i\dot{x}^j \right) = 0 \,. \tag{1.72}$$

The final equation can be written as

$$\ddot{x}^i + \Gamma^i_{jk} \dot{x}^j \dot{x}^k = 0, \tag{1.73}$$

which is called the *geodesic equations*. Γ^i_{jk}s are the *Christoffel symbols*

$$\Gamma^i_{jk} = \frac{1}{2} g^{il} \left(\frac{\partial g_{lk}}{\partial x^j} + \frac{\partial g_{jl}}{\partial x^k} - \frac{\partial g_{jk}}{\partial x^l} \right). \tag{1.74}$$

As we will see in Sect. 5.2.1, the Christoffel symbols are not the components of a tensor.

Note that the geodesic equations can be obtained even if we apply the Least Action Principle to the length of the trajectory of the particle

$$\ell = \int_\Gamma dl = \int_\Gamma \sqrt{g_{ij} \dot{x}^i \dot{x}^j} \, dt. \tag{1.75}$$

Indeed, modulo a constant, the new Lagrangian is $L' = \sqrt{L}$ and the Euler–Lagrange equations for L' are

$$\frac{d}{dt} \left(\frac{1}{2\sqrt{L}} \frac{\partial L}{\partial \dot{x}^k} \right) - \frac{1}{2\sqrt{L}} \frac{\partial L}{\partial x^k} = 0,$$

$$-\frac{1}{4L^{3/2}} \frac{dL}{dt} \frac{\partial L}{\partial \dot{x}^k} + \frac{1}{2\sqrt{L}} \frac{d}{dt} \frac{\partial L}{\partial \dot{x}^k} - \frac{1}{2\sqrt{L}} \frac{\partial L}{\partial x^k} = 0. \tag{1.76}$$

Since L is (modulo a constant) the particle kinetic energy and is conserved for a free particle, $dL/dt = 0$, and we recover the Euler–Lagrange equations for L.[4]

If we have a metric tensor g_{ij}, we can compute the Christoffel symbols from their definition (1.74). Nevertheless, it is usually faster to write the Euler–Lagrange equations for the motion of a free point-like particle and identify the non-vanishing Christoffel symbols by comparing the result with the geodesic equations. As an example, let us consider the metric tensor in (1.15). The Lagrangian of a free point-like particle is

$$L = \frac{1}{2} m \left(\dot{r}^2 + r^2 \dot{\theta}^2 + r^2 \sin^2\theta \dot{\phi}^2 \right). \tag{1.77}$$

The Euler–Lagrange equation for the r coordinate is

$$\ddot{r} - r\dot{\theta}^2 - r \sin^2\theta \dot{\phi}^2 = 0. \tag{1.78}$$

For the θ coordinate, we have

---

[4]If we think of t as the parameter that parametrizes the curve (rather than the time coordinate), we can always choose t such that L is constant and thus recover the geodesic equations. Of course, the choice of the parametrization does not affect the solution of the equations. It only simplifies the equations to solve.

$$\frac{d}{dt}\left(r^2\dot\theta\right) - r^2\sin\theta\cos\theta\dot\phi^2 = 0\,,$$

$$2r\dot r\dot\theta + r^2\ddot\theta - r^2\sin\theta\cos\theta\dot\phi^2 = 0\,, \tag{1.79}$$

which can be rewritten as

$$\ddot\theta + \frac{2}{r}\dot r\dot\theta - \sin\theta\cos\theta\dot\phi^2 = 0\,. \tag{1.80}$$

Lastly, for the ϕ coordinate we have

$$\frac{d}{dt}\left(r^2\sin^2\theta\dot\phi\right) = 0\,,$$

$$2r\dot r\sin^2\theta\dot\phi + 2r^2\sin\theta\cos\theta\dot\theta\dot\phi + r^2\sin^2\theta\ddot\phi = 0\,, \tag{1.81}$$

which we rewrite as

$$\ddot\phi + \frac{2}{r}\dot r\dot\phi + 2\cot\theta\dot\theta\dot\phi = 0\,. \tag{1.82}$$

If we compare Eqs. (1.78), (1.80), and (1.82) with the geodesic equations (1.73), we see that the non-vanishing Christoffel symbols are

$$\Gamma^r_{\theta\theta} = -r\,, \quad \Gamma^r_{\phi\phi} = -r\sin^2\theta\,,$$

$$\Gamma^\theta_{r\theta} = \Gamma^\theta_{\theta r} = \frac{1}{r}\,, \quad \Gamma^\theta_{\phi\phi} = -\sin\theta\cos\theta\,,$$

$$\Gamma^\phi_{r\phi} = \Gamma^\phi_{\phi r} = \frac{1}{r}\,, \quad \Gamma^\phi_{\theta\phi} = \Gamma^\phi_{\phi\theta} = \cot\theta\,. \tag{1.83}$$

1.8 Newton's Gravity

Let us consider a point-like test-particle of mass m moving in the gravitational field of a point-like massive body of mass M ($M \gg m$). The Lagrangian of the point-like test-particle is

$$L = \frac{1}{2}m\dot{\mathbf{x}}^2 - m\Phi = \frac{1}{2}m\dot{\mathbf{x}}^2 + \frac{G_N M m}{r}\,, \tag{1.84}$$

where $\dot{\mathbf{x}}$ is the velocity of the particle, Φ is the gravitational potential, and r is the distance of the particle from the massive body. In Cartesian coordinates, we have

$$\dot{\mathbf{x}}^2 = \delta_{ij}\dot x^i\dot x^j\,, \quad r = \sqrt{x^2 + y^2 + z^2}\,. \tag{1.85}$$

Cartesian coordinates are not convenient for such a system with spherical symmetry. While the physics is independent of the choice of coordinates, it is easier to study the system in spherical coordinates. In spherical coordinates we have

$$L = \frac{1}{2}m\left(\dot{r}^2 + r^2\dot{\theta}^2 + r^2\sin^2\theta\dot{\phi}^2\right) + \frac{G_N M m}{r}. \tag{1.86}$$

The Euler–Lagrange equation for the θ coordinate is

$$\frac{d}{dt}\left(mr^2\dot{\theta}\right) - mr^2\sin\theta\cos\theta\dot{\phi}^2 = 0. \tag{1.87}$$

If the motion of the particle is initially in the equatorial plane, namely $\theta(t_0) = \pi/2$ and $\dot{\theta}(t_0) = 0$, where t_0 is some initial time, it remains in the equatorial plane. Without loss of generality, we can thus study the case of a particle moving in the equatorial plane (if this were not the case, we can always perform a proper rotation of the coordinate system to meet such a condition). The Lagrangian of the particle can thus be simplified to the form

$$L = \frac{1}{2}m\left(\dot{r}^2 + r^2\dot{\phi}^2\right) + \frac{G_N M m}{r}. \tag{1.88}$$

Since the Lagrangian does not explicitly depend on the coordinate ϕ and the time t, we have two constants of motion, which are, respectively, the axial component of the angular momentum L_z and the energy E. Following the approach of Sect. 1.6, we have[5]

$$\frac{d}{dt}\left(mr^2\dot{\phi}\right) = 0 \quad \Rightarrow \quad mr^2\dot{\phi} = \text{const.} = L_z, \tag{1.89}$$

and

$$\begin{aligned} E &= \frac{1}{2}m\dot{\mathbf{x}}^2 - \frac{G_N M m}{r} \\ &= \frac{1}{2}m\left(\dot{r}^2 + r^2\dot{\phi}^2\right) - \frac{G_N M m}{r} \\ &= \frac{1}{2}m\dot{r}^2 + \frac{L_z^2}{2mr^2} - \frac{G_N M m}{r}. \end{aligned} \tag{1.90}$$

If we define $\tilde{E} \equiv E/m$ and $\tilde{L}_z \equiv L_z/m$, we can write the following equation of motion

$$\frac{1}{2}\dot{r}^2 = \tilde{E} - V_{\text{eff}}, \tag{1.91}$$

---

[5]We use the notation L_z because this is the axial component of the angular momentum and we do not want to call it L because it may generate confusion with the Lagrangian.

Fig. 1.2 Plot of the effective potential V_{eff} in Eq. (1.92) for $G_N M = 1$ and $\tilde{L}_z = 3.9$

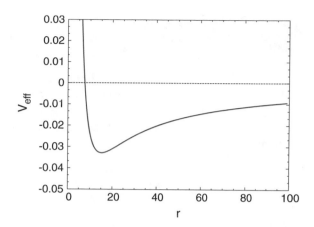

where V_{eff} is the effective potential

$$V_{eff} = -\frac{G_N M}{r} + \frac{\tilde{L}_z^2}{2r^2}. \tag{1.92}$$

Figure 1.2 shows V_{eff} as a function of the radial coordinate r. The first term on the right hand side in (1.92) is the (attractive) gravitational potential of a point-like body of mass M and is dominant when the particle is at large radii. The second term is the (repulsive) centrifugal potential and is dominant when the particle is at small radii. In Newton's gravity, the centrifugal potential prevents a point-like particle with non-vanishing angular momentum from falling onto a point-like massive body. As shown in Sect. 8.4, this is not true in Einstein's gravity.

1.9 Kepler's Laws

Kepler's Laws were empirically discovered by Johannes Kepler at the beginning of the 17th century by studying astronomical data of the planets in the Solar System. It was later shown by Isaac Newton that these laws are a direct consequence of Newtonian mechanics and Newton's Law of Universal Gravitation. Kepler's Laws read as follows:

Kepler's First Law. The orbit of a planet is an ellipse with the Sun at one of the two foci.

Kepler's Second Law. The line connecting a planet to the Sun sweeps out equal areas in equal times.

Kepler's Third Law. The square of the period of any planet is proportional to the cube of the semi-major axis of its orbit.

The Lagrangian describing the motion of a planet of mass m orbiting the Sun with mass M ($m \ll M$) is given in Eq. (1.88). Kepler's Second Law is the direct consequence of the fact that the angular momentum of the planet is a constant of motion. The area swept out by the line connecting the planet to the Sun is indeed given by

$$A(\phi_1, \phi_2) = \frac{1}{2} \int_{\phi_1}^{\phi_2} r^2 d\phi = \frac{1}{2} \int_{t_1}^{t_2} r^2 \dot{\phi} dt = \frac{L_z}{2m} \int_{t_1}^{t_2} dt = \frac{L_z}{2m} (t_2 - t_1) ,$$

(1.93)

where ϕ_1 and ϕ_2 are the values of the ϕ coordinate at the time t_1 and t_2, respectively.

Kepler's First Law can be derived as follows. We rewrite Eq. (1.90) by replacing $\dot{r}$ with

$$\dot{r} = \frac{dr}{d\phi} \frac{d\phi}{dt} = \frac{L_z}{mr^2} \frac{dr}{d\phi} ,$$

(1.94)

and we find

$$E = \frac{L_z^2}{2mr^4} \left(\frac{dr}{d\phi} \right)^2 + \frac{L_z^2}{2mr^2} - \frac{G_N M m}{r} .$$

(1.95)

We define

$$r = \frac{1}{u} , \quad u' = \frac{du}{d\phi} ,$$

(1.96)

and Eq. (1.95) becomes

$$E = \frac{L_z^2}{2m} u'^2 + \frac{L_z^2}{2m} u^2 - G_N M m u ,$$

(1.97)

and then

$$u'^2 - \frac{2 G_N M m^2}{L_z^2} u + u^2 = \frac{2mE}{L_z^2} .$$

(1.98)

We derive Eq. (1.98) with respect to ϕ and we find

$$2u' \left(u'' - \frac{G_N M m^2}{L_z^2} + u \right) = 0 .$$

(1.99)

We have thus two equations:

$$u' = 0, \tag{1.100}$$

$$u'' - \frac{G_N M m^2}{L_z^2} + u = 0. \tag{1.101}$$

The solutions of the differential equations in (1.100) and (1.101) are, respectively, the equations of a circle and of a conic

$$\frac{1}{r} = u = \text{constant}, \tag{1.102}$$

$$\frac{1}{r} = u = \frac{G_N M m^2}{L_z^2} + A \cos \phi, \tag{1.103}$$

where A is a constant. Eq. (1.103) describes an ellipse if $0 < A < G_N M m^2 / L_z^2$ [see Eqs. (D.6) and (D.7) in Appendix D], a parabola if $A = G_N M m^2 / L_z^2$, and a hyperbola if $A > G_N M m^2 / L_z^2$.

From Eq. (1.93), we can write the area of the ellipse described by the orbit of the planet as

$$\pi a^2 \sqrt{1 - e^2} = \frac{1}{2} \int_0^{2\pi} r^2 d\phi = \frac{L_z}{2m} \int_0^T dt = \frac{L_z T}{2m}, \tag{1.104}$$

where a is the ellipse semi-major axis, e is the ellipse eccentricity, and T is the orbital period of the planet. From Eqs. (D.6) and (D.7) in Appendix D

$$\frac{G_N M m^2}{L_z^2} = \frac{1}{a \left(1 - e^2\right)} \quad \Rightarrow \quad L_z^2 = a \left(1 - e^2\right) G_N M m^2. \tag{1.105}$$

Combining Eq. (1.104) with Eq. (1.105), we find Kepler's Third Law

$$\frac{a^3}{T^2} = \frac{G_N M}{4\pi^2} = \text{constant} \quad \Rightarrow \quad a^3 \propto T^2. \tag{1.106}$$

1.10 Maxwell's Equations

The physics of the 17th and 18th centuries was in perfect agreement with the Principle of Relativity, according to which there is no preferred reference frame and all inertial reference frames are equivalent and related to each other by the transformations of the Galilean group. However, the situation changed in the 19th century with the study of electromagnetic phenomena. Maxwell's equations in vacuum are (in Gaussian units)

$$\nabla \cdot \mathbf{E} = 0, \tag{1.107}$$

$$\nabla \cdot \mathbf{B} = 0, \tag{1.108}$$

$$\nabla \times \mathbf{E} = -\frac{1}{c}\frac{\partial \mathbf{B}}{\partial t}, \tag{1.109}$$

$$\nabla \times \mathbf{B} = \frac{1}{c}\frac{\partial \mathbf{E}}{\partial t}, \tag{1.110}$$

and they look to be inconsistent with the Principle of Relativity. Here c is the speed of light. In the context of Galileo's relativity, this is a physical quantity that changes according to Eq. (1.38) if we change inertial reference frame, which suggests that Maxwell's equations must hold in some special inertial reference frame and not in the others. Moreover, Maxwell's equations are not invariant under Galilean transformations. There are thus two possibilities:

1. The Principle of Relativity does not hold for electromagnetic phenomena, which require a preferred reference frame.
2. The laws of physics are the same in all inertial reference frames, but the Galilean transformations are wrong.

It is easy to see from Maxwell's equations in vacuum that electromagnetic phenomena have wave properties. For a generic vector $\mathbf{V}$, we have the following identity (see Appendix B.3)

$$\nabla \times (\nabla \times \mathbf{V}) = \nabla (\nabla \cdot \mathbf{V}) - \nabla^2 \mathbf{V}. \tag{1.111}$$

From Maxwell's third equation (1.109), we can write

$$\nabla \times (\nabla \times \mathbf{E}) = \nabla \times \left(-\frac{1}{c}\frac{\partial \mathbf{B}}{\partial t} \right). \tag{1.112}$$

We rewrite the left hand side of this equation by employing Eq. (1.111) and Maxwell's first equation (1.107)

$$\nabla^2 \mathbf{E} = \frac{1}{c}\frac{\partial}{\partial t} (\nabla \times \mathbf{B}), \tag{1.113}$$

where the right hand side has been rewritten by exploiting the fact that $\partial/\partial t$ and $\nabla\times$ commute. We then use Eq. (1.110) and we obtain

$$\left(\nabla^2 - \frac{1}{c^2}\frac{\partial^2}{\partial t^2} \right) \mathbf{E} = \mathbf{0}. \tag{1.114}$$

With a similar procedure, we can also write

$$\left(\nabla^2 - \frac{1}{c^2}\frac{\partial^2}{\partial t^2} \right) \mathbf{B} = \mathbf{0}. \tag{1.115}$$

Equations (1.114) and (1.115) are wave equations for the fields $\mathbf{E}$ and $\mathbf{B}$. In Newtonian mechanics, waves arise from the displacement of a portion of an elastic medium from its natural position, which then starts oscillating about its equilibrium position. A mechanical wave is a perturbation of the elastic medium and propagates with velocity

$$v = \sqrt{\frac{T}{\rho}}, \tag{1.116}$$

where T and ρ are, respectively, the tension and the density of the elastic medium.

At the end of the 19th century, it was natural to postulate the existence of a medium for the propagation of electromagnetic phenomena. Such a medium was called the *aether* and the speed of the electromagnetic phenomena was c. The aether had to be the preferred reference frame for the description of electromagnetic phenomena. At this point, it was necessary to find direct or indirect evidence for the existence of the aether.

1.11 Michelson–Morley Experiment

The Michelson-Morley experiment is an interferometer used to measure the difference of the speed of light between two orthogonal directions as the result of the possible motion of the set-up with respect to the aether. The set-up is sketched in Fig. 1.3. The source S emits a beam of light, which is separated into two beams by a beamsplitter B. One of the light beams goes to branch 1, is reflected by mirror M_1, and returns to beamsplitter B. The second light beam goes to branch 2, is reflected by mirror M_2, and returns to beamsplitter B. The two beams then go to a detector D, where we can observe the interference pattern. As shown in Fig. 1.3, we indicate with d_1 and d_2 the lengths of branch 1 and branch 2, respectively.

Let us assume that the aether is moving with the speed v parallel to branch 1 (configuration A). According to the Galilean transformations, the travel time of the light beam in branch 1, namely the time that light takes to go from beamsplitter B to mirror M_1 and return to beamsplitter B, would be

$$\Delta t_1^A = \frac{d_1}{c-v} + \frac{d_1}{c+v} = \frac{2d_1 c}{c^2 - v^2} = \frac{2d_1}{c}\left[1 + \frac{v^2}{c^2} + O\left(\frac{v^4}{c^4}\right)\right], \tag{1.117}$$

where c is the speed of light in the reference frame of the aether. In the reference frame of the aether, the travel distance of the light beam in branch 2 would be

$$D = \left[4d_2^2 + v^2\left(\Delta t_2^A\right)^2\right]^{1/2}, \tag{1.118}$$

Fig. 1.3 Sketch of the
Michelson-Morley
experiment. S is the source,
B is the beamsplitter, M_1
and M_2 are the mirrors in
branches 1 and 2,
respectively, and D is the
detector where we can
observe the interference
pattern. d_1 and d_2 are the
lengths of branches 1 and 2,
respectively

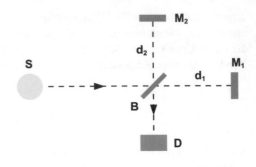

where Δt_2^A is the time that the light beam takes to go from B to M_2 and return to B.
Since the speed of light in the reference frame of the aether is c, Δt_2^A is given by

$$\Delta t_2^A = \frac{D}{c}. \tag{1.119}$$

If we combine Eq. (1.118) with Eq. (1.119), we get Δt_2^A

$$\Delta t_2^A = \frac{2d_2}{\left(c^2 - v^2\right)^{1/2}} = \frac{2d_2}{c}\left[1 + \frac{1}{2}\frac{v^2}{c^2} + O\left(\frac{v^4}{c^4}\right)\right]. \tag{1.120}$$

The time difference of the two signals is

$$\delta t^A = \Delta t_1^A - \Delta t_2^A = 2\frac{d_1 - d_2}{c} + (2d_1 - d_2)\frac{v^2}{c^3} + O\left(\frac{v^4}{c^5}\right). \tag{1.121}$$

Let us now rotate the set-up by 90° (configuration B) and evaluate the time difference in the new configuration. Now the aether should move with the speed v parallel
to branch 2. We find

$$\delta t^B = \Delta t_1^B - \Delta t_2^B = 2\frac{d_1 - d_2}{c} + (d_1 - 2d_2)\frac{v^2}{c^3} + O\left(\frac{v^4}{c^5}\right). \tag{1.122}$$

The hypothesis of the existence of the aether together with the Galilean transformations predict the following fringe shift of the interference pattern between configurations A and B

$$\delta n = \frac{\delta T}{T} = \frac{c}{\lambda}\left(\delta t^A - \delta t^B\right) = \frac{d_1 + d_2}{\lambda}\frac{v^2}{c^2} + O\left(\frac{v^4}{c^4}\right), \tag{1.123}$$

where $T = \lambda/c$ is the period of the radiation and λ is its wavelength.

The first attempt to measure the speed of the aether with respect to Earth was done
in 1881 by Albert Michelson, but the result was not conclusive. The experiment was
repeated in 1887 by Albert Michelson and Edward Morley. In the 1887 experiment,

$d_1 = d_2 = 11$ m and $\lambda \approx 600$ nm. If we assume that the rest frame of the aether is that of the Solar System and we consider that the orbital velocity of Earth is about 30 km/s, we should expect $\delta n = 0.4$. If we consider the velocity of the Solar System around the galactic center, which is about 220 km/s, we should expect $\delta n = 3$. The resolution of the 1887 experiment was $\delta n = 0.01$ and no difference in the interference pattern was observed between the two orientations of the set-up. The experiment was repeated by other physicists with the same negative result. Strictly speaking, these experiments do not rule out the hypothesis of the existence of the aether if we postulate that the space is not isotropic with respect to length measurements, and such a possibility was the first solution explored by physicists to save the aether and explain the experimental results. However, similar attempts eventually failed.

1.12 Towards the Theory of Special Relativity

In 1887 Woldemar Voigt wrote the following coordinate transformation relating an inertial reference frame with Cartesian coordinates (x, y, z) to another inertial reference frame with Cartesian coordinates (x', y', z') and moving in the positive x-direction at constant speed v relative to the former reference frame

$$x \to x' = x - vt \, ,$$

$$y \to y' = y \left(1 - \frac{v^2}{c^2}\right)^{1/2} ,$$

$$z \to z' = z \left(1 - \frac{v^2}{c^2}\right)^{1/2} ,$$

$$t \to t' = t - \frac{vx}{c^2} \, . \tag{1.124}$$

Such a coordinate transformation is a modification of the Galilean transformation in Eq. (1.36). Voigt showed that some electrodynamics equations were invariant under the transformation in (1.124). For instance, the electromagnetic wave Eqs. (1.114) and (1.115) become

$$\left(1 - \frac{v^2}{c^2}\right) \left(\nabla'^2 - \frac{1}{c^2}\frac{\partial^2}{\partial t'^2}\right) \mathbf{E} = 0 \, ,$$

$$\left(1 - \frac{v^2}{c^2}\right) \left(\nabla'^2 - \frac{1}{c^2}\frac{\partial^2}{\partial t'^2}\right) \mathbf{B} = 0 \, . \tag{1.125}$$

However, it seems he did not attribute any particular physical meaning to this coordinate transformation.

In 1887, Michelson and Morley announced the results of their experiment. Unlike that in 1881, the result was convincing. However, the hypothesis of the existence of

the aether was not immediately abandoned. In 1889, George FitzGerald and, independently, in 1892, Hendrik Lorentz showed that the Michelson-Morley experiment
could be explained postulating the contraction of the length L of an object moving
parallel to the aether (*FitzGerald-Lorentz contraction*)

$$L \to L\left(1 - \frac{v^2}{c^2}\right)^{1/2}. \tag{1.126}$$

In 1897, Joseph Larmor extended Lorentz's work, writing the coordinate transformation of special relativity between two inertial reference frames differing by a
constant relative speed, and showed that the FitzGerald-Lorentz contraction was a
consequence of this coordinate transformation. In 1905, Henri Poincaré gave these
coordinate transformations their modern form and called them the Lorentz transformations. Poincaré also showed that these transformations together with the rotations
form a group, which was called the Lorentz group. In the same year, Albert Einstein
showed that these transformations can be derived assuming the Principle of Relativity and the constancy of the speed of light.

Problems

1.1 Verify Eq. (1.14).

1.2 The transformation between spherical coordinates (r, θ, ϕ) and cylindrical coordinates (ρ, z, ϕ') is

$$\rho = r\sin\theta, \quad z = r\cos\theta, \quad \phi' = \phi, \tag{1.127}$$

with inverse

$$r = \sqrt{\rho^2 + z^2}, \quad \theta = \arctan\left(\frac{\rho}{z}\right), \quad \phi = \phi'. \tag{1.128}$$

Write the metric tensor g_{ij} and then the line element dl in cylindrical coordinates.

1.3 Consider the transformation $x^i \to x'^i$ described by the Galilean transformation
in Eq. (1.36). Show that the expression of the Euclidean metric δ_{ij} does not change.

1.4 Consider the transformation $x^i \to x'^i$ described by the rotation in the xy plane
R_{xy} in Eq. (1.42). Show that the expression of the Euclidean metric δ_{ij} does not
change.

1.5 The Lagrangian of a free point-like particle in spherical coordinates is in Eq. (1.77) and the transformations between spherical coordinates (r, θ, ϕ) and cylindrical coordinates (ρ, z, ϕ') are given by Eqs. (1.127) and (1.128). Write the Lagrangian in cylindrical coordinates and then the corresponding Euler–Lagrange equations.

1.6 From the Euler–Lagrange equations obtained in the previous exercise, write the Christoffel symbols in cylindrical coordinates.

1.7 The Lagrangian of a free point-like particle of mass m moving on a spherical surface of radius R is

$$L = \frac{1}{2}mR^2 \left(\dot{\theta}^2 + \sin^2\theta\dot{\phi}^2\right) . \qquad (1.129)$$

Note that here the Lagrangian coordinates are (θ, ϕ), while R is a constant. Write the Euler–Lagrange equations for the Lagrangian in (1.129).

1.8 Let us consider the following Lagrangian

$$L = \frac{1}{2}m \left(\dot{x}^2 + \dot{y}^2\right) - \frac{1}{2}k \left(x^2 + y^2\right) , \qquad (1.130)$$

where x and y are the Lagrangian coordinates. It is the Lagrangian of a particle moving in a 2-dimensional space and subject to the potential $V = k(x^2 + y^2)/2$. Find the constant(s) of motion and then write the corresponding Euler–Lagrange equations.

1.9 Show that the Maxwell equations are not invariant under Galilean transformations.

Chapter 2
Special Relativity

The theory of *special relativity* refers to the theoretical framework based on the Einstein Principle of Relativity. In Newtonian mechanics, we have space and time as two distinct entities and time is the same for all observers. Such a set-up breaks down when we impose that interactions do not propagate instantaneously with infinite velocity (an implicit assumption in the theory of Galilean relativity). In special relativity we have spacetime as the natural stage for the description of physical phenomena.

2.1 Einstein's Principle of Relativity

Experiments support the validity of the Special Principle of Relativity, namely that the laws of physics have the same form in all inertial reference frames. A key assumption in Newtonian mechanics is that interactions propagate with infinite velocity. For example, Newton's Law of Universal Gravitation can be written as

$$\mathbf{F}_{12} = G_N \frac{M_1 M_2}{r^2} \hat{\mathbf{r}}_{12} \,, \tag{2.1}$$

where $\mathbf{F}_{12}$ is the force acting on body 1 by body 2, M_1 (M_2) is the mass of body 1 (2), r is the distance between the two bodies, and $\hat{\mathbf{r}}_{12}$ is the unit vector located at the position of body 1 and pointing towards the direction of body 2. If body 2 changes position, the force acting on body 1 changes immediately without any time delay. However, experiments show that there are no interactions propagating with infinite velocity. The speed of light is very high in comparison with the typical velocities of objects around us, but still it is finite. Assuming instantaneous interactions, Newtonian mechanics inevitably introduces some approximations.

© Springer Nature Singapore Pte Ltd. 2018
C. Bambi, *Introduction to General Relativity*, Undergraduate Lecture Notes
in Physics, https://doi.org/10.1007/978-981-13-1090-4_2

The crucial ingredient to go beyond Newtonian mechanics is to note that there exists a maximum velocity for the propagation of interactions. *Einstein's Principle of Relativity* reads as follows:

Einstein's Principle of Relativity.

1. The Special Principle of Relativity holds.
2. The speed of light in vacuum is the maximum velocity for the propagation of interactions.

Newtonian mechanics is clearly inconsistent with the Einstein Principle of Relativity. In Newtonian mechanics, if the velocity of the propagation of interactions were finite, its value should change in different reference frames. If we measure the velocity $\mathbf{w}$ in the reference frame (x, y, z), in the reference frame (x', y', z'), moving with constant velocity $\mathbf{v}$ with respect to the reference frame (x, y, z), the velocity of propagation of the interaction should be $\mathbf{w}' = \mathbf{w} - \mathbf{v}$ and may exceed the speed of light in vacuum.

In Newtonian mechanics, there is an absolute time, which is valid for all reference frames. However, the existence of an absolute time valid for all reference frames is inconsistent with the Einstein Principle of Relativity. For instance, let us consider an electromagnetic source at rest at the origin in the reference frame (x, y, z). The source emits a flash of light propagating in all directions with the same velocity. The electromagnetic signal reaches all the points equidistant from the origin at the same time. Let us now consider the reference frame (x', y', z') moving with velocity $\mathbf{v}$ with respect to the reference frame (x, y, z). According to the reference frame (x', y', z'), the electromagnetic signal emitted by the source reaches all the points equidistant from the emission point at the same time, because the electromagnetic signal moves at the speed of light in all directions. It is clear that the reference frames (x, y, z) and (x', y', z') have a different notion of "same time". Two events that are considered to occur at the same time in one reference frame are not at the same time in the other reference frame. This simple example also shows that the line element in (1.1) cannot be an invariant any longer.

2.2   Minkowski Spacetime

In Newtonian mechanics, we have the space coordinates, e.g. (x, y, z) if we consider a 3-dimensional space with a system of Cartesian coordinates, and an absolute time, t. We have thus "space" and "time" as two distinct entities. In special relativity it is useful to introduce the concept of *spacetime*, where the space coordinates and the time coordinate become the coordinates of the spacetime. Every point of the spacetime is an *event*, because it indeed represents an "event" occurring at a particular point of the space and at a certain time.

First, we want to find the counterpart of the line element dl in Eq. (1.1). Indeed, if we assume the Einstein Principle of Relativity, dl cannot be an invariant any longer: if we consider a light signal moving from the point A to the point B in two different reference frames, the distance between the two points cannot be the same in the two different reference frames because the speed of light is the same but the travel time is not, in general.

Let us consider the trajectory of a photon in a $(3 + 1)$-dimensional spacetime (3 spatial dimensions + 1 temporal dimension). If the trajectory is parametrized by the coordinate time t, we can write[1]

$$||x^\mu|| = \begin{pmatrix} x^0(t) \\ x^1(t) \\ x^2(t) \\ x^3(t) \end{pmatrix} = \begin{pmatrix} ct \\ x(t) \\ y(t) \\ z(t) \end{pmatrix} = \begin{pmatrix} ct \\ \mathbf{x}(t) \end{pmatrix}. \qquad (2.2)$$

Note that we have defined $x^0 = ct$, where c is the speed of light, and x^0 thus has the dimensions of a length same as the space coordinates x^1, x^2, and x^3. Since the speed of light is c in all inertial reference frames, the quantity

$$ds^2 = -c^2 dt^2 + dx^2 + dy^2 + dz^2 = -c^2 dt^2 + d\mathbf{x}^2$$
$$= \left[-c^2 + \left(\frac{dx}{dt} \right)^2 + \left(\frac{dy}{dt} \right)^2 + \left(\frac{dz}{dt} \right)^2 \right] dt^2 \qquad (2.3)$$

must vanish along the photon trajectory in all inertial reference frames (the case of non-inertial reference frame will be discussed later; $ds^2 = 0$ still holds, but the expression of ds^2 is different). ds is the line element of the spacetime.

If $ds^2 = 0$ in an inertial reference frame, it vanishes in all inertial reference frames. However, this is not yet enough to say it is an invariant. Indeed we may have

$$ds^2 = k\, ds'^2, \qquad (2.4)$$

where ds'^2 is the line element of another inertial reference frame and k some coefficient. If we believe that the spacetime is homogeneous (no preferred points) and isotropic (no preferred directions), k cannot depend on the spacetime coordinates x^μ. So it can at most be a function of the relative velocity between the two reference frames. However, this is not the case. Let us consider reference frames 0, 1, and 2, and let $\mathbf{v}_{01}$ and $\mathbf{v}_{02}$ be the velocities of, respectively, coordinate systems 1 and 2 relative to coordinate system 0. We can write

$$ds_0^2 = k(v_{01})\, ds_1^2, \quad ds_0^2 = k(v_{02})\, ds_2^2, \qquad (2.5)$$

---

[1] Some authors use a different convention. For an $(n + 1)$-dimensional spacetime (n spatial dimensions + 1 temporal dimension), they write the coordinates of the spacetime as $(x^1, x^2, \dots, x^{n+1})$, where x^{n+1} is the time coordinate. In such a case, the Minkowski metric in Eq. (2.9) becomes $\eta_{\mu\nu} = \text{diag}(1, 1, \dots, 1, -1)$.

where v_{01} and v_{02} are the absolute velocities of $\mathbf{v}_{01}$ and $\mathbf{v}_{02}$ because k cannot depend on their direction given the isotropy of the spacetime. Let $\mathbf{v}_{12}$ be the velocity of reference frame 2 relative to reference frame 1. We have

$$ds_1^2 = k(v_{12})\, ds_2^2 \,. \tag{2.6}$$

Combining Eqs. (2.5) and (2.6), we have

$$k(v_{12}) = \frac{ds_1^2}{ds_2^2} = \frac{k(v_{02})}{k(v_{01})} \,. \tag{2.7}$$

However, v_{12} depends on the directions of $\mathbf{v}_{01}$ and $\mathbf{v}_{02}$, while the right hand side in (2.7) does not. Equation (2.7) thus implies that k is a constant and since it is independent of the reference frame it must be 1. We can now conclude that ds^2 is an invariant.

The line element (2.3) can be rewritten as

$$ds^2 = \eta_{\mu\nu} dx^\mu dx^\nu \,, \tag{2.8}$$

where $\eta_{\mu\nu}$ is the *Minkowski metric* and reads[2]

$$||\eta_{\mu\nu}|| = \begin{pmatrix} -1 & 0 & 0 & 0 \\ 0 & 1 & 0 & 0 \\ 0 & 0 & 1 & 0 \\ 0 & 0 & 0 & 1 \end{pmatrix} \,. \tag{2.9}$$

The n-dimensional *Minkowski spacetime* can be identified with $\mathbb{R}^n$ with the metric $\eta_{\mu\nu}$.[3] It is the counterpart of the n-dimensional Euclidean space with the metric δ_{ij} reviewed in Sect. 1.2. The Minkowski spacetime is the space of relativistic mechanics as the Euclidean space is the space of Newtonian mechanics. In general, we can write the metric of the spacetime as $g_{\mu\nu}$. For example, in spherical coordinates we have

---

[2] The convention of a metric with signature $(-+++)$ is common in the gravity community. In the particle physics community it is more common the convention of a metric with signature $(+---)$.

[3] Throughout the book, we use Latin letters $i, j, k, \ldots$ for space indices $(1, 2, \ldots, n)$, where n is the number of spatial dimensions, and Greek letters $\mu, \nu, \rho, \ldots$ for spacetimes indices $(0, 1, 2, \ldots, n)$. Such a convention is also used when we sum over repeated indices. For instance, for $n = 3$ we have

$$ds^2 = \eta_{\mu\nu} dx^\mu dx^\nu \equiv -\left(dx^0\right)^2 + \left(dx^1\right)^2 + \left(dx^2\right)^2 + \left(dx^3\right)^2 \,. \tag{2.10}$$

If we wrote $\eta_{ij} dx^j dx^i$, we would mean

$$\eta_{ij} dx^i dx^j \equiv \left(dx^1\right)^2 + \left(dx^2\right)^2 + \left(dx^3\right)^2 \,, \tag{2.11}$$

because i and j can run from 1 to n.

$$
\|g_{\mu\nu}\| = \begin{pmatrix} -1 & 0 & 0 & 0 \\ 0 & 1 & 0 & 0 \\ 0 & 0 & r^2 & 0 \\ 0 & 0 & 0 & r^2 \sin^2\theta \end{pmatrix}.
\tag{2.12}
$$

If we consider a transformation of spatial coordinates only, such as from Cartesian to spherical coordinates or from spherical to cylindrical coordinates, we have $g_{0i} = 0$. However, in some cases it may be useful to mix temporal and spatial coordinates and g_{0i} may be non-vanishing.

The results reviewed in Sect. 1.3 still hold, but space indices are replaced by spacetime indices. Under the coordinate transformation $x^\mu \to x'^\mu$, vectors and dual vectors transform as

$$
V^\mu \to V'^\mu = \frac{\partial x'^\mu}{\partial x^\nu} V^\nu, \quad V_\mu \to V'_\mu = \frac{\partial x^\nu}{\partial x'^\mu} V_\nu.
\tag{2.13}
$$

Upper indices are now lowered by $g_{\mu\nu}$ and lower indices are raised by $g^{\mu\nu}$, for example

$$
V_\mu = g_{\mu\nu} V^\nu, \quad V^\mu = g^{\mu\nu} V_\nu.
\tag{2.14}
$$

The generalization to tensors of any type and order is straightforward.

The trajectory of a point-like particle in the spacetime is a curve called the *world line*. Particles moving at the speed of light follow trajectories with $ds^2 = 0$ by definition, while for particles moving at lower (higher) velocities the line element along their trajectories is $ds^2 < 0$ ($ds^2 > 0$). Curves with $ds^2 < 0$ are called *time-like*, those with $ds^2 = 0$ are called *light-like* or *null*, and those with $ds^2 > 0$ are called *space-like*.[4] According to the Einstein Principle of Relativity, there are no particles following space-like trajectories, but in some contexts it is necessary to consider space-like curves.

Figure 2.1 shows a diagram of the Minkowski spacetime. For simplicity and graphical reasons, we consider a spacetime in $1 + 1$ dimensions, so we have the coordinates t and x only. Here the line element is $ds^2 = -c^2 dt^2 + dx^2$. P is an event, namely a point in the spacetime. Without loss of generality, we can put P at the origin of the coordinate system. The set of points in the spacetime that are connected to P by a light-like trajectory is called the *light-cone*. If we consider a $(2 + 1)$-dimensional spacetime (ct, x, y), the light-cone is indeed the surface of a double cone. In $1 + 1$ dimensions, like in Fig. 2.1, the light-cone is defined by $ct = \pm x$ and we have two straight lines. In the case of an $(n + 1)$-dimensional spacetime with $n \geq 3$, the light-cone is a hyper-surface. We can also distinguish the *future light-cone* (the set of points of the light-cone with $t > 0$) and the *past light-cone* (the set of points of the light-cone with $t < 0$).

---

[4]With the convention $\eta_{\mu\nu} = \mathrm{diag}(1, -1, -1, -1)$ common in particle physics, the line element along particle trajectories is $ds^2 > 0$ ($ds^2 < 0$) if the particle moves at a speed lower (higher) than c. In that context, time-like curves have $ds^2 > 0$ and space-like curves have $ds^2 < 0$.

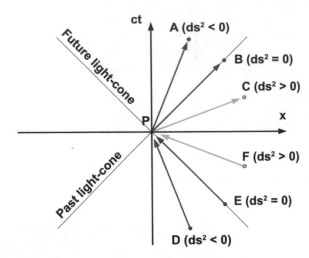

Fig. 2.1 Diagram of the Minkowski spacetime. Event P is at the origin of the coordinate system (ct, x). The future light-cone consists of the two half straight lines $ct = \pm x$ with $t > 0$. Event B is on the future light-cone and is connected to P by a light-like trajectory. Event A is inside the future light-cone: it is connected to P by a time-like trajectory. Events inside the future light-cone are causally connected to point P, i.e. they can be affected by what happens in P. Event C is outside the future light-cone: it is connected to P by a space-like trajectory. Events outside the future light-cone are causally disconnected to point P, i.e. they cannot be affected by what happens in P. The past light-cone consists of the two half straight lines $ct = \pm x$ with $t < 0$. Event D is connected to P by a time-like trajectory, event E is connected to P by a light-like trajectory, and event F is connected to P by a space-like trajectory. D and E can affect P, while F cannot. See the text for more details

According to the Einstein Principle of Relativity, there are no interactions propagating faster than light in vacuum. The light cone thus determines the causally connected and disconnected regions with respect to a certain point. With reference to Fig. 2.1, an event in P can influence an event in A and in B, but cannot influence an event in C. An event in D or in E can influence an event in P, while one in F cannot.

2.3 Lorentz Transformations

Now we want to find the relativistic counterpart of the Galilean transformations, namely those transformations connecting the coordinates of two inertial reference frames which differ only by constant relative motion and are consistent with the Einstein Principle of Relativity. Let $\Lambda^\mu{}_\nu$ be the matrix connecting the two inertial reference frames, that is

$$dx^\mu \to dx'^\mu = \Lambda^\mu{}_\nu dx^\nu . \tag{2.15}$$

We have

$$ds^2 = \eta_{\mu\nu}dx^\mu dx^\nu \rightarrow ds'^2 = \eta_{\mu\nu}dx'^\mu dx'^\nu , \tag{2.16}$$

because the form of the metric tensor cannot change, and therefore we find the following condition that the transformation $\Lambda^\mu{}_\nu$ must satisfy

$$\eta_{\mu\nu} = \Lambda^\rho{}_\mu \Lambda^\sigma{}_\nu \eta_{\rho\sigma} . \tag{2.17}$$

There are several ways to obtain the expression of the transformation $\Lambda^\mu{}_\nu$. In our case, we proceed as follows. First, we consider the transformation

$$t \rightarrow \tilde{t} = i\,ct , \tag{2.18}$$

where i is the imaginary unit, i.e. $i^2 = -1$. The line element now reads

$$ds^2 = d\tilde{t}^2 + dx^2 + dy^2 + dz^2 , \tag{2.19}$$

which is the line element in the 4-dimensional Euclidean space $\mathbb{R}^4$. The transformations that leave the line element unchanged are the rotations and the translations. As mentioned in Sect. 1.4, in $\mathbb{R}^3$ every rotation can be written as the combination of elementary rotations in the planes yz, xz, and xy. Now in $\mathbb{R}^4$ we have six "elementary" rotations because there are also the rotations in the planes $\tilde{t}x$, $\tilde{t}y$, and $\tilde{t}z$. It is clear that the rotations yz, xz, and xy are the standard spatial rotations that do not link reference frames moving with respect to each other. The generalization of the Galilean transformations should thus come from the rotations in the planes $\tilde{t}x$, $\tilde{t}y$, and $\tilde{t}z$.

Let us consider a rotation in the plane $\tilde{t}x$. The rotation matrix is

$$R_{\tilde{t}x}(\psi) = \begin{pmatrix} \cos\psi & \sin\psi & 0 & 0 \\ -\sin\psi & \cos\psi & 0 & 0 \\ 0 & 0 & 1 & 0 \\ 0 & 0 & 0 & 1 \end{pmatrix} , \tag{2.20}$$

where ψ is the rotation angle. We have

$$\begin{pmatrix} d\tilde{t} \\ dx \\ dy \\ dz \end{pmatrix} \rightarrow \begin{pmatrix} d\tilde{t}' \\ dx' \\ dy' \\ dz' \end{pmatrix} = \begin{pmatrix} \cos\psi\,d\tilde{t} + \sin\psi\,dx \\ -\sin\psi\,d\tilde{t} + \cos\psi\,dx \\ dy \\ dz \end{pmatrix} . \tag{2.21}$$

If we consider an event that is not changing space coordinates in the reference frame $(\tilde{t}, x, y, z)$, we have $dx = 0$ and therefore

$$\frac{dx'}{d\tilde{t}'} = -\frac{\sin\psi \, d\tilde{t}}{\cos\psi \, d\tilde{t}} = -\tan\psi \, . \tag{2.22}$$

If we reintroduce the variable t and t', we have

$$\frac{dx'}{dt'} = -ic\tan\psi \, . \tag{2.23}$$

Note that $dx'/dt' = -v$, where $-v$ is the velocity of the reference frame (ct, x, y, z) relative to the reference frame (ct', x', y', z'), and therefore

$$\tan\psi = -i\beta \, , \tag{2.24}$$

where we have defined $\beta = v/c$. Since

$$\cos\psi = \frac{1}{\sqrt{1 + \tan^2\psi}} = \frac{1}{\sqrt{1 - \beta^2}} \, ,$$

$$\sin\psi = \tan\psi \cos\psi = -\frac{i\beta}{\sqrt{1 - \beta^2}} \, , \tag{2.25}$$

Equation (2.21) becomes

$$\begin{pmatrix} cdt \\ dx \\ dy \\ dz \end{pmatrix} \rightarrow \begin{pmatrix} cdt' \\ dx' \\ dy' \\ dz' \end{pmatrix} = \begin{pmatrix} \gamma cdt - \gamma\beta dx \\ -\gamma\beta cdt + \gamma dx \\ dy \\ dz \end{pmatrix} , \tag{2.26}$$

where we have defined the *Lorentz factor*

$$\gamma = \frac{1}{\sqrt{1 - \beta^2}} \, . \tag{2.27}$$

The expression for $\Lambda^\mu{}_\nu$ is thus

$$||\Lambda_x{}^\mu{}_\nu|| = \begin{pmatrix} \gamma & -\gamma\beta & 0 & 0 \\ -\gamma\beta & \gamma & 0 & 0 \\ 0 & 0 & 1 & 0 \\ 0 & 0 & 0 & 1 \end{pmatrix} , \tag{2.28}$$

where we have introduced a subindex x to indicate that the relative velocity of the two reference frames is along the x-axis. We have similar expressions if the relative motion between the two reference frames is along the y and z axes

$$||\Lambda_y{}^{\mu}{}_{\nu}|| = \begin{pmatrix} \gamma & 0 & -\gamma\beta & 0 \\ 0 & 1 & 0 & 0 \\ -\gamma\beta & 0 & \gamma & 0 \\ 0 & 0 & 0 & 1 \end{pmatrix}, \quad ||\Lambda_z{}^{\mu}{}_{\nu}|| = \begin{pmatrix} \gamma & 0 & 0 & -\gamma\beta \\ 0 & 1 & 0 & 0 \\ 0 & 0 & 1 & 0 \\ -\gamma\beta & 0 & 0 & \gamma \end{pmatrix}. \quad (2.29)$$

The inverse transformations can be easily obtained with the substitution $\mathbf{v} \to -\mathbf{v}$. For example, the inverse transformation of (2.28) is

$$||\Lambda_x^{-1\,\mu}{}_{\nu}|| = \begin{pmatrix} \gamma & \gamma\beta & 0 & 0 \\ \gamma\beta & \gamma & 0 & 0 \\ 0 & 0 & 1 & 0 \\ 0 & 0 & 0 & 1 \end{pmatrix}, \quad (2.30)$$

and it is easy to verify that $\Lambda_x{}^{\mu}{}_{\nu}\Lambda_x^{-1\,\nu}{}_{\rho} = \delta^{\mu}_{\rho}$.

The transformations connecting two inertial reference frames which differ only by constant relative motion are called the *Lorentz transformations* or *Lorentz boosts*. They are the relativistic generalization of the Galilean transformations. The Lorentz transformations with the spatial rotations form a group, which is called the *Lorentz group*. If we add the translations to the Lorentz group, we have the *Poincaré group*, which is the counterpart in special relativity of the Galilean group in Newtonian mechanics. If we explicitly show the speed of light c, the Lorentz factor is

$$\gamma = \frac{1}{\sqrt{1 - \frac{v^2}{c^2}}}. \quad (2.31)$$

For $c \to \infty$, $\gamma \to 1$, and the Lorentz transformations reduce to the Galilean ones.

Note that

$$\Lambda^{\mu}{}_{\nu} = \frac{\partial x'^{\mu}}{\partial x^{\nu}}, \quad (2.32)$$

and therefore vectors, dual vectors, etc. transform as[5]

$$V^{\mu} \to V'^{\mu} = \Lambda^{\mu}{}_{\nu}V^{\nu}, \quad V_{\mu} \to V'_{\mu} = \Lambda_{\mu}{}^{\nu}V_{\nu}, \quad \text{etc.} \quad (2.34)$$

Any combination between a vector and a dual vector like $V^{\mu}W_{\mu}$ is a scalar

$$V^{\mu}W_{\mu} \to V'^{\mu}W'_{\mu} = \left(\Lambda^{\mu}{}_{\nu}V^{\nu}\right)\left[\Lambda_{\mu}{}^{\rho}W_{\rho}\right] = \delta^{\rho}_{\nu}V^{\nu}W_{\rho} = V^{\nu}W_{\nu}. \quad (2.35)$$

---

[5]Note the difference of the position of the indices μ and ν in $\Lambda^{\mu}{}_{\nu}$ and $\Lambda_{\mu}{}^{\nu}$. Indeed

$$\Lambda^{\mu}{}_{\nu} = \frac{\partial x'^{\mu}}{\partial x^{\nu}}, \quad \Lambda_{\mu}{}^{\nu} = \frac{\partial x^{\nu}}{\partial x'^{\mu}}. \quad (2.33)$$

However, since the matrices of the Lorentz transformations are symmetric, sometimes the notation Λ^{μ}_{ν} and Λ^{ν}_{μ} is used when it is clear the initial and the final coordinate systems.

2.4 Proper Time

Let us consider an inertial reference frame (ct, x, y, z). In this 4-dimensional coordinate system, the trajectory of a particle moving along a time-like curve is described by four coordinates as in Eq. (2.2). The line element along the curve is

$$ds^2 = -c^2 dt^2 + dx^2 + dy^2 + dz^2$$
$$= -\left[c^2 - \left(\frac{dx}{dt} \right)^2 - \left(\frac{dy}{dt} \right)^2 - \left(\frac{dz}{dt} \right)^2 \right] dt^2$$
$$= -\left(1 - \frac{\dot{\mathbf{x}}^2}{c^2} \right) c^2 dt^2 = -\frac{c^2 dt^2}{\gamma^2}, \tag{2.36}$$

where $\dot{\mathbf{x}}^2$ is the square of the velocity of the particle. Note that we are not assuming that the particle is moving at a constant speed in a straight line.

Let us now consider the reference frame in which the particle is at rest $(c\tau, X, Y, Z)$. Here τ is called the *proper time* of the particle, because it is the time measured by a clock moving with the particle. In such a reference frame, the motion of the particle is simply

$$||x'^{\mu}|| = \begin{pmatrix} c\tau \\ 0 \\ 0 \\ 0 \end{pmatrix}, \tag{2.37}$$

because the particle is at rest at the origin. The line element is thus $ds^2 = -c^2 d\tau^2$ and, since it is an invariant, it must be equal to (2.36). We thus find that

$$dt = \gamma \, d\tau. \tag{2.38}$$

Integrating over a finite time and assuming γ constant for simplicity, we have

$$\Delta t = \gamma \, \Delta \tau. \tag{2.39}$$

Since $\gamma > 1$, we find that the clock of the reference frame (ct, x, y, z) is faster than the clock of the moving particle. The phenomenon is called *time dilation*. For $c \to \infty$, $\gamma \to 1$, and we recover the Newtonian result that all clocks measure the same time.

If the particle is another observer, we have two reference frames, (ct, x, y, z) and (ct', x', y', z'). If they are both inertial reference frames, the observer with the coordinate system (ct, x, y, z) will see that the clock of the observer with the coordinate system (ct', x', y', z') is slower, because $\Delta t = \gamma \Delta t'$. At the same time, the observer with the coordinate system (ct', x', y', z') will see that the clock of the observer with the coordinate system (ct, x, y, z) is slower, $\Delta t' = \gamma \Delta t$. There is no contradiction, because the clocks are in different places.

On the contrary, if we consider the scenario in which the two observers are initially at the same point and they meet again at the same point later, one of the two observers cannot be in an inertial reference frame. In such a case, ds^2 is still an invariant and we find the result in (2.39), but we cannot invert the argument for the observer with proper time τ and the clock in the non-inertial frame is indeed slower. This is the picture of the so-called *twin paradox*. We can consider two twins that at time t_1 are on Earth and, as twins, have the same age. One of the two twins travels through space on a spacecraft and, at the end of the trip, he/she returns to Earth. The twin of the spacecraft will now be younger than the twin that remained on Earth. However, even the twin of the spacecraft sees the other twin moving. So each twin should paradoxically expect the other to have aged less. The explanation of the paradox is that there is not a symmetry between the two twins. The one that remained on Earth is in a (quasi) inertial reference frame (ct, x, y, z), and can apply Eq. (2.36) to describe the trajectory of the twin on the spacecraft. The twin of the spacecraft is not in an inertial reference frame, because otherwise he/she could not come back to Earth. In such a case, he/she cannot describe the twin on Earth with Eq. (2.36). As we will see later in the context of general relativity, the line element ds is an invariant in any (inertial and non-inertial) reference frame, but the expression of $g_{\mu\nu}$ is not, and this is why the twin on the spacecraft cannot use Eq. (2.36).

Note that for $v \to c, \gamma \to \infty$, and $\Delta\tau \to 0$. Particles moving at the speed of light do not have a proper time. It is as if their clock were completely frozen.

2.5 Transformation Rules

Now we want to figure out how the values of some physical quantities measured in a reference frame change when they are measured in another reference frame. The issue of time intervals has already been addressed in the previous section.

First, we want to find the transformation rule for the Cartesian coordinates, namely the relativistic generalization of the Galilean transformation in (1.35). Let $\{x^\mu\}$ be the Cartesian coordinates of the first inertial reference frame and let $\{x'^\mu\}$ be the Cartesian coordinates of the second inertial reference frame moving with constant velocity $\mathbf{v} = (v, 0, 0)$ with respect to the former. If the two Cartesian coordinate systems coincide at the time $t = t' = 0$, we have

$$x^\mu \to x'^\mu = \Lambda_x{}^\mu{}_\nu x^\nu, \tag{2.40}$$

as follows by the simple integration of Eq. (2.15) and where $\Lambda_x{}^\mu{}_\nu$ is given by Eq. (2.28). The extension to the case in which the two reference frames have a different relative velocity is straightforward.

Let us now consider the case of length measurements. We have again two inertial reference systems, the first has Cartesian coordinates (ct, x, y, z) and the second, moving with constant velocity $\mathbf{v} = (v, 0, 0)$ with respect to the former, has Cartesian coordinates (ct', x', y', z'). We have a rod at rest in the reference frame (ct', x', y', z') and parallel to the x'-axis. If the two ends of the rod have coordinates x'_1 and x'_2, the *proper length* of the rod, i.e. the length in the rest frame of the rod, is $l_0 = x'_2 - x'_1$. From the coordinate transform in Eq. (2.40), we find that the two ends of the rod have coordinates x'_1 and x'_2

$$x'_1 = -\gamma v t + \gamma x_1, \quad x'_2 = -\gamma v t + \gamma x_2. \tag{2.41}$$

Since the observer with Cartesian coordinates (ct, x, y, z) must measure the length of the rod by measuring the two ends at the same time t, we have

$$l_0 = x'_2 - x'_1 = \gamma (x_2 - x_1) = \gamma l, \tag{2.42}$$

where l is the length measured in the system (ct, x, y, z). We thus see that the proper length is larger than the length measured in another reference frame

$$l = \frac{l_0}{\gamma}. \tag{2.43}$$

This phenomenon is called *Lorentz contraction*. If we consider a volume and the fact that $y = y'$ and $z = z'$, we find the relation between the proper volume V_0 and the volume in another reference frame

$$V = \frac{V_0}{\gamma}. \tag{2.44}$$

Let us now consider the trajectory of a particle. The Cartesian coordinates of the particle in the first inertial reference frame are (ct, x, y, z). In the second inertial reference frame, the Cartesian coordinates of the particle are (ct', x', y', z'). The infinitesimal displacements of the particle in the two coordinate systems are related by the following equations

$$
\begin{aligned}
dt' &= \gamma dt - \frac{\gamma v dx}{c^2}, \\
dx' &= -\gamma v dt + \gamma dx, \\
dy' &= dy, \\
dz' &= dz,
\end{aligned}
\tag{2.45}
$$

The velocity of the particle in reference frames (ct, x, y, z) and (ct', x', y', z') is, respectively,

$$\mathbf{w} = (dx/dt, dy/dt, dz/dt), \quad \mathbf{w}' = (dx'/dt', dy'/dt', dz'/dt'). \tag{2.46}$$

We thus have

$$\dot{x}' = \frac{dx'}{dt'} = \frac{-\gamma v dt + \gamma dx}{\gamma dt - \gamma v dx/c^2} = \frac{-v + \dot{x}}{1 - v\dot{x}/c^2},$$

$$\dot{y}' = \frac{dy'}{dt'} = \frac{dy}{\gamma dt - \gamma v dx/c^2} = \frac{\dot{y}}{\gamma \left(1 - v\dot{x}/c^2\right)},$$

$$\dot{z}' = \frac{dz'}{dt'} = \frac{dz}{\gamma dt - \gamma v dx/c^2} = \frac{\dot{z}}{\gamma \left(1 - v\dot{x}/c^2\right)}. \tag{2.47}$$

Let us consider a particle moving at the speed of light in the inertial reference frame (ct, x, y, z). Without loss of generality, we can rotate the reference frame to have the motion of the particle along the x axis. We thus have $\dot{x} = c$ and $\dot{y} = \dot{z} = 0$. In the inertial reference frame (ct', x', y', z') moving with constant velocity $\mathbf{v} = (v, 0, 0)$ with respect to the former, we have

$$\dot{x}' = \frac{-v + c}{1 - v/c} = c, \quad \dot{y}' = 0, \quad \dot{z}' = 0. \tag{2.48}$$

If $\dot{x} = -c$ and $\dot{y} = \dot{z} = 0$, then $\dot{x}' = -c$. The particle moves at the speed of light in both reference frames, as is expected because ds^2 is an invariant. If $|\dot{x}| < c$, then $|\dot{x}'| < c$ as no particle can exceed the speed of light just because we are changing reference frame.

We now have a particle moving in the xy plane in the reference frame with coordinates (ct, x, y, z). Its velocity is $\dot{x} = w \cos \Theta$ and $\dot{y} = w \sin \Theta$, so the absolute velocity is w and Θ is the angle between the x axis and the particle velocity. In the reference frame with coordinates (ct', x', y', z') moving with constant velocity $\mathbf{v} = (v, 0, 0)$ with respect to the former, the particle has the velocity $\dot{x}' = w' \cos \Theta'$ and $\dot{y}' = w' \sin \Theta'$, where w' and Θ' are, respectively, the absolute velocity and the angle as measured in the reference frame with coordinates (ct', x', y', z'). The relation between Θ and Θ' can be obtained by calculating the ratio between $\dot{y}'$ and $\dot{x}'$ from Eq. (2.47)

$$\frac{\dot{y}'}{\dot{x}'} = \tan \Theta' = \frac{\dot{y}}{\gamma(\dot{x} - v)} = \frac{w \sin \Theta}{\gamma(w \cos \Theta - v)}. \tag{2.49}$$

In the case where the particle moves at the speed of light ($w = c$), we have

$$\tan \Theta' = \frac{\sin \Theta}{\gamma(\cos \Theta - v/c)}. \tag{2.50}$$

The fact that $\Theta \neq \Theta'$ is the phenomenon known as *aberration of light*.

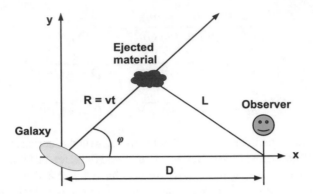

Fig. 2.2 The distance between the observer and the galaxy is D. At time $t = 0$, some material is ejected with velocity v from the center of the galaxy, and thus the distance between the ejected material and the center of the galaxy at t is $R = vt$. The radiation emitted by the ejected material at t and detected by the observer at time t' travels a distance L. φ is the angle between the trajectory of the ejected material and the line of sight of the galaxy with respect to the distant observer

2.5.1 Superluminal Motion

Superluminal motion refers to the phenomenon of observation of some material ejected by certain galactic nuclei that apparently moves faster than the speed of light. Actually there is no violation of the Einstein Principle of Relativity. As illustrated in Fig. 2.2, we have an observer at the distance D from a galaxy. Relativistic material is ejected at time $t = 0$ from the galactic nucleus with velocity v and φ is the angle between the trajectory of the ejected material and the line of sight of the observer. At the time t, the ejected material is at a distance $R = vt$ from the galactic nucleus, and emits some radiation. The latter travels a distance L and reaches the distant observer at time t'.

Considering that $R \ll D$, we can calculate the distance L between the position of the ejected material at time t and the observer

$$L = \sqrt{(D - vt\cos\varphi)^2 + v^2 t^2 \sin^2\varphi} = \sqrt{D^2 - 2Dvt\cos\varphi + v^2 t^2}$$
$$= D - vt\cos\varphi + O\left(\frac{v^2 t^2}{D^2}\right). \tag{2.51}$$

Since the radiation emitted by the ejected material moves at the speed of light c, we also have

$$L = c\left(t' - t\right). \tag{2.52}$$

Combining Eqs. (2.51) and (2.52), we can write

$$t' = \frac{D}{c} + t\,(1 - \beta \cos\varphi)\,, \tag{2.53}$$

where $\beta = v/c$. The apparent velocity of the ejected material along the y axis as measured by the distant observer is

$$v'_y = \frac{dy}{dt'} = \frac{dy}{dt}\frac{dt}{dt'} = \frac{v\sin\varphi}{1 - \beta\cos\varphi}\,. \tag{2.54}$$

The angle φ_{max} for which we have the highest apparent velocity v'_y is given by

$$\frac{\partial v'_y}{\partial\varphi} = 0\,,$$

$$v\cos\varphi_{max}\,(1 - \beta\cos\varphi_{max}) - v\beta\sin^2\varphi_{max} = 0\,,$$

$$v\cos\varphi_{max} - v\beta = 0\,,$$

$$\cos\varphi_{max} = \beta\,. \tag{2.55}$$

$\sin\varphi_{max}$ is

$$\sin\varphi_{max} = \sqrt{1 - \beta^2} = \frac{1}{\gamma}\,. \tag{2.56}$$

If we plug $\cos\varphi_{max}$ and $\sin\varphi_{max}$ into Eq. (2.54), we find

$$v'_y = v\gamma\,, \tag{2.57}$$

and we can see that v'_y can easily exceed c for relativistic matter even if there is no violation of the Einstein Principle of Relativity.

2.6 Example: Cosmic Ray Muons

Primary cosmic rays are high-energy particles (mostly protons and helium-4 nuclei) created outside of Earth's atmosphere and mainly outside of the Solar System. When they enter Earth's atmosphere, they collide with air molecules and produce showers of particles. The latter are called secondary cosmic rays as they are created in Earth's atmosphere. Some cosmic rays can reach the surface of Earth.

Muons (μ^-) and anti-muons (μ^+) are secondary products, mainly created by the decay of pions around 15 km above the surface of Earth. Muons/anti-muons are highly penetrating particles able to reach the ground. However, they are unstable and decay into electrons/positrons (e^-/e^+), electron anti-neutrinos/neutrinos ($\bar{\nu}_e/\nu_e$), and muon neutrinos/anti-neutrinos ($\nu_\mu/\bar{\nu}_\mu$):

$$\mu^- \to e^- + \bar{\nu}_e + \nu_\mu ,$$
$$\mu^+ \to e^+ + \nu_e + \bar{\nu}_\mu . \tag{2.58}$$

The mean lifetime of muons/anti-muons is $\tau_\mu \approx 2.2\,\mu s$.

Let us consider the cosmic ray muons traveling near normal to Earth's surface. Their average velocity is roughly $v_\mu = 0.995\,c$ and the flux at 15 km is $\Phi_0 \sim 0.1\,\text{muons/s/sr/cm}^2$. For simplicity, we can assume that the muon flux only decreases due to muon decay (i.e. we neglect the possibility of collisions producing other particles). In Newtonian mechanics, the muon flux at sea level should be

$$\Phi = \Phi_0 e^{-\Delta t/\tau_\mu} \sim 10^{-11}\,\text{muons/s/sr/cm}^2 , \tag{2.59}$$

where $\Delta t = 15\,\text{km}/v_\mu \approx 50\,\mu s$ is the travel time of muons to reach the surface of Earth as measured by an observer on Earth, $\Delta t/\tau_\mu \approx 23$, and $e^{-\Delta t/\tau_\mu} \sim 10^{-10}$. This is not what we observe. The flux of muons traveling near normal to Earth's surface at sea level is $\Phi \sim 0.01\,\text{muons/s/sr/cm}^2$, in agreement with the predictions of special relativity.

For $v_\mu = 0.995\,c$, the Lorentz factor is $\gamma = 10$. In the rest frame of the muon, the time the particle takes to reach the surface of Earth is $\Delta t' = \Delta t/\gamma \approx 5\,\mu s$. Now $e^{-\Delta t'/\tau_\mu} \sim 0.1$ and we recover the measured flux $\Phi \sim 0.01\,\text{muons/s/sr/cm}^2$.

Alternatively, we can interpret the phenomenon as Lorentz contraction of the travel distance. In the rest-frame of the muon, the Earth is moving at velocity $v = 0.995\,c$ and has Lorentz factor $\gamma = 10$. The distance of 15 km becomes $L = 15\,\text{km}/\gamma = 1.5\,\text{km}$. The muon takes about $5\,\mu s$ to travel such a distance and, again, we find that the flux is only decreased by an order of magnitude.

Problems

2.1 The energy-momentum tensor of a perfect fluid in Cartesian coordinates and in the rest-frame of the fluid has the following form

$$||T^{\mu\nu}|| = \begin{pmatrix} \varepsilon & 0 & 0 & 0 \\ 0 & P & 0 & 0 \\ 0 & 0 & P & 0 \\ 0 & 0 & 0 & P \end{pmatrix} , \tag{2.60}$$

where ε and P are, respectively, the energy density and the pressure of the fluid. Write $T^{\mu\nu}$, $T^\mu{}_\nu$, and $T_{\mu\nu}$ in spherical coordinates.

2.2 Consider the coordinate transformation $x^\mu \to x'^\mu$ in Eq. (2.28). Write the energy-momentum tensor of a perfect fluid in Eq. (2.60) in the new reference frame moving with velocity v along the x axis.

2.3 Let us consider three inertial reference frames with, respectively, Cartesian coordinates (ct, x, y, z), (ct', x', y', z'), and (ct'', x'', y'', z''). The reference frame (ct', x', y', z') moves with the velocity $\mathbf{v} = (v, 0, 0)$ with respect to the reference frame (ct, x, y, z), and the reference frame (ct'', x'', y'', z'') moves with the velocity $\mathbf{v}' = (v', 0, 0)$ with respect to the reference frame (ct', x', y', z'). For $t = t' = t'' = 0$ the three reference frames coincide. Write the Lorentz boost connecting the reference frames (ct, x, y, z) and (ct'', x'', y'', z'').

2.4 Show that Lorentz transformations do not commute in general.

2.5 The GPS Navigation System consists of a network of satellites in high orbits around Earth. Each satellite has an orbital speed of about 14,000 km/hour. What is the relation between the time measured by a clock on one of these satellites and by a clock on Earth due to the orbital motion of the satellite?[6]

---

[6] As we will see in Sect. 6.6, there is also a contribution of opposite sign due to the difference in the gravitational field between the two points.

Chapter 3
Relativistic Mechanics

Relativistic mechanics is the mechanics based on the Einstein Principle of Relativity. It reduces to Newtonian mechanics in the limit $c \to \infty$.

3.1 Action for a Free Particle

As we have seen in Sect. 1.7, in Newtonian mechanics the equations of motion of a free point-like particle can be obtained by either requiring that the Lagrangian is the kinetic energy of the particle, Eq. (1.70),

$$L = \frac{1}{2} m g_{ij} \dot{x}^i \dot{x}^j \, , \qquad (3.1)$$

or imposing that the trajectory of the particle is the one that makes its length stationary, Eq. (1.75),

$$\ell = \int_\Gamma dl \, . \qquad (3.2)$$

In this section we want to find the relativistic generalization for the case of massive particles. Massless particles will be discussed in Sect. 3.3.

As we have already pointed out in Sect. 1.5, there is no fundamental rule or standard recipe to write the Lagrangian of a physical system. One can construct the Lagrangian on the basis of the symmetries of the system and other physical requirements, but observations are always the final arbiter: eventually it is necessary to compare the predictions of the theoretical model with experimental data.

© Springer Nature Singapore Pte Ltd. 2018
C. Bambi, *Introduction to General Relativity*, Undergraduate Lecture Notes in Physics, https://doi.org/10.1007/978-981-13-1090-4_3

The natural relativistic generalization of Eq. (3.1) is

$$L = \frac{1}{2}mg_{\mu\nu}\dot{x}^{\mu}\dot{x}^{\nu} .$$ (3.3)

However, in general it is not convenient to parametrize the particle trajectory with the coordinate time t, i.e. $x^{\mu} = x^{\mu}(t)$. In the case of massive particles, we can always choose the proper time τ, so $x^{\mu} = x^{\mu}(\tau)$. By definition of proper time

$$c^2 d\tau^2 = -g_{\mu\nu}dx^{\mu}dx^{\nu} ,$$ (3.4)

and therefore

$$g_{\mu\nu}\dot{x}^{\mu}\dot{x}^{\nu} = -c^2 .$$ (3.5)

$\dot{x}^{\mu} = u^{\mu} = dx^{\mu}/d\tau$ is the particle *4-velocity*. If we write the Euler-Lagrange equations for the Lagrangian in (3.3), we find the geodesic equations

$$\ddot{x}^{\mu} + \Gamma^{\mu}_{\nu\rho}\dot{x}^{\nu}\dot{x}^{\rho} = 0 ,$$ (3.6)

where $\Gamma^{\mu}_{\nu\rho}$s are the Christoffel symbols

$$\Gamma^{\mu}_{\nu\rho} = \frac{1}{2}g^{\mu\sigma}\left(\frac{\partial g_{\sigma\rho}}{\partial x^{\nu}} + \frac{\partial g_{\nu\sigma}}{\partial x^{\rho}} - \frac{\partial g_{\nu\rho}}{\partial x^{\sigma}}\right) .$$ (3.7)

Note that the only difference between these geodesic equations and those in Sect. 1.7 is that now we have spacetime indices $(\mu, \nu, \rho, \ldots)$ running from 0 to n instead of space indices $(i, j, k, \ldots)$ running from 1 to n.

Let us now find the counterpart of Eq. (3.2). Its natural generalization is[1]

$$\ell = \int_{\Gamma} \sqrt{-ds^2} .$$ (3.9)

Now Γ is a curve in the $(n+1)$-dimensional spacetime, the initial and the final points are two spacetime events, and ds is the spacetime line element. When we apply the Least Action Principle, we obtain the counterpart of Eq. (1.76). Since we are

---

[1]The minus sign in front of ds^2 is because we are using a metric with signature $(-+++)$ and therefore time-like trajectories have $ds^2 < 0$. With a metric with signature $(+---)$, as is the common convention among the particle physics community, we could write Eq. (3.9) as

$$\ell = \int_{\Gamma} ds .$$ (3.8)

It is clear that the convention of the signature of the metric does not change the solution of the particle trajectory.

parametrizing the particle's trajectory with the proper time τ, $L = -mc^2/2$ is a constant, $dL/d\tau = 0$, and we recover the geodesic equation in its standard form.

Let us now parametrize the particle trajectory with the time coordinate t. From Eq. (3.9), the action of the free particle should have the following form

$$S = K \int_\Gamma \sqrt{-g_{\mu\nu}\dot{x}^\mu\dot{x}^\nu}\, dt \,, \qquad (3.10)$$

where K is some constant, $\dot{x}^\mu = dx^\mu/dt$, and t is the time. It is easy to see that $K = -mc$ in order to recover the correct Newtonian limit, where m is the particle mass and c is the speed of light. With $K = -mc$, the Lagrangian is

$$\begin{aligned}
L &= -mc\sqrt{-g_{\mu\nu}\dot{x}^\mu\dot{x}^\nu} = -mc\sqrt{c^2 - \dot{\mathbf{x}}^2} = -mc^2\sqrt{1 - \frac{\dot{\mathbf{x}}^2}{c^2}} \\
&= -mc^2\left[1 - \frac{1}{2}\frac{\dot{\mathbf{x}}^2}{c^2} + O\left(\frac{\dot{\mathbf{x}}^4}{c^4}\right)\right] \\
&= -mc^2 + \frac{1}{2}m\dot{\mathbf{x}}^2 \,. \qquad (3.11)
\end{aligned}$$

Remember that $x^0 = ct$, see Eq. (2.2), and therefore $\dot{x}^0 = c$. The term $-mc^2$ is a constant and therefore does not play any role in the equations of motion. The second term is the Newtonian kinetic energy and then we have higher orders terms in $\dot{\mathbf{x}}^2/c^2$.

In the next sections of the present chapter, we will derive a number of predictions from these Lagrangians that turn out to be consistent with experiments and thus support these Lagrangians as the correct model.

3.2 Momentum and Energy

Now we want to derive some basic physical properties from the Lagrangians in Eqs. (3.3) and (3.10). For simplicity, we assume three space dimensions, but all results can be easily generalized to a number n of space dimensions. There are two slightly different approaches. We can write the Lagrangian and the action as if we were in Newtonian mechanics, where time and space are two different entities (3-dimensional formalism). In this case we work with an effective Lagrangian in Newtonian mechanics. Alternatively, we treat the spacetime coordinates all in the same way and we have the proper time as our "time" (4-dimensional formalism).

3.2.1 3-Dimensional Formalism

We employ Cartesian coordinates and we have the Minkowski metric $\eta_{\mu\nu}$. The Lagrangian in Eq. (3.10) becomes

$$L = -mc\sqrt{c^2 - \dot{x}^2 - \dot{y}^2 - \dot{z}^2} = -mc\sqrt{c^2 - \dot{\mathbf{x}}^2}\,. \qquad (3.12)$$

Here the dot ˙ stands for the derivative with respect to the time t. The 3-momentum of the particle is[2]

$$\mathbf{p} = \frac{\partial L}{\partial \dot{\mathbf{x}}} = \frac{mc\dot{\mathbf{x}}}{\sqrt{c^2 - \dot{\mathbf{x}}^2}}\,, \qquad (3.15)$$

which can be rewritten in the more compact form

$$\mathbf{p} = m\gamma\dot{\mathbf{x}}\,, \qquad (3.16)$$

where γ is the Lorentz factor of the particle. As we can see from Eq. (3.16), the 3-momentum is the momentum vector of Newtonian mechanics multiplied by γ. As we will see in Eq. (3.32), the 3-momentum is the spatial part of the 4-momentum of the particle.

The equations of motion are obtained from the 3-dimensional Euler-Lagrange equations, where the time t is not a Lagrangian coordinate but the temporal coordinate of the system

$$\frac{d}{dt}\frac{\partial L}{\partial \dot{\mathbf{x}}} - \frac{\partial L}{\partial \mathbf{x}} = 0\,. \qquad (3.17)$$

The equations of motion are

$$\frac{d\mathbf{p}}{dt} = 0\,. \qquad (3.18)$$

With the component notation, Eqs. (3.17) and (3.18) are, respectively,

$$\frac{d}{dt}\frac{\partial L}{\partial \dot{x}^i} - \frac{\partial L}{\partial x^i} = 0\,, \quad \frac{dp_i}{dt} = 0\,. \qquad (3.19)$$

The solution is that the particle moves at a constant speed along a straight line, as in Newtonian mechanics.

The energy of the particle (the Hamiltonian) is obtained from the Legendre transform of the Lagrangian

---

[2]Note that we are using Cartesian coordinates. In non-Cartesian coordinates, we have to be more careful. The conjugate momentum is

$$\mathbf{p}^* = \frac{\partial L}{\partial \dot{\mathbf{x}}}\,, \qquad (3.13)$$

or, with the component notation,

$$p_i = \frac{\partial L}{\partial \dot{x}^i}\,. \qquad (3.14)$$

The 3-momentum is $p^i = g^{ij} p_j$. In Cartesian coordinates, $g^{ij} = \delta^{ij}$, and therefore $\mathbf{p}^* = \mathbf{p}$.

$$E = \frac{\partial L}{\partial \dot{\mathbf{x}}} \cdot \dot{\mathbf{x}} - L = \frac{mc^3}{\sqrt{c^2 - \dot{\mathbf{x}}^2}}. \tag{3.20}$$

Even in this case, we can use the Lorentz factor of the particle and write

$$E = m\gamma c^2. \tag{3.21}$$

We thus see that the particle energy diverges when its velocity approaches the speed of light. Massive particles can never reach the speed of light because this would require providing them with infinite energy.

3.2.2 4-Dimensional Formalism

Let us now consider the Lagrangian in Eq. (3.3). Now the Lagrangian coordinates are $\{x^\mu(\tau)\}$, i.e. the temporal coordinate is also a Lagrangian coordinate, and the particle trajectory is parametrized by the particle proper time τ. The action now reads

$$S = \frac{1}{2}m \int_\Gamma g_{\mu\nu} \dot{x}^\mu \dot{x}^\nu \, d\tau, \tag{3.22}$$

where $\dot{x}^\mu = dx^\mu/d\tau$.

The conjugate momentum of the particle is

$$p_\mu = \frac{\partial L}{\partial \dot{x}^\mu} = m g_{\mu\nu} \dot{x}^\nu = m\dot{x}_\mu. \tag{3.23}$$

The *4-momentum* is $p^\mu = g^{\mu\nu} p_\nu$. From Eq. (3.5), we have the conservation of the particle mass

$$p^\mu p_\mu = -m^2 c^2. \tag{3.24}$$

If we employ Cartesian coordinates, $g_{\mu\nu} = \eta_{\mu\nu}$, and the Euler-Lagrange equations for the massive particle read

$$\frac{dp_\mu}{d\tau} = 0. \tag{3.25}$$

Note that the normalization of the 4-velocity

$$\eta_{\mu\nu} \dot{x}^\mu \dot{x}^\nu = -c^2 \dot{t}^2 + \dot{x}^2 + \dot{y}^2 + \dot{z}^2 = -c^2, \tag{3.26}$$

can be rewritten as

$$\dot{t}^2 \left[c^2 - \left(\frac{dx}{dt} \right)^2 - \left(\frac{dy}{dt} \right)^2 - \left(\frac{dz}{dt} \right)^2 \right] = c^2 , \tag{3.27}$$

and therefore

$$\dot{t} = \frac{c}{\sqrt{c^2 - \mathbf{v}^2}} = \gamma , \tag{3.28}$$

where $\mathbf{v} = d\mathbf{x}/dt$ is the particle 3-velocity. The x component of the 4-velocity is

$$\dot{x} = \frac{dx}{d\tau} = \frac{dx}{dt} \frac{dt}{d\tau} = \gamma v_x , \tag{3.29}$$

where v_x is the x component of the particle 3-velocity and there are similar expressions for the y and z components, namely $\dot{y} = \gamma v_y$ and $\dot{z} = \gamma v_z$. The spatial components of the 4-velocity are thus the velocity vector of Newtonian mechanics multiplied by γ. The particle 4-velocity is given by (remember that $x^0 = ct$)

$$\dot{x}^\mu = (\gamma c, \gamma \mathbf{v}) . \tag{3.30}$$

The components of the conjugate momentum are

$$p_t = -m\gamma c , \quad p_x = m\gamma v_x , \quad p_y = m\gamma v_y , \quad p_z = m\gamma v_z . \tag{3.31}$$

Note that p_t is negative because $p_t = m\dot{x}_t = m\eta_{t\mu}\dot{x}^\mu = -mc\dot{t}$.

Comparing Eq. (3.31) with Eqs. (3.16) and (3.21), we see that $p^t c$ is the particle energy and p^x, p^y, and p^z are, respectively, the x, y, and z components of the particle 3-momentum. We can thus write

$$p^\mu = \left(\frac{E}{c}, \mathbf{p} \right) = (m\gamma c, m\gamma \mathbf{v}) . \tag{3.32}$$

If we write the 4-momentum with lower indices, we have $p_\mu = (-E/c, \mathbf{p})$. The conservation of the particle mass in Eq. (3.24) becomes

$$E^2 = m^2 c^4 + \mathbf{p}^2 c^2 . \tag{3.33}$$

If a particle is at rest, $\mathbf{p} = \mathbf{0}$, and we find the very celebrated formula

$$E = mc^2 . \tag{3.34}$$

From Eq. (3.25) we see that the system has four constants of motion, the four components of p_μ.

3.3 Massless Particles

Generally speaking, the results in Sects. 3.1 and 3.2 can be extended to massless particles by considering the limit $m/E \to 0$. Indeed every particle can be regarded as a massless particle when its mass is much smaller than its energy. However, some concepts are not defined. For $m/E \to 0$ we have

$$\frac{E}{m} = \gamma c^2 \to \infty \,, \tag{3.35}$$

that is, massless particles moves at the speed of light and therefore we cannot define their Lorentz factor. $m\gamma c^2$ is well defined as being equal to the energy and therefore we can replace this quantity with E in the expressions in the previous sections. The conservation of the particle mass (3.24) becomes

$$p^\mu p_\mu = 0 \,, \tag{3.36}$$

or, equivalently,

$$E^2 = \mathbf{p}^2 c^2 \,. \tag{3.37}$$

From the consideration above, it is clear that the Euler-Lagrange equations provide the geodesic equations for massless particles as well, because the mass m appears just as a constant in front of the Lagrangian. If we employ the Lagrangian in (3.3), the trajectory of the particle cannot be parametrized with the particle proper time because the latter is not defined for massless particles (see Sect. 2.4). However, whatever the parameter of the trajectory is, we can write the Euler-Lagrange equations and get the geodesic equations.

If we want to find the equations of motion by minimizing/maximizing the length of the particle trajectory, the Lagrangian is

$$L = -mc\sqrt{-g_{\mu\nu}\dot{x}^\mu \dot{x}^\nu} \,, \tag{3.38}$$

where m has to be interpreted as some constant with the dimensions of a mass. In general, we do not obtain the geodesic equations in the form in Eq. (3.6), but as in (1.76). The geodesic equations in (3.6) are obtained when the trajectory is parametrized with an *affine parameter* λ, i.e. $x^\mu = x^\mu(\lambda)$. If we consider a different parametrization, say $\lambda \to \lambda' = \lambda'(\lambda)$, the geodesic equations become

$$\left(\ddot{x}^\mu + \Gamma^\mu_{\nu\rho}\dot{x}^\nu \dot{x}^\rho\right)\left(\frac{d\lambda'}{d\lambda}\right)^2 = -\dot{x}^\mu \frac{d^2\lambda'}{d\lambda^2} \tag{3.39}$$

The physical solution of these equations is the same, because it is independent of the parametrization, but the form of the equations is slightly different because there is the extra term on the right hand side of Eq. (3.39). Such an extra term can always be removed by a suitable reparametrization, which is the case when we employ an affine parameter. Affine parameters are related to each other by relations of the form

$$\lambda' = a\lambda + b,\tag{3.40}$$

where a and b are constants. In such a case $d^2\lambda'/d\lambda^2 = 0$.

3.4 Particle Collisions

In Sects. 3.2 and 3.3 we have inferred some fundamental predictions of our Lagrangian for relativistic free point-like particles. Now we want to study possible applications of those results.

An important example is the study of collisions of particles like protons, electrons, etc., where the velocities can indeed be very high and Newtonian mechanics does not work. For instance, a typical problem in particle physics is to find the *threshold energy* for a certain reaction. We have a target particle A with mass m_A and we fire a particle B with mass m_B. We want to know the minimum energy necessary for particle B to produce particles C and D, respectively with masses m_C and m_D. The reaction is

$$A + B \rightarrow C + D.\tag{3.41}$$

In Cartesian coordinates,[3] the 4-momentum before the collision must be equal to the 4-momentum after the collision or, in other words, the total 4-momentum of the initial state is the same as the total 4-momentum of the final state

$$p_i^\mu = p_f^\mu,\tag{3.42}$$

where $p_i^\mu = p_A^\mu + p_B^\mu$, $p_f^\mu = p_C^\mu + p_D^\mu$, and p_X^μ is the 4-momentum of the particle $X = A$, B, C, and D. Note that in Eq. (3.42) we are evaluating p_i^μ and p_f^μ in the same reference frame.

Let us consider the reference frame of the laboratory, where particle A is at rest. For simplicity and without loss of generality, we assume that particle B moves along the x axis. We have

$$||p_A^\mu|| = (m_A c, 0, 0, 0), \quad ||p_B^\mu|| = (m_B \gamma c, m_B \gamma v, 0, 0),\tag{3.43}$$

where v is the velocity of particle B in the reference frame of the laboratory.

---

[3]In a different coordinate system, the total 4-momentum may not be a conserved quantity. Remember, for instance, that p^μs and p_μs are not all constants of motion for a free particle in spherical coordinates.

Let us now consider the reference frame in which the total 3-momentum of the system vanishes, namely $\mathbf{p}_i = \mathbf{p}_f = \mathbf{0}$. The 4-momenta of particles C and D are

$$||p_C^\mu|| = \left(\sqrt{m_C^2 c^2 + \mathbf{p}_C^2}, \mathbf{p}_C\right), \quad ||p_D^\mu|| = \left(\sqrt{m_D^2 c^2 + \mathbf{p}_C^2}, -\mathbf{p}_C\right), \quad (3.44)$$

where $\mathbf{p}_C$ is the 3-momentum of particle C and therefore $-\mathbf{p}_C$ is that of particle D because we are in the rest frame of the system. The minimum energy of the system is when $\mathbf{p}_C = \mathbf{0}$ and particles C and D are created at rest

$$||p_C^\mu|| = (m_C c, 0, 0, 0), \quad ||p_D^\mu|| = (m_D c, 0, 0, 0). \quad (3.45)$$

The 4-momenta in (3.43) are evaluated in the rest frame of the laboratory, while those in (3.45) are in the rest frame of the system. They cannot be directly compared.

The *invariant mass* M of a system of particles with total 4-momentum p_{tot}^μ is defined as

$$-M^2 c^2 = p_{\text{tot}}^\mu p_\mu^{\text{tot}}, \quad (3.46)$$

and is a scalar by construction, i.e. it does not depend on the choice of the reference frame. For the reaction in (3.41) we can thus write

$$p_\mu^i p^{i\mu} = -M^2 c^2 = p_\mu^f p^{f\mu}, \quad (3.47)$$

and evaluate the left hand side in the reference frame of the laboratory and the right hand side in the rest frame of the system. Eq. (3.47) becomes

$$(m_A c + m_B \gamma c)^2 - m_B^2 \gamma^2 v^2 = (m_C c + m_D c)^2,$$
$$m_A^2 c^2 + m_B^2 \gamma^2 c^2 + 2m_A m_B \gamma c^2 - m_B^2 \gamma^2 v^2 = m_C^2 c^2 + m_D^2 c^2 + 2m_C m_D c^2. \quad (3.48)$$

Since $m_B^2 \gamma^2 c^2 - m_B^2 \gamma^2 v^2 = m_B^2 c^2$, we have

$$2m_A m_B \gamma = m_C^2 + m_D^2 + 2m_C m_D - m_A^2 - m_B^2. \quad (3.49)$$

The threshold energy of particle B is

$$E_{\text{th}}^B = m_B \gamma c^2 = \frac{\left(m_C^2 + m_D^2 + 2m_C m_D - m_A^2 - m_B^2\right) c^2}{2m_A}. \quad (3.50)$$

3.5 Example: Colliders Versus Fixed-Target Accelerators

Let us consider the production of a Z^0-boson from the collision of a positron and an electron

$$e^+ + e^- \to Z^0 . \tag{3.51}$$

The mass of the Z^0-boson is $m_Z c^2 = 91$ GeV, while those of the positron and electron are $m_{e^+} c^2 = m_{e^-} c^2 = 0.5$ MeV.

In a particle collider, we can make the positron and electron collide head-on, so their 4-momenta are (assuming they are moving along the x axis)

$$||p_{e^+}^\mu|| = (E/c, p, 0, 0) , \quad ||p_{e^-}^\mu|| = (E/c, -p, 0, 0) , \tag{3.52}$$

where $E = \sqrt{m_e^2 c^4 + p^2 c^2}$ is their energy. The threshold energy for this reaction is

$$-\left(p_{e^+}^\mu + p_{e^-}^\mu\right)\left(p_\mu^{e^+} + p_\mu^{e^-}\right) = \frac{4E_{th}^2}{c^2} = m_Z^2 c^2 , \tag{3.53}$$

and therefore $E_{th} = 45.5$ GeV.

Let us now assume that we want to have the same reaction in a fixed-targed accelerator, namely we accelerate the positron and we fire it against an electron which is at rest in the reference frame of the laboratory. Now the 4-momenta of the two particles is (assuming the positron moves along the x axis)

$$||p_{e^+}^\mu|| = (E/c, p, 0, 0) , \quad ||p_{e^-}^\mu|| = (m_e c, 0, 0, 0) . \tag{3.54}$$

where $E = \sqrt{m_e^2 c^4 + p^2 c^2}$ is now the energy of the positron only. The mass invariant is

$$\frac{E^2}{c^2} + 2m_e E + m_e^2 c^2 - p^2 = 2m_e^2 c^2 + 2m_e E \approx 2m_e E , \tag{3.55}$$

and the threshold energy is now

$$2m_e E_{th} = m_Z^2 c^2 \quad \Rightarrow \quad E_{th} = \frac{m_Z^2 c^2}{2m_e} = 8,000 \text{ TeV} . \tag{3.56}$$

We can thus see that particle colliders are much more efficient than fixed-target accelerators. In the former case, we have to accelerate the two particles to 45.5 GeV to create a Z^0-boson. With a fixed-target experiment, we should accelerate one of the two particles to 8000 TeV to have the same reaction. Big particle physics accelerators are indeed colliders and not fixed-target accelerators.

3.6 Example: The GZK Cut-Off

Cosmic rays of very high energy cannot travel for very long distances because they interact with the photons of the cosmic microwave background (CMB) and lose energy. The CMB is the leftover of the electromagnetic radiation of the primordial plasma. After recombination (the epoch in which electrons and protons first became bound to form hydrogen atoms about 400,000 yrs after the Big Bang), photons decoupled from matter (the photon cross section with neutral hydrogen is much smaller than the Thomson cross section with free electrons), and photons started traveling freely through the Universe without interacting with matter. These are the CMB photons observed today around us with a density of about 400 photons/cm^3.

Protons represent about 90% of the cosmic rays. The Universe becomes opaque for them when the following reaction becomes energetically allowed

$$p + \gamma_{CMB} \rightarrow \Delta^+ \rightarrow N + \pi \,, \tag{3.57}$$

where γ_{CMB} is a CMB photon, Δ^+ is the resonant intermediate state, and the final product is either a proton and neutral pion $(p + \pi^0)$ or a neutron and a charged pion $(n + \pi^+)$.

Let us consider the cosmic reference frame, where the CMB is isotropic and which is not too different from our reference frame (Earth is moving with a velocity of about 370 km/s with respect to the cosmic reference frame). Without loss of generality, we assume that the CMB photon moves along the x axis and the high-energy proton moves in the xy plane. Their 4-momenta are

$$||p_p^\mu|| = (\sqrt{m_p^2 c^2 + p^2},\, p \cos\theta,\, p \sin\theta, 0)\,, \quad ||p_\gamma^\mu|| = (q, q, 0, 0)\,, \tag{3.58}$$

where θ is the angle between the proton propagation direction and the x axis. The total 4-momentum is

$$p_i^\mu = p_p^\mu + p_\gamma^\mu \,, \tag{3.59}$$

and the reaction (3.57) is energetically allowed when the invariant mass is not lower than the minimum invariant mass of the final state, which happens when the two final particles are created at rest with vanishing 3-momenta,

$$- p_i^\mu p_\mu^i \geq (m_N c + m_\pi c)^2 \,. \tag{3.60}$$

Plugging the expressions in Eq. (3.58) for p_p^μ and p_γ^μ, we find

$$m_p^2 c^2 + p^2 + q^2 + 2q\sqrt{m_p^2 c^2 + p^2} - p^2 \cos^2\theta - q^2$$
$$-2qp\cos\theta - p^2 \sin^2\theta \geq m_N^2 c^2 + m_\pi^2 c^2 + 2m_N m_\pi c^2 \,,$$
$$m_p^2 c^2 + 2q\left(\sqrt{m_p^2 c^2 + p^2} - p\cos\theta\right) \geq m_N^2 c^2 + m_\pi^2 c^2 + 2m_N m_\pi c^2 \,. \tag{3.61}$$

$E_p/c = \sqrt{m_p^2 c^2 + p^2}$, where E_p is the proton energy, and we can write

$$E_p - pc\cos\theta \geq \frac{m_N^2 c^3 + m_\pi^2 c^3 + 2m_N m_\pi c^3 - m_p^2 c^3}{2q} \approx \frac{m_p m_\pi c^3}{q}, \quad (3.62)$$

where in the last passage we have neglected the pion mass because it is much smaller than the proton/neutron mass as well as the difference between the masses of the proton and neutron ($m_p c^2 = 938$ MeV, $m_n c^2 = 940$ MeV, $m_{\pi^0} c^2 = 135$ MeV, $m_{\pi^+} c^2 = 140$ MeV).

The average energy of CMB photons is $\langle qc \rangle = 2 \cdot 10^{-4}$ eV. Ignoring the term $pc\cos\theta$ in Eq. (3.62), we find

$$E_p \sim 10^{20} \text{ eV}. \quad (3.63)$$

Protons with energies exceeding 10^{20} eV interact with the CMB photons and lose energy, so they cannot travel for long distances in the Universe. This is called the *Greisen-Zatsepin-Kuzmin cut-off* (GZK cut-off).

The cross section for the reaction (3.57) is $\sigma \sim 10^{-28}$ cm^{-2} and the number density of CMB photons in the Universe today is $n_{CMB} \sim 400$ photons/cm^3. The mean free path for high-energy protons, namely the average distance that a high-energy proton can travel before having the reaction (3.57), is

$$l = \frac{1}{\sigma n_{CMB}} \sim 10 \text{ Mpc}, \quad (3.64)$$

where we have used the relation 1 Mpc $\approx 3 \cdot 10^{24}$ cm. Assuming that a proton loses $\sim 20\%$ of its energy in every collision, one can estimate that almost all energy is lost in about 100 Mpc. So protons with an energy exceeding the GZK limit cannot travel for more than about 100 Mpc.

3.7 Multi-body Systems

In Sect. 3.2, we saw that the energy of a particle is $E = m\gamma c^2$, which is a well-defined positive quantity. This is in contrast to Newtonian mechanics, where the absolute value of the energy has no physical meaning because it cannot be measured and only energy differences are relevant.

The results found in the previous sections of this chapter were obtained for free point-like particles, but we never required that the particle had to be "elementary". Those results can thus be applied even to bound states like an atomic nucleus, which can be made of several protons and neutrons. In such a case, we see that the energy of the system at rest is $E = mc^2$, where m is the mass of the bound state, not the sum of the masses of its constituents.

If we consider the single constituents of a bound state, the total action should look like

$$S = \sum_i -mc \int_{\Gamma_i} \sqrt{-g_{\mu\nu}\dot{x}_i^\mu \dot{x}_i^\nu} \, dt + S_{\text{int}} \, , \tag{3.65}$$

where S_{int} is the contribution from the interactions between particles[4]

$$S_{\text{int}} = \sum_{i,j} \Phi(x_i, x_j) \, . \tag{3.66}$$

If we could turn the interactions off, we would have free particles and the total action would be the sum of the actions of the free particles. Taking particle interactions into account, we have to add the interaction terms.

The total energy of the system seen as a single body and as a bound state made of a number of more elementary constituents should be the same. The total energy of the bound state includes the rest energy of the single constituents, their kinetic energies, and all the interaction terms. The *binding energy* is defined as the difference between the sum of the rest energies of the single constituents and the rest energy of the composite body

$$E_{\text{B}} = \sum_i m_i - E \, . \tag{3.67}$$

For example, if we consider an α particle, namely a nucleus of helium-4 which is made of two protons and two neutrons, its mass is $m_\alpha c^2 = 3.727\,\text{GeV}$. The masses of a proton and of a neutron are, respectively, $m_p c^2 = 0.938\,\text{GeV}$ and $m_n c^2 = 0.940\,\text{GeV}$. The binding energy of a helium-4 nucleus is thus

$$E_{\text{B}} = (2 \cdot 0.938 + 2 \cdot 0.940 - 3.727) \,\text{GeV} = 29 \,\text{MeV} \, . \tag{3.68}$$

If we have two protons and two neutrons and we combine them to have a nucleus of helium-4, we get 29 MeV of energy from the reaction. For example, in the case of nuclear fusion in stars, we have something similar, and light nuclei can bound together to form heavier nuclei. The process releases energy. If we have a nucleus of helium-4 and we want to create two free protons and two free neutrons, we have to provide 29 MeV of energy.

3.8 Lagrangian Formalism for Fields

Up to now, we have considered systems whose action and Lagrangian had the form

---

[4]In Eq. (3.66) we assume that there are only two-body interactions, namely that interactions are only between two particles. In principle, there may exist even three-body and, more in general, multi-body interaction terms.

$$S = \int_{t_1}^{t_2} L dt , \quad L = L [\mathbf{q}, \dot{\mathbf{q}}, t] , \tag{3.69}$$

and the Lagrangian coordinates defining the state of the system had the form

$$\mathbf{q} = [q^1(t), q^2(t), \dots, q^n(t)] , \tag{3.70}$$

where t is the time coordinate. For example, in the case of a free point-like particle moving in a 3-dimensional space in Newtonian mechanics, the Lagrangian is the kinetic energy of the particle and the Lagrangian coordinates can be the Cartesian coordinates of the particle

$$\mathbf{q} = [x(t), y(t), z(t)] . \tag{3.71}$$

The Lagrangian formalism and the Least Action Principle can be extended to the description of fields (e.g. scalar fields, electromagnetic field, gravitational field, etc.). Now the variable describing the configuration of the field is a quantity (scalar, vector, or tensor) that has some value at every point of the spacetime. For example, if the system is described by r scalar fields, the Lagrangian coordinates will be like

$$\Phi = [\Phi^1(x), \Phi^2(x), \dots, \Phi^r(x)] , \tag{3.72}$$

where $x = (x^0, x^1, x^2, x^3)$ and therefore each Lagrangian coordinate, in general, will be a function of the point of the spacetime, not just of time. In the case of vector or tensor fields, the Lagrangian coordinates will also have spacetime indices.

The natural generalization of the action S in (3.69) is[5]

$$S = \frac{1}{c} \int \mathscr{L} d^4 \Omega , \tag{3.73}$$

where the integral is over the whole 4-dimensional volume. If we do not mix time and space coordinates, in special relativity we have $g_{0i} = 0$ (see the discussion in Sect. 2.2), and therefore $d^4 \Omega = c dt d^3 V$, where $d^3 V$ is the 3-dimensional volume. In Cartesian coordinates, $d^3 V = dx\, dy\, dz$. If we move from Cartesian to non-Cartesian coordinates, we have to calculate the determinant of the Jacobian, J, and $d^3 V = |J| d^3 x$.

In Eq. (3.73), $\mathscr{L}$ is the *Lagrangian density* of the system, which will depend on the Lagrangian coordinates and on their first derivative; that is,

$$\mathscr{L} = \mathscr{L} \left(\Phi^i, \frac{\partial \Phi^i}{\partial x^\mu} \right) . \tag{3.74}$$

---

[5]The factor $1/c$ in front of the integral is introduced because we define $d^4 \Omega = c dt d^3 V$, not $d^4 \Omega = dt d^3 V$.

Eq. (3.73) can also be rewritten as

$$S = \int L \, dt \quad \text{with} \quad L = \int \mathscr{L} \, d^3V \, . \tag{3.75}$$

Applying the Least Action Principle, the equations of motion (now called the *field equations*) should be obtained by varying the action S with respect to Φ

$$\Phi(x) \to \tilde{\Phi}(x) = \Phi(x) + \delta\Phi(x) \, , \tag{3.76}$$

where $\delta\Phi$ must vanish at the boundary of the integration region. We have

$$\delta S = \frac{1}{c} \int \left(\frac{\partial \mathscr{L}}{\partial \Phi^i} \delta\Phi^i + \frac{\partial \mathscr{L}}{\partial \frac{\partial \Phi^i}{\partial x^\mu}} \delta \frac{\partial \Phi^i}{\partial x^\mu} \right) d^4\Omega \, . \tag{3.77}$$

As we will see in Sect. 6.7, $d^4\Omega = \sqrt{-g} \, d^4x$ where g is the determinant of the metric tensor, and the second term on the right hand side can be written as

$$\frac{1}{c} \int \frac{\partial \mathscr{L}}{\partial \frac{\partial \Phi^i}{\partial x^\mu}} \delta \frac{\partial \Phi^i}{\partial x^\mu} \sqrt{-g} \, d^4x$$

$$= \frac{1}{c} \int \left\{ \frac{\partial}{\partial x^\mu} \left(\sqrt{-g} \frac{\partial \mathscr{L}}{\partial \frac{\partial \Phi^i}{\partial x^\mu}} \delta\Phi^i \right) - \left[\frac{\partial}{\partial x^\mu} \left(\sqrt{-g} \frac{\partial \mathscr{L}}{\partial \frac{\partial \Phi^i}{\partial x^\mu}} \right) \right] \delta\Phi^i \right\} d^4x \, . \tag{3.78}$$

Since $\delta\Phi$ vanishes at the boundary of the integration region, the first term in the second line does not provide any contribution to δS, so we can write

$$\delta S = \frac{1}{c} \int \left[\frac{\partial \mathscr{L}}{\partial \Phi^i} - \frac{1}{\sqrt{-g}} \frac{\partial}{\partial x^\mu} \left(\sqrt{-g} \frac{\partial \mathscr{L}}{\partial \frac{\partial \Phi^i}{\partial x^\mu}} \right) \right] \delta\Phi^i \, d^4\Omega \, . \tag{3.79}$$

Since $\delta S = 0$ for any arbitrary variation $\delta\Phi$, we have

$$\frac{1}{\sqrt{-g}} \frac{\partial}{\partial x^\mu} \left(\sqrt{-g} \frac{\partial \mathscr{L}}{\partial \frac{\partial \Phi^i}{\partial x^\mu}} \right) - \frac{\partial \mathscr{L}}{\partial \Phi^i} = 0 \, . \tag{3.80}$$

These are the field equations. It is sometimes more convenient to define the Lagrangian density as $\mathscr{L}' = \sqrt{-g}\mathscr{L}$. In such a case, the field equations are

$$\frac{\partial}{\partial x^\mu} \frac{\partial \mathscr{L}'}{\partial \frac{\partial \Phi^i}{\partial x^\mu}} - \frac{\partial \mathscr{L}'}{\partial \Phi^i} = 0 \, . \tag{3.81}$$

If we work in Cartesian coordinates, which is often the case in these chapters on special relativity, $\sqrt{-g} = 1$ and the two conventions provides the same Lagrangian density.

As in the case of the action in (3.69), it is not guaranteed that every field has an action such that we can derive its field equations after applying the Least Action Principle. However, for the known fields this is the case. In the same way, there is not a clear recipe to find the action and the Lagrangian density of a field. Every Lagrangian density is a theory and makes some predictions. If the predictions agree well with the observations of a certain system, then that theory describes that system well.

3.9 Energy-Momentum Tensor

In Sect. 1.6, we saw that, if a Lagrangian does not explicitly depend on time, there is a constant of motion, which is the energy of the system. If the Lagrangian does not depend on one of the Lagrangian coordinates, we have the conservation of the corresponding conjugate momentum. Now we want to extend those results. In this and next sections, we will employ Cartesian coordinates; the generalization to an arbitrary coordinate system will be clear after Chaps. 6 and 7.

Let us consider a system described by the Lagrangian density $\mathscr{L}$ that does not explicitly depend on the spacetime coordinates x^μ. We have[6]

$$\frac{\partial \mathscr{L}}{\partial x^\mu} = \frac{\partial \mathscr{L}}{\partial \Phi^i} \frac{\partial \Phi^i}{\partial x^\mu} + \frac{\partial \mathscr{L}}{\partial \frac{\partial \Phi^i}{\partial x^\nu}} \frac{\partial^2 \Phi^i}{\partial x^\mu \partial x^\nu} + \left(\frac{\partial \mathscr{L}}{\partial x^\mu}\right)_{\Phi,\partial\Phi}$$

$$= \frac{\partial \mathscr{L}}{\partial \Phi^i} \frac{\partial \Phi^i}{\partial x^\mu} + \frac{\partial \mathscr{L}}{\partial \frac{\partial \Phi^i}{\partial x^\nu}} \frac{\partial^2 \Phi^i}{\partial x^\mu \partial x^\nu}. \tag{3.84}$$

---

[6]Note the confusion in the notation. In Sect. 1.6, the derivative of t was

$$\frac{dL}{dt} = \frac{\partial L}{\partial q^i} \frac{dq^i}{dt} + \frac{\partial L}{\partial \frac{dq^i}{dt}} \frac{d^2 q^i}{dt^2} + \frac{\partial L}{\partial t}. \tag{3.82}$$

The counterpart here is

$$\frac{\partial \mathscr{L}}{\partial x^\mu} = \frac{\partial \mathscr{L}}{\partial \Phi^i} \frac{\partial \Phi^i}{\partial x^\mu} + \frac{\partial \mathscr{L}}{\partial \frac{\partial \Phi^i}{\partial x^\nu}} \frac{\partial^2 \Phi^i}{\partial x^\mu \partial x^\nu} + \left(\frac{\partial \mathscr{L}}{\partial x^\mu}\right)_{\Phi,\partial\Phi}, \tag{3.83}$$

where the last term $\left(\frac{\partial \mathscr{L}}{\partial x^\mu}\right)_{\Phi,\partial\Phi}$ is the possible contribution from an explicit dependence of $\mathscr{L}$ on x^μ (which we assume to vanish in Eq. (3.84)). In this section, we employ the partial derivatives because $x = (x^0, x^1, x^2, x^3)$ and $\frac{\partial \mathscr{L}}{\partial x^\mu}$ is not the counterpart of $\partial L/\partial t$. The equivalent of $\partial L/\partial t$ is now $\left(\frac{\partial \mathscr{L}}{\partial x^\mu}\right)_{\Phi,\partial\Phi}$.

Employing the equations of motion (3.80) (remember $\sqrt{-g} = 1$ in Cartesian coordinates), we have

$$\frac{\partial \mathcal{L}}{\partial x^\mu} = \left(\frac{\partial}{\partial x^\nu} \frac{\partial \mathcal{L}}{\partial \frac{\partial \Phi^i}{\partial x^\nu}} \right) \frac{\partial \Phi^i}{\partial x^\mu} + \frac{\partial \mathcal{L}}{\partial \frac{\partial \Phi^i}{\partial x^\nu}} \frac{\partial^2 \Phi^i}{\partial x^\mu \partial x^\nu} , \tag{3.85}$$

which we can rewrite as

$$\frac{\partial}{\partial x^\nu} \left(\frac{\partial \mathcal{L}}{\partial \frac{\partial \Phi^i}{\partial x^\nu}} \frac{\partial \Phi^i}{\partial x^\mu} - \delta^\nu_\mu \mathcal{L} \right) = 0 . \tag{3.86}$$

If we define the *energy-momentum tensor* T^ν_μ as[7]

$$T^\nu_\mu = - \frac{\partial \mathcal{L}}{\partial \frac{\partial \Phi^i}{\partial x^\nu}} \frac{\partial \Phi^i}{\partial x^\mu} + \delta^\nu_\mu \mathcal{L} , \tag{3.88}$$

Eq. (3.86) can be written as

$$\partial_\nu T^\nu_\mu = 0 . \tag{3.89}$$

Note that $T^{\mu\nu} = \eta^{\mu\sigma} T^\nu_\sigma$ is not unique. If $T^{\mu\nu}$ is given by Eq. (3.88), then even $T'^{\mu\nu}$ defined as

$$T'^{\mu\nu} = T^{\mu\nu} + \frac{\partial t^{\mu\nu\rho}}{\partial x^\rho} \quad \text{with} \quad t^{\mu\nu\rho} = -t^{\mu\rho\nu} , \tag{3.90}$$

satisfies Eq. (3.89)

$$\partial_\nu T'^\nu_\mu = 0 , \tag{3.91}$$

because

$$\frac{\partial^2 t^{\mu\nu\rho}}{\partial x^\nu \partial x^\rho} = 0 . \tag{3.92}$$

Let us assume that the energy-momentum tensor is symmetric. Eq. (3.89) has the form of a conservation equation. For $\mu = t$, we can write

---

[7] There are different conventions in the literature. If we change sign in Eq. (3.88), then T^{tt} is minus the energy density of the system [see later after Eq. (3.97)]. If the metric has signature $(+ - - -)$, we define

$$T^\nu_\mu = \frac{\partial \mathcal{L}}{\partial \frac{\partial \Phi^i}{\partial x^\nu}} \frac{\partial \Phi^i}{\partial x^\mu} - \delta^\nu_\mu \mathcal{L} , \tag{3.87}$$

and T^{tt} is the energy density of the system.

$$\frac{1}{c}\frac{\partial T^{tt}}{\partial t} = -\frac{\partial T^{ti}}{\partial x^i}.$$ (3.93)

If we integrate over the 3-dimensional space and we apply Gauss's theorem, we find

$$\frac{1}{c}\frac{\partial}{\partial t}\int_V T^{tt}d^3V = -\int_V \frac{\partial T^{ti}}{\partial x^i}d^3V = -\int_\Sigma T^{ti}d^2\sigma_i = 0,$$ (3.94)

where Σ is the 2-dimensional boundary of the 3-dimensional volume V and we assume that at large distances $T^{\mu\nu} = 0$. Proceeding in the same way, we can start from Eq. (3.89), consider the case $\mu = i$, and find

$$\frac{1}{c}\frac{\partial}{\partial t}\int_V T^{ti}d^3V = -\int_V \frac{\partial T^{ij}}{\partial x^j}d^3V = -\int_\Sigma T^{ij}d^2\sigma_j = 0.$$ (3.95)

We can define as the 4-momentum of the system the quantity

$$P^\mu = \frac{1}{c}\int_V T^{t\mu}d^3V.$$ (3.96)

Note also that T^{tt} is given by

$$T^{tt} = \dot{\phi}^i\frac{\partial \mathscr{L}}{\partial \dot{\phi}^i} - \mathscr{L},$$ (3.97)

and can thus be naturally associated to the energy density of the system. T^{0i}s can be naturally associated (modulo a factor $1/c$) to the momentum density of the system. T^{0i}s are thus the three components of the energy flux density. T^{ij}s can be interpreted as the components of the momentum flux density.

As we will see in Sect. 7.4, it is possible to define the energy-momentum tensor without ambiguities and in a way that it is automatically symmetric in the indices.

3.10   Examples

3.10.1   Energy-Momentum Tensor of a Free Point-Like Particle

We have seen in the previous section that $T^{0\mu}$s can be interpreted as the components of the 4-momentum density of the system. For a free point-like particle, the 4-momentum is $p^\mu = mu^\mu$, where m is the particle mass and u^μ is the particle 4-velocity. For a point-like particle the mass density is

$$\rho = m\delta^3\left[\mathbf{x} - \tilde{\mathbf{x}}(\tau)\right],$$ (3.98)

where δ^3 is the 3-dimensional Dirac delta function, $\tilde{\mathbf{x}}(\tau) = (x(\tau), y(\tau), z(\tau))$ are the space coordinates of the particle, and we assume Cartesian coordinates. We can thus write

$$T^{0\mu} = m\delta^3 \left[\mathbf{x} - \tilde{\mathbf{x}}(\tau) \right] c u^\mu . \tag{3.99}$$

T^{ij}s are proportional to the components of the momentum flux density and therefore we should add one more velocity. Since $u^0 = \gamma c$, Eq. (3.99) can be rewritten as

$$T^{0\mu} = m\delta^3 \left[\mathbf{x} - \tilde{\mathbf{x}}(\tau) \right] \frac{u^0 u^\mu}{\gamma} . \tag{3.100}$$

Eventually we have the following expression for the energy-momentum tensor of a free point-like particle in Cartesian coordinates

$$T^{\mu\nu} = m\delta^3 \left[\mathbf{x} - \tilde{\mathbf{x}}(\tau) \right] \frac{u^\mu u^\nu}{\gamma} . \tag{3.101}$$

As we will see in Sect. 7.4, Eq. (3.101) can be derived from the Lagrangian density in a straightforward way.

3.10.2 Energy-Momentum Tensor of a Perfect Fluid

A perfect fluid can be described as a gas of quasi-free point-like particles in which the fluid pressure comes from the particle collisions. Assuming that all particles have the same mass m for simplicity, the energy-momentum tensor can be written as

$$T^{\mu\nu} = \sum_{k=1}^{N} m\delta^3 \left[\mathbf{x}_k - \tilde{\mathbf{x}}_k(\tau) \right] \frac{u_k^\mu u_k^\nu}{\gamma} \tag{3.102}$$

where $k = 1, \ldots, N$ is the label of every particle. In the rest-frame of the fluid, $T^{00} = \varepsilon$, where ε is the energy density (not the mass density!) of the fluid. $T^{0i} = 0$, because $\langle u_k^\mu \rangle = 0$, where $\langle \cdot \rangle$ indicates the average over the sample and we assume that $N \to \infty$. $T^{ij} = 0$ for $i \neq j$, because $\langle u_k^i u_k^j \rangle = 0$. $T^{ij} = P$ for $i = j$, where P is the pressure. Eventually, the energy-momentum tensor of a perfect fluid in its rest-frame in Cartesian coordinates should be

$$||T^{\mu\nu}|| = ||T_{\mu\nu}|| = \begin{pmatrix} \varepsilon & 0 & 0 & 0 \\ 0 & P & 0 & 0 \\ 0 & 0 & P & 0 \\ 0 & 0 & 0 & P \end{pmatrix} . \tag{3.103}$$

At this point we have to write $T^{\mu\nu}$ in terms of vectors/tensors and such that we recover the expression in Eq. (3.103) when we are in the fluid rest-frame. It turns out that the correct form is

$$T^{\mu\nu} = (\varepsilon + P)\frac{U^\mu U^\nu}{c^2} + P\eta^{\mu\nu}. \tag{3.104}$$

where U^μ is the 4-velocity of the fluid (u_i^μ was the 4-velocity of the particle i) and in the fluid rest-frame $U^\mu = (c, 0, 0, 0)$.

Problems

3.1 Write the 4-momentum p^μ and p_μ for a free point-like particle in spherical coordinates.

3.2 Find the constants of motion for a free point-like particles in spherical coordinates.

3.3 Repeat Problems 3.1 and 3.2 for cylindrical coordinates.

3.4 The counterpart of reaction (3.57) for cosmic ray photons is

$$\gamma + \gamma_{CMB} \rightarrow e^+ + e^- \tag{3.105}$$

namely a high-energy photon (γ) collides with a CMB photon (γ_{CMB}) and there is the production of a pair electron-positron ($e^+ e^-$). Calculate the threshold energy of the high-energy photon that permits the reaction. $m_{e^+}c^2 = m_{e^-}c^2 = 0.5$ MeV.

3.5 Iron-56 is the most common isotope of iron. Its nucleus is made of 26 protons and 30 neutrons. Calculate the binding energy per nucleon, namely the binding energy divided the number of protons and neutrons. The mass of a nucleus of iron-56 is $Mc^2 = 52.103$ GeV. The masses of a proton and of a neutron are, respectively, $m_p c^2 = 0.938$ GeV and $m_n c^2 = 0.940$ GeV.

3.6 Write the field equations of the following Lagrangian density

$$\mathscr{L}(\phi, \partial_\mu\phi) = -\frac{\hbar}{2}g^{\mu\nu}(\partial_\mu\phi)(\partial_\nu\phi) - \frac{1}{2}\frac{m^2 c^2}{\hbar}\phi^2. \tag{3.106}$$

3.7 Write the field equations in Problem 3.6 in Cartesian and spherical coordinates.

3.8 Let us consider a Cartesian coordinate system. Write the energy-momentum tensor associated with the Lagrangian in Problem 3.6.

3.9 Eqs. (3.103) and (3.104) are the expressions in Cartesian coordinates. Find their expressions in spherical coordinates and compare the result with that found in Problem 2.1.

Chapter 4
Electromagnetism

In this chapter, we will revise the pre-relativistic theory of electromagnetic phenomena, namely the theory of electromagnetism formulated before the advent of special relativity that the reader is expected to know, and we will write a fully relativistic and manifestly covariant theory. For *manifestly covariant* (or, equivalently, *manifestly Lorentz-invariant*) formulation, we mean that the theory is written in terms of tensors; that is, the laws of physics are written as equalities between two tensors or equalities to zero of a tensor. Since we know how tensors change when we move from a Cartesian inertial reference frame to another Cartesian inertial reference frame, it is straightforward to realize that the equations are invariant under Lorentz transformations. In this and the following chapters, we will always employ Gaussian units to describe electromagnetic phenomena. In the present chapter, unless stated otherwise, we will use Cartesian coordinates; the generalization to arbitrary coordinate systems will be discussed in Sect. 6.7.

The basic equations governing electromagnetic phenomena are the Lorentz force law and Maxwell's equations. The equation of motion of a particle of mass m and electric charge e moving in an electromagnetic field is

$$m\ddot{\mathbf{x}} = e\mathbf{E} + \frac{e}{c}\dot{\mathbf{x}} \times \mathbf{B}\,, \tag{4.1}$$

where c is the speed of light and $\mathbf{E}$ and $\mathbf{B}$ are, respectively, the electric and magnetic fields. Equation (4.1) is just Newton's Second Law, $m\ddot{\mathbf{x}} = \mathbf{F}$, when $\mathbf{F}$ is the Lorentz force. The equations of motion (field equations) for the electric and magnetic fields are the Maxwell equations

$$\nabla \cdot \mathbf{E} = 4\pi\rho\,, \tag{4.2}$$

$$\nabla \cdot \mathbf{B} = 0\,, \tag{4.3}$$

$$\nabla \times \mathbf{E} = -\frac{1}{c}\frac{\partial \mathbf{B}}{\partial t}\,, \tag{4.4}$$

$$\nabla \times \mathbf{B} = \frac{4\pi}{c}\mathbf{J} + \frac{1}{c}\frac{\partial \mathbf{E}}{\partial t}\,, \tag{4.5}$$

© Springer Nature Singapore Pte Ltd. 2018, corrected publication 2020
C. Bambi, *Introduction to General Relativity*, Undergraduate Lecture Notes in Physics, https://doi.org/10.1007/978-981-13-1090-4_4

where ρ is the electric charge density and $\mathbf{J}$ is the electric current density.

From Eq. (4.3), we see that we can introduce the *vector potential* $\mathbf{A}$ defined as

$$\mathbf{B} = \nabla \times \mathbf{A}. \tag{4.6}$$

Indeed the divergence of the curl of any vector field vanishes (see Appendix B.3). If we plug Eq. (4.6) into Eq. (4.4), we find

$$\nabla \times \left(\mathbf{E} + \frac{1}{c} \frac{\partial \mathbf{A}}{\partial t} \right) = 0. \tag{4.7}$$

We can thus introduce the *scalar potential* ϕ as

$$\mathbf{E} + \frac{1}{c} \frac{\partial \mathbf{A}}{\partial t} = -\nabla \phi, \tag{4.8}$$

because $\nabla \times (\nabla \phi) = 0$ (see Appendix B.3). The electric field can thus be written in terms of the scalar potential ϕ and the vector potential $\mathbf{A}$

$$\mathbf{E} = -\nabla \phi - \frac{1}{c} \frac{\partial \mathbf{A}}{\partial t}. \tag{4.9}$$

4.1   Action

Now we want to write the action of a system composed of a point-like particle of mass m and electric charge e and an electromagnetic field. The action should be made of three parts: the action of the free point-like particle, S_m, the action of the electromagnetic field, S_{em}, and the action describing the interaction between the particle and the electromagnetic field, S_{int}. The total action should thus have the following form

$$S = S_m + S_{int} + S_{em}. \tag{4.10}$$

In the absence of the particle, we would be left with S_{em}. In the absence of the electromagnetic field, we would be left with S_m. If we turn the value of the electric charge of the particle off, we eliminate S_{int}, and the particle and the electromagnetic field do not interact, so the equations of motion of the particle are independent of the electromagnetic field and, vice versa, those of the electromagnetic field do not depend on the particle.

Once we apply the Least Action Principle to the action in (4.10) we have to recover the Maxwell equations and the relativistic version of the Lorentz force law (4.1). As we will see in the next sections, this can be achieved with S_m, S_{int}, and S_{em} given, respectively, by

$$S_{\mathrm{m}} = -mc \int_{\Gamma} \sqrt{-ds^2} \, ,$$

$$S_{\mathrm{int}} = \frac{e}{c} \int_{\Gamma} A_{\mu} dx^{\mu} \, ,$$

$$S_{\mathrm{em}} = -\frac{1}{16\pi c} \int_{\Omega} F^{\mu\nu} F_{\mu\nu} \, d^4 \Omega \, . \tag{4.11}$$

$F_{\mu\nu}$ is the *Faraday tensor* defined as[1]

$$F_{\mu\nu} = \partial_{\mu} A_{\nu} - \partial_{\nu} A_{\mu} \, , \tag{4.13}$$

and A_{μ} is the *4-vector potential* of the electromagnetic field. The latter is a 4-vector in which the time component is the scalar potential ϕ and the space components are the components of the 3-vector potential **A**

$$A^{\mu} = (\phi, \mathbf{A}) \, . \tag{4.14}$$

From Eqs. (4.6), (4.9), (4.13), and (4.14), we can write the Faraday tensor in terms of the electric and magnetic fields. In Cartesian coordinates (ct, x, y, z), we have

$$||F_{\mu\nu}|| = \begin{pmatrix} 0 & -E_x & -E_y & -E_z \\ E_x & 0 & B_z & -B_y \\ E_y & -B_z & 0 & B_x \\ E_z & B_y & -B_x & 0 \end{pmatrix} , \quad ||F^{\mu\nu}|| = \begin{pmatrix} 0 & E_x & E_y & E_z \\ -E_x & 0 & B_z & -B_y \\ -E_y & -B_z & 0 & B_x \\ -E_z & B_y & -B_x & 0 \end{pmatrix} , \tag{4.15}$$

where E_i and B_i are the i components of, respectively, the electric and the magnetic fields. We write them with lower indices, but they are not the spatial components of a dual vector. We remind that $F^{\mu\nu} = \eta^{\mu\rho} \eta^{\nu\sigma} F_{\rho\sigma}$ and note that the ti and it components change sign, while the ij components are unchanged, when we pass from $F_{\mu\nu}$ to $F^{\mu\nu}$.

The electric and magnetic fields can be written in terms of the Faraday tensor as follows

$$E_i = F_{it} \, , \quad B_i = \frac{1}{2} \varepsilon_{ijk} F^{jk} \, , \tag{4.16}$$

---

[1] As it will be more clear later (see Sect. 6.7), the definition (4.13) is valid in any coordinate system, i.e. both Cartesian and non-Cartesian systems.

If the spacetime metric has signature $(+ - --)$, the Faraday tensor is still defined as in Eq. (4.13), but now the expression of $F_{\mu\nu}$ in Cartesian coordinates is

$$||F_{\mu\nu}|| = \begin{pmatrix} 0 & E_x & E_y & E_z \\ -E_x & 0 & -B_z & B_y \\ -E_y & B_z & 0 & -B_x \\ -E_z & -B_y & B_x & 0 \end{pmatrix} . \tag{4.12}$$

where ε_{ijk} is the *Levi–Civita symbol* (see Appendix B.2). The space-space components of the Faraday tensor can be written in terms of the magnetic field

$$F^{ij} = \varepsilon^{ijk} B_k \,. \tag{4.17}$$

Indeed

$$\varepsilon^{ijk} B_k = \varepsilon^{kij} B_k = \frac{1}{2} \varepsilon^{kij} \varepsilon_{klm} F^{lm} = \frac{1}{2} \left(\delta^i_l \delta^j_m - \delta^i_m \delta^j_l \right) F^{lm}$$

$$= \frac{1}{2} \left(F^{ij} - F^{ji} \right) = F^{ij} \,. \tag{4.18}$$

If we write the electric charge e as the integral of the electric charge density over the 3-volume, we can write the interaction term in the action, S_{int}, as follows (note that $d^4 \Omega = c dt d^3 V$)

$$S_{\text{int}} = \frac{1}{c} \int_V \rho d^3 V \int_\Gamma A_\mu dx^\mu = \frac{1}{c} \int_V \rho d^3 V \int_{t_1}^{t_2} \frac{dx^\mu}{dt} A_\mu \, dt$$

$$= \frac{1}{c^2} \int_\Omega J^\mu A_\mu \, d^4 \Omega \,, \tag{4.19}$$

where J^μ is the *current 4-vector*[2]

$$J^\mu = (\rho c, \rho \mathbf{v}) \,. \tag{4.21}$$

In the case of a system with many electrically charged particles, the total action can be written as

$$S = -\sum_i m_i c \int_{\Gamma_i} \sqrt{-ds^2} - \int_\Omega \left(\frac{1}{16\pi c} F^{\mu\nu} F_{\mu\nu} - \frac{1}{c^2} J^\mu A_\mu \right) d^4 \Omega \,. \tag{4.22}$$

In the next sections, we will see that the action in (4.10), with S_{m}, S_{int}, and S_{em} given by Eq. (4.11) and A^μ given by Eq. (4.14), provides the relativistic version of the Lorentz force law (4.1) and the correct Maxwell equations. As we have already emphasized in the previous chapters, there is no fundamental recipe to obtain the action of a certain physical system. A specific action represents a specific theory. If the theoretical predictions agree well with observations, we have the right theory

---

[2]Note that the electric charge density ρ can be written as $\rho = \delta Q / \delta V$, where δQ is the electric charge in the infinitesimal volume δV. The latter depends on the reference frame and can be written in terms of the proper infinitesimal volume $\delta V = \delta V_0 / \gamma$, see Eq. (2.44). Now $\delta Q / \delta V_0$ is an invariant and the current 4-vector can be written as

$$J^\mu = \frac{\delta Q}{\delta V_0} (\gamma c, \gamma \mathbf{v}) = \frac{\delta Q}{\delta V_0} u^\mu \,, \tag{4.20}$$

where u^μ is the 4-velocity. So $J^\mu \propto u^\mu$ and is clearly a 4-vector.

up to when we find a phenomenon that cannot be explained within our model. For the moment, the action above is our current theory for the description of a charged particle in an electromagnetic field.

4.2 Motion of a Charged Particle

As we did in Sect. 3.2, we can either employ the standard Lagrangian formalism of Newtonian mechanics with a time coordinate and a space, or a fully 4-dimensional approach in which all the spacetime coordinates are treated in the same way and we have some parameter (e.g. the particle proper time) to parametrize the particle trajectory.

4.2.1 3-Dimensional Formalism

The motion of the particle in a (pre-determined) electromagnetic field is described by the action of the free particle and by the interaction term. If the trajectory of the particle is parametrized by the time coordinate t, we have

$$S = \int_\Gamma (L_m + L_{int})\, dt \,, \tag{4.23}$$

where the Lagrangians of the free particle and of the interaction term between the particle and the electromagnetic field are, respectively,

$$L_m = -mc\sqrt{1 - \frac{\dot{\mathbf{x}}^2}{c^2}}\,,$$

$$L_{int} = -e\phi + \frac{e}{c}\mathbf{A} \cdot \dot{\mathbf{x}}\,. \tag{4.24}$$

Here the dot $\cdot$ indicates the derivative with respect to the coordinate t, so $\dot{\mathbf{x}} = d\mathbf{x}/dt$ where $\mathbf{x} = (x, y, z)$.

The 3-momentum is[3]

$$\mathbf{P} = \frac{\partial L}{\partial \dot{\mathbf{x}}} = \frac{m\dot{\mathbf{x}}}{\sqrt{1 - \dot{\mathbf{x}}^2/c^2}} + \frac{e}{c}\mathbf{A}\,, \tag{4.26}$$

---

[3] As already pointed out in Sect. 3.2, the conjugate 3-momentum is

$$\mathbf{P}^* = \frac{\partial L}{\partial \dot{\mathbf{x}}}\,. \tag{4.25}$$

In Cartesian coordinates, the metric tensor is δ_{ij}, and therefore $\mathbf{P}^* = \mathbf{P}$, where $\mathbf{P}$ is the 3-momentum.

which can also be written as

$$\mathbf{P} = \mathbf{p} + \frac{e}{c}\mathbf{A}\,, \qquad (4.27)$$

where $\mathbf{p} = m\gamma\dot{\mathbf{x}}$ is the 3-momentum of the free particle (see Sect. 3.2). The energy
of the particle is

$$E = \frac{\partial L}{\partial \dot{\mathbf{x}}}\dot{\mathbf{x}} - L = \frac{mc^2}{\sqrt{1 - \dot{\mathbf{x}}^2/c^2}} + e\phi\,. \qquad (4.28)$$

Let us now derive the equations of motion. The first term in the Euler-Lagrange
equation is

$$\frac{d}{dt}\frac{\partial L}{\partial \dot{\mathbf{x}}} = \frac{d}{dt}\left(\mathbf{p} + \frac{e}{c}\mathbf{A}\right) = \frac{d\mathbf{p}}{dt} + \frac{e}{c}\frac{\partial \mathbf{A}}{\partial t} + \frac{e}{c}(\dot{\mathbf{x}}\cdot\nabla)\mathbf{A}\,. \qquad (4.29)$$

The second term in the Euler-Lagrange equation is

$$\frac{\partial L}{\partial \mathbf{x}} = -e\nabla\phi + \frac{e}{c}\nabla(\mathbf{A}\cdot\dot{\mathbf{x}})\,. \qquad (4.30)$$

From the identity

$$\nabla(\mathbf{V}\cdot\mathbf{W}) = (\mathbf{W}\cdot\nabla)\mathbf{V} + (\mathbf{V}\cdot\nabla)\mathbf{W} + \mathbf{W}\times(\nabla\times\mathbf{V}) + \mathbf{V}\times(\nabla\times\mathbf{W})\,, \qquad (4.31)$$

we can rewrite Eq. (4.30) as

$$\frac{\partial L}{\partial \mathbf{x}} = -e\nabla\phi + \frac{e}{c}(\dot{\mathbf{x}}\cdot\nabla)\mathbf{A} + \frac{e}{c}\dot{\mathbf{x}}\times(\nabla\times\mathbf{A})\,, \qquad (4.32)$$

because the differentiation with respect to $\mathbf{x}$ is carried out for constant $\dot{\mathbf{x}}$. Eventually,
we get the following equation of motion

$$\frac{d\mathbf{p}}{dt} = -\frac{e}{c}\frac{\partial \mathbf{A}}{\partial t} - e\nabla\phi + \frac{e}{c}\dot{\mathbf{x}}\times(\nabla\times\mathbf{A})\,. \qquad (4.33)$$

In terms of the electric and magnetic fields, we have

$$\frac{d\mathbf{p}}{dt} = e\mathbf{E} + \frac{e}{c}\dot{\mathbf{x}}\times\mathbf{B}\,, \qquad (4.34)$$

In the non-relativistic limit, $\mathbf{p} = m\dot{\mathbf{x}}$, and we recover Eq. (4.1).

4.2.2 4-Dimensional Formalism

Let us now parametrize the particle trajectory with the particle proper time τ. The action reads

$$S = \int_\Gamma (L_{\rm m} + L_{\rm int})\, d\tau \,, \tag{4.35}$$

where τ is the proper time of the particle, the Lagrangian terms are

$$L_{\rm m} = \frac{1}{2} m \eta_{\mu\nu} \dot{x}^\mu \dot{x}^\nu \,,$$

$$L_{\rm int} = \frac{e}{c} A_\mu \dot{x}^\mu \,, \tag{4.36}$$

and now the dot indicates the derivative with respect to τ, i.e. $\dot{x}^\mu = dx^\mu/d\tau$. From the Euler-Lagrange equations, we find

$$\frac{d}{d\tau} \frac{\partial L_{\rm m}}{\partial \dot{x}^\mu} + \frac{d}{d\tau} \frac{\partial L_{\rm int}}{\partial \dot{x}^\mu} - \frac{\partial L_{\rm int}}{\partial x^\mu} = 0 \,,$$

$$\frac{d}{d\tau} \left(m \eta_{\mu\nu} \dot{x}^\nu \right) + \frac{d}{d\tau} \left(\frac{e}{c} A_\mu \right) - \frac{e}{c} \frac{\partial A_\nu}{\partial x^\mu} \dot{x}^\nu = 0 \,,$$

$$m \ddot{x}_\mu + \frac{e}{c} \frac{\partial A_\mu}{\partial x^\nu} \dot{x}^\nu - \frac{e}{c} \frac{\partial A_\nu}{\partial x^\mu} \dot{x}^\nu = 0 \,,$$

$$m \ddot{x}_\mu - \frac{e}{c} F_{\mu\nu} \dot{x}^\nu = 0 \,. \tag{4.37}$$

The equations of motion of the particle are

$$\ddot{x}^\mu = \frac{e}{mc} F^{\mu\nu} \dot{x}_\nu \,. \tag{4.38}$$

If we do not use Cartesian coordinates, in Eq. (4.36) we have $g_{\mu\nu}$ instead of $\eta_{\mu\nu}$, $L_{\rm m}$ provides the geodesic equations, and Eq. (4.38) reads

$$\ddot{x}^\mu + \Gamma^\mu_{\nu\rho} \dot{x}^\nu \dot{x}^\rho = \frac{e}{mc} F^{\mu\nu} \dot{x}_\nu \,. \tag{4.39}$$

4.3 Maxwell's Equations in Covariant Form

4.3.1 Homogeneous Maxwell's Equations

The covariant form of the homogeneous Maxwell equations, Eqs. (4.3) and (4.4), is

$$\partial_\mu F_{\nu\rho} + \partial_\nu F_{\rho\mu} + \partial_\rho F_{\mu\nu} = 0 \,. \tag{4.40}$$

Equation (4.40) directly follows from the definition of $F_{\mu\nu}$. If we plug Eq. (4.13) into Eq. (4.40) we find

$$\partial_\mu \left(\partial_\nu A_\rho - \partial_\rho A_\nu \right) + \partial_\nu \left(\partial_\rho A_\mu - \partial_\mu A_\rho \right) + \partial_\rho \left(\partial_\mu A_\nu - \partial_\nu A_\mu \right) = 0 \,. \tag{4.41}$$

Let us now check that Eq. (4.40) is equivalent to Eqs. (4.3) and (4.4). Equation (4.40) is non-trivial only when there are no repeated indices. So we have four independent equations, which correspond to the cases in which (μ, ν, ρ) is (t, x, y), (t, x, z), (t, y, z), and (x, y, z). Permutations of these indices provide the same equations.

For $(\mu, \nu, \rho) = (t, x, y)$ we have

$$\frac{1}{c} \frac{\partial F_{xy}}{\partial t} + \frac{\partial F_{yt}}{\partial x} + \frac{\partial F_{tx}}{\partial y} = 0 \,. \tag{4.42}$$

Employing the relations in (4.16), we replace the components of the Faraday tensor with those of the electric and magnetic fields

$$\frac{1}{c} \frac{\partial B_z}{\partial t} + \frac{\partial E_y}{\partial x} - \frac{\partial E_x}{\partial y} = 0 \,. \tag{4.43}$$

Equation (4.43) can be rewritten with the Levi–Civita symbol ε_{ijk} as

$$\frac{1}{c} \frac{\partial B_z}{\partial t} + \varepsilon_{zjk} \frac{\partial E_k}{\partial x^j} = 0 \,. \tag{4.44}$$

This is the z component of Eq. (4.4). For $(\mu, \nu, \rho) = (t, x, z)$, we recover the y component of Eq. (4.4), and for $(\mu, \nu, \rho) = (t, y, z)$ we get the x component.

For $(\mu, \nu, \rho) = (x, y, z)$ (no t component), we have

$$\frac{\partial F_{yz}}{\partial x} + \frac{\partial F_{zx}}{\partial y} + \frac{\partial F_{xy}}{\partial z} = 0 \,. \tag{4.45}$$

Employing Eq. (4.16), we find

$$\frac{\partial B_x}{\partial x} + \frac{\partial B_y}{\partial y} + \frac{\partial B_z}{\partial z} = \frac{\partial B_i}{\partial x^i} = 0 \,, \tag{4.46}$$

and we recover Eq. (4.3).

4.3.2 Inhomogeneous Maxwell's Equations

The covariant form of the inhomogeneous Maxwell equations, Eqs. (4.2) and (4.5), is[4]

$$\partial_\mu F^{\mu\nu} = -\frac{4\pi}{c} J^\nu \,. \tag{4.48}$$

Equation (4.48) is the field equation from the Least Action Principle, so it follows from

$$\frac{\partial}{\partial x^\mu} \frac{\partial \mathscr{L}}{\partial \left(\partial_\mu A_\nu \right)} - \frac{\partial \mathscr{L}}{\partial A_\nu} = 0 \,. \tag{4.49}$$

$S_{\rm m}$ does not provide any contribution, so we only have to consider

$$S_{\rm em} + S_{\rm int} = \frac{1}{c} \int_\Omega \left(\mathscr{L}_{\rm em} + \mathscr{L}_{\rm in} \right) d^4\Omega \,, \tag{4.50}$$

where

$$\mathscr{L}_{\rm em} = -\frac{1}{16\pi} F^{\mu\nu} F_{\mu\nu} \,, \quad \mathscr{L}_{\rm int} = \frac{1}{c} J^\mu A_\mu \,. \tag{4.51}$$

$\mathscr{L}_{\rm em}$ only depends on $\partial_\mu A_\nu$ and we have

$$
\begin{aligned}
\frac{\partial \mathscr{L}_{\rm em}}{\partial \left(\partial_\mu A_\nu \right)} &= -\frac{\partial}{\partial \left(\partial_\mu A_\nu \right)} \frac{1}{16\pi} \left[\left(\partial_\rho A_\sigma - \partial_\sigma A_\rho \right) \left(\partial_\tau A_\nu - \partial_\nu A_\tau \right) \eta^{\rho\tau} \eta^{\sigma\nu} \right] \\
&= -\frac{1}{16\pi} \left[\left(\delta^\mu_\rho \delta^\nu_\sigma - \delta^\mu_\sigma \delta^\nu_\rho \right) F^{\rho\sigma} + F^{\tau\nu} \left(\delta^\mu_\tau \delta^\nu_\nu - \delta^\mu_\nu \delta^\nu_\tau \right) \right] \\
&= -\frac{1}{16\pi} \left(F^{\mu\nu} - F^{\nu\mu} + F^{\mu\nu} - F^{\nu\mu} \right) \,.
\end{aligned} \tag{4.52}
$$

Since the Faraday tensor is antisymmetric, i.e. $F^{\mu\nu} = -F^{\nu\mu}$, we have

$$\frac{\partial \mathscr{L}_{\rm em}}{\partial \left(\partial_\mu A_\nu \right)} = -\frac{1}{4\pi} F^{\mu\nu} \,. \tag{4.53}$$

---

[4]Since $F^{\mu\nu}$ is antisymmetric, we can also write

$$\partial_\nu F^{\mu\nu} = \frac{4\pi}{c} J^\mu \,. \tag{4.47}$$

$\mathscr{L}_{\text{int}}$ depends on A_ν and is independent of $\partial_\mu A_\nu$. We have

$$\frac{\partial \mathscr{L}_{\text{int}}}{\partial A_\nu} = \frac{1}{c} J^\nu . \tag{4.54}$$

When we combine Eqs. (4.53) and (4.54), we find Eq. (4.48).

Let us now check that we recover the inhomogeneous Maxwell equations (4.2) and (4.5). For $\nu = t$ we have

$$\partial_\mu F^{\mu t} = \partial_i F^{it} = -\partial_i E_i = -\frac{4\pi}{c} J^t , \tag{4.55}$$

and it is Eq. (4.2) because $J^t = \rho c$.

For $\nu = i$ we have

$$\partial_\mu F^{\mu i} = \frac{1}{c} \partial_t F^{ti} + \partial_j F^{ji} = -\frac{4\pi}{c} J^i . \tag{4.56}$$

We replace the components of the Faraday tensor with those of the electric and magnetic fields and we find

$$\frac{1}{c} \frac{\partial E_i}{\partial t} - \varepsilon^{ijk} \frac{\partial B_k}{\partial x^j} = -\frac{4\pi}{c} J^i . \tag{4.57}$$

For a 3-dimensional vector $\mathbf{V}$, we have

$$(\nabla \times \mathbf{V})^i = \varepsilon^{ijk} \partial_j V_k , \tag{4.58}$$

and we thus recover Eq. (4.5) for $i = x$, y, and z.

Let us note that Eq. (4.48) implies the conservation of the electric current. If we apply the differential operator ∂_ν to both sides of this equation, we find

$$\partial_\nu \partial_\mu F^{\mu\nu} = -\frac{4\pi}{c} \partial_\nu J^\nu . \tag{4.59}$$

Since $F^{\mu\nu}$ is an antisymmetric tensor, $\partial_\nu \partial_\mu F^{\mu\nu} = 0$, and thus we find the conservation of the electric current

$$\partial_\mu J^\mu = 0 . \tag{4.60}$$

Equation (4.60) is a continuity equation, as we can easily see if we rewrite it as

$$\frac{1}{c} \frac{\partial J^t}{\partial t} = -\frac{\partial J^i}{\partial x^i} , \tag{4.61}$$

and we integrate both sides over the 3-dimensional space volume

$$\frac{1}{c}\frac{d}{dt}\int_V J^t\, d^3V = -\int_V \left(\partial_i J^i\right) d^3V = -\int_\Sigma J^i d^2\sigma_i\,, \tag{4.62}$$

where in the last passage we applied Gauss's theorem. The total electric charge in the volume V is

$$Q = \int_V J^t\, d^3V\,. \tag{4.63}$$

Equation (4.62) thus tells us that any variation of the total electric charge can be calculated from the outflow/inflow of the electric current at the boundary of the region. If there is no outflow/inflow of the electric current, we have the conservation of the electric charge in that region

$$\frac{d}{dt}Q = 0\,. \tag{4.64}$$

4.4 Gauge Invariance

If we plug Eq. (4.13) into Eq. (4.48), we find

$$\Box A^\mu - \partial^\mu\left(\partial_\nu A^\nu\right) = -\frac{4\pi}{c}J^\mu\,. \tag{4.65}$$

where $\Box = \partial_\mu\partial^\mu$ is the d'Alembertian. Equation (4.65) is invariant under the following transformation

$$A_\mu \to A'_\mu = A_\mu + \partial_\mu\Lambda\,, \tag{4.66}$$

where $\Lambda = \Lambda(x)$ is a generic function. Indeed we have

$$\Box A'^\mu - \Box\partial^\mu\Lambda - \partial^\mu\left(\partial_\nu A'^\nu - \Box\Lambda\right) = -\frac{4\pi}{c}J^\mu\,,$$
$$\Box A'^\mu - \partial^\mu\left(\partial_\nu A'^\nu\right) = -\frac{4\pi}{c}J^\mu\,, \tag{4.67}$$

which is the same as Eq. (4.65) with A'_μ replacing A_μ. We can thus always choose A_μ such that it satisfies the following condition

$$\partial_\mu A^\mu = 0\,. \tag{4.68}$$

If our initial A_μ does not meet this condition, we can perform the transformation (4.66) such that

$$\Box \Lambda = \partial_\mu A^\mu , \tag{4.69}$$

and our new equation is

$$\Box A^\mu = -\frac{4\pi}{c} J^\mu . \tag{4.70}$$

The condition (4.68) is called the *Lorentz gauge* and it is quite a common choice because it sometimes simplifies the calculations. The formal solution of Eq. (4.70) is

$$A^\mu = \frac{1}{c} \int d^3\mathbf{x}' \, \frac{J^\mu(t - |\mathbf{x} - \mathbf{x}'|/c, \mathbf{x}')}{|\mathbf{x} - \mathbf{x}'|} . \tag{4.71}$$

4.5   Energy-Momentum Tensor of the Electromagnetic Field

From Eq. (3.88), we can compute the energy-momentum tensor of the electromagnetic field

$$\begin{aligned}
T^\nu_\mu &= -\frac{\partial \mathcal{L}_{\text{em}}}{\partial (\partial_\nu A_\rho)} \left(\partial_\mu A_\rho \right) + \delta^\nu_\mu \mathcal{L}_{\text{em}} \\
&= \frac{1}{4\pi} F^{\nu\rho} \left(\partial_\mu A_\rho \right) - \frac{1}{16\pi} \delta^\nu_\mu F^{\rho\sigma} F_{\rho\sigma} ,
\end{aligned} \tag{4.72}$$

which we can rewrite as

$$T^{\mu\nu} = \frac{1}{4\pi} F^{\nu\rho} \left(\partial^\mu A_\rho \right) - \frac{1}{16\pi} \eta^{\mu\nu} F^{\rho\sigma} F_{\rho\sigma} . \tag{4.73}$$

This tensor is not symmetric. It can be made symmetric as shown in Sect. 3.9

$$T^{\mu\nu} \to T^{\mu\nu} - \frac{1}{4\pi} \frac{\partial}{\partial x^\rho} \left(A^\mu F^{\nu\rho} \right) , \tag{4.74}$$

where $A^\mu F^{\nu\rho} = -A^\mu F^{\rho\nu}$. Let us note that

$$\frac{1}{4\pi} \frac{\partial}{\partial x^\rho} \left(A^\mu F^{\nu\rho} \right) = \frac{1}{4\pi} F^{\nu\rho} \left(\partial_\rho A^\mu \right) , \tag{4.75}$$

because $\partial_\rho F^{\nu\rho} = 0$ in the absence of electric currents. The energy-momentum tensor of the electromagnetic field thus becomes

$$T^{\mu\nu} = \frac{1}{4\pi} F^{\nu\rho} F^\mu{}_\rho - \frac{1}{16\pi} \eta^{\mu\nu} F^{\rho\sigma} F_{\rho\sigma} , \tag{4.76}$$

and it is symmetric, so we can also write

$$T^{\mu\nu} = \frac{1}{4\pi} F^{\mu\rho} F^{\nu}{}_{\rho} - \frac{1}{16\pi} \eta^{\mu\nu} F^{\rho\sigma} F_{\rho\sigma} . \qquad (4.77)$$

For the moment, Eq. (4.74) may look an *ad hoc* recipe. We will see in Sect. 7.4 that it is possible to define the energy-momentum tensor without ambiguities.

Lastly, it is worth noting that the energy-momentum tensor of the electromagnetic field has vanishing trace

$$\begin{aligned} T^{\mu}{}_{\mu} &= \frac{1}{4\pi} F^{\mu\rho} F_{\mu\rho} - \frac{1}{16\pi} \delta^{\mu}_{\mu} F^{\rho\sigma} F_{\rho\sigma} \\ &= \frac{1}{4\pi} F^{\mu\rho} F_{\mu\rho} - \frac{1}{4\pi} F^{\rho\sigma} F_{\rho\sigma} = 0 . \end{aligned} \qquad (4.78)$$

4.6 Examples

4.6.1 Motion of a Charged Particle in a Constant Uniform Electric Field

Let us consider a particle of mass m and electric charge e in a constant uniform electric field $\mathbf{E} = (E, 0, 0)$. At the time $t = 0$, the position of the particle is $\mathbf{x} = (x_0, y_0, z_0)$ and its velocity is $\dot{\mathbf{x}} = (\dot{x}_0, \dot{y}_0, 0)$.

In the non-relativistic theory, the equation of motion is given by Eq. (4.1). For the x component we have

$$m\ddot{x} = eE \quad \rightarrow \quad \dot{x}(t) = \dot{x}_0 + \frac{eE}{m}t , \quad x(t) = x_0 + \dot{x}_0 t + \frac{eE}{2m}t^2 . \qquad (4.79)$$

For the y component we have

$$m\ddot{y} = 0 \quad \rightarrow \quad \dot{y}(t) = \dot{y}_0 , \quad y(t) = y_0 + \dot{y}_0 t . \qquad (4.80)$$

From Eq. (4.80) we can write t in terms of y, y_0, and $\dot{y}_0$. We plug this expression into Eq. (4.79) and we find

$$x(y) = \left(\frac{eE}{2m\dot{y}_0^2} \right) y^2 + \left(\frac{\dot{x}_0}{\dot{y}_0} - \frac{eEy_0}{m\dot{y}_0^2} \right) y + \left(x_0 - \frac{\dot{x}_0 y_0}{\dot{y}_0} + \frac{eEy_0^2}{2m\dot{y}_0^2} \right) . \qquad (4.81)$$

This is the equation of a parabola.

In the relativistic theory, the equation governing the motion of the particle is (4.34). Let us assume that at the time $t = 0$ the position of the particle is $\mathbf{x} = (x_0, y_0, z_0)$ and its 3-momentum is $\mathbf{p} = (q_x, q_y, 0)$. From Eq. (4.34) we have

$$\dot{p}_x = eE \rightarrow p_x = q_x + eEt, \quad \dot{p}_y = 0 \rightarrow p_y = q_y. \tag{4.82}$$

The particle energy is thus (here we use ε to indicate the particle energy because E is the electric field)

$$\varepsilon = \sqrt{m^2 c^4 + p_x^2 c^2 + p_y^2 c^2}, \tag{4.83}$$

as found in Sect. 3.2. The 3-velocity is $\dot{\mathbf{x}} = \mathbf{p}c^2/\varepsilon$. For the x component we have

$$\dot{x} = \frac{p_x c^2}{\varepsilon} = \frac{q_x c + eEct}{\sqrt{m^2 c^2 + (q_x + eEt)^2 + q_y^2}}$$

$$\rightarrow \quad x(t) = x_0 + \frac{\sqrt{m^2 c^4 + (q_x + eEt)^2 c^2 + q_y^2 c^2}}{eE}$$

$$- \frac{\sqrt{m^2 c^4 + q_x^2 c^2 + q_y^2 c^2}}{eE}. \tag{4.84}$$

For the y component we have

$$\dot{y} = \frac{p_y c^2}{\varepsilon} = \frac{q_y c}{\sqrt{m^2 c^2 + (q_x + eEt)^2 + q_y^2}}$$

$$\rightarrow y(t) = y_0 + \frac{q_y c}{eE} \ln \left[q_x + eEt + \sqrt{m^2 c^2 + (q_x + eEt)^2 + q_y^2} \right]$$

$$- \frac{q_y c}{eE} \ln \left(q_x + \sqrt{m^2 c^2 + q_x^2 + q_y^2} \right). \tag{4.85}$$

Let us note the difference between the non-relativistic and relativistic theories. In the non-relativistic theory, the velocity diverges

$$\lim_{t \to \infty} \dot{x}(t) = \infty, \quad \lim_{t \to \infty} \dot{y}(t) = \dot{y}_0. \tag{4.86}$$

In the relativistic theory, the velocity of the charged particle asymptotically tends to the speed of light

$$\lim_{t \to \infty} \dot{x}(t) = c, \quad \lim_{t \to \infty} \dot{y}(t) = 0, \tag{4.87}$$

but it never reaches it

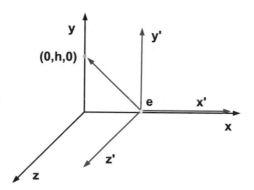

Fig. 4.1 The particle with electric charge e is moving with 3-velocity $\dot{\mathbf{x}} = (v, 0, 0)$ with respect to the reference frame with the Cartesian coordinates (x, y, z). In the reference frame with the Cartesian coordinates (x', y', z'), the particle is at rest at the point $\mathbf{x}' = (0, 0, 0)$. We want to calculate the electric and magnetic fields at the point $\mathbf{x} = (0, h, 0)$ measured by the observer with the Cartesian coordinates (x, y, z)

$$\dot{x}^2 + \dot{y}^2 = c^2 \left[\frac{(q_x + eEt)^2 + q_y^2}{m^2 c^2 + (q_x + eEt)^2 + q_y^2} \right] < c^2. \tag{4.88}$$

4.6.2 Electromagnetic Field Generated by a Charged Particle

Let us consider a particle with electric charge e. In the reference frame with the Cartesian coordinates (x, y, z), the particle has 3-velocity $\dot{\mathbf{x}} = (v, 0, 0)$. We want to calculate the electric and magnetic fields at the point $\mathbf{x} = (0, h, 0)$. The system is sketched in Fig. 4.1.

In the system with the Cartesian coordinates (x', y', z'), the particle is at rest at the point $\mathbf{x}' = (0, 0, 0)$. In this coordinate system, it is straightforward to write the intensities of the electric and magnetic fields. The magnetic field vanishes, because there is no electric current. The electric field at every point is simply that of a static charge. So we have

$$\mathbf{E}' = \frac{e}{r'^3} \mathbf{r}' = \frac{e}{(h^2 + v^2 t'^2)^{3/2}} \begin{pmatrix} -vt' \\ h \\ 0 \end{pmatrix}, \quad \mathbf{B}' = \begin{pmatrix} 0 \\ 0 \\ 0 \end{pmatrix}. \tag{4.89}$$

where $r' = \sqrt{h^2 + v^2 t'^2}$ is the distance between the particle and the point where we want to evaluate the electric and magnetic fields. The Faraday tensor in the reference frame (x', y', z') is thus

$$||F'^{\mu\nu}|| = \begin{pmatrix} 0 & E'_x & E'_y & 0 \\ -E'_x & 0 & 0 & 0 \\ -E'_y & 0 & 0 & 0 \\ 0 & 0 & 0 & 0 \end{pmatrix}. \tag{4.90}$$

The Faraday tensor in the reference frame (x, y, z) can be obtained with the coordinate transformation

$$F^{\mu\nu} = \Lambda^\mu{}_\rho \Lambda^\nu{}_\sigma F'^{\rho\sigma}, \tag{4.91}$$

where $\Lambda^\mu{}_\rho$ is the Lorentz boost

$$||\Lambda^\mu{}_\nu|| = \begin{pmatrix} \gamma & \gamma\beta & 0 & 0 \\ \gamma\beta & \gamma & 0 & 0 \\ 0 & 0 & 1 & 0 \\ 0 & 0 & 0 & 1 \end{pmatrix}. \tag{4.92}$$

Since $F^{\mu\nu}$ is antisymmetric, the diagonal components vanish and we have only six independent components. Let us start from F^{tx}. We have

$$\begin{aligned} F^{tx} &= \Lambda^t{}_\rho \Lambda^x{}_\sigma F'^{\rho\sigma} = \Lambda^t{}_t \Lambda^x{}_x F'^{tx} + \Lambda^t{}_x \Lambda^x{}_t F'^{xt} \\ &= \gamma^2 E'_x - \gamma^2 \beta^2 E'_x = E'_x, \end{aligned} \tag{4.93}$$

because $1 - \beta^2 = 1/\gamma^2$. For F^{ty}, we have

$$F^{ty} = \Lambda^t{}_\rho \Lambda^y{}_\sigma F'^{\rho\sigma} = \Lambda^t{}_\rho \Lambda^y{}_y F'^{\rho y} = \Lambda^t{}_t \Lambda^y{}_y F'^{ty} = \gamma E'_y, \tag{4.94}$$

because the only non-vanishing $\Lambda^y{}_\rho$ is $\Lambda^y{}_y = 1$, and then the only non-vanishing $F'^{\rho y}$ is F'^{ty}. For F^{xy}, we have

$$F^{xy} = \Lambda^x{}_\rho \Lambda^y{}_\sigma F'^{\rho\sigma} = \Lambda^x{}_\rho \Lambda^y{}_y F'^{\rho y} = \Lambda^x{}_t \Lambda^y{}_y F'^{ty} = \gamma\beta E'_y, \tag{4.95}$$

because, again, the only non-vanishing $\Lambda^y{}_\rho$ is $\Lambda^y{}_y = 1$, and then the only non-vanishing $F'^{\rho y}$ is F'^{ty}. Lastly, all $F^{\mu z}$s vanish, because $F^{\mu z} = \Lambda^\mu{}_\rho \Lambda^z{}_\sigma F'^{\rho\sigma}$, the only non-vanishing $\Lambda^z{}_\sigma$ is $\Lambda^z{}_z = 1$, but $F'^{\rho z} = 0$.

In the end, we have

$$||F^{\mu\nu}|| = \begin{pmatrix} 0 & E'_x & \gamma E'_y & 0 \\ -E'_x & 0 & \gamma\beta E'_y & 0 \\ -\gamma E'_y & -\gamma\beta E'_y & 0 & 0 \\ 0 & 0 & 0 & 0 \end{pmatrix}. \tag{4.96}$$

Since we are considering the point $\mathbf{x} = (0, h, 0)$, $t' = \gamma t$ (because $x = 0$), and the electric and magnetic fields measured in the reference frame (x, y, z) can be written

Fig. 4.2 E_x as a function of time at the position $(0, h, 0)$ for $\beta = 0.95$. Units in which $e = h = 1$ are used. See the text for more details

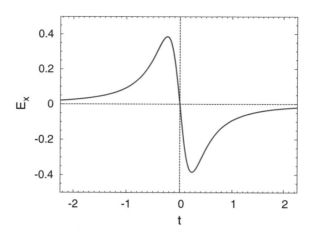

as

$$\mathbf{E} = \frac{e\gamma}{\left(h^2 + v^2\gamma^2t^2\right)^{3/2}} \begin{pmatrix} -vt \\ h \\ 0 \end{pmatrix}, \quad \mathbf{B} = \frac{e\gamma\beta h}{\left(h^2 + v^2\gamma^2t^2\right)^{3/2}} \begin{pmatrix} 0 \\ 0 \\ 1 \end{pmatrix}. \quad (4.97)$$

The x component of the electric field in the reference frame (x, y, z) at the point $\mathbf{x} = (0, h, 0)$ is shown in Fig. 4.2 as a function of the time t. The maximum value of E_x is

$$(E_x)_{\text{max}} = \frac{2}{3\sqrt{3}} \frac{e}{h^2}, \quad (4.98)$$

which is reached at the time $t = h/(\sqrt{2}v\gamma)$. As shown in Fig. 4.3, the y component of the electric field looks like a pulse. The maximum intensity and the time width of the pulse are

$$\left(E_y\right)_{\text{max}} = \frac{e\gamma}{h^2}, \quad (\Delta t)_{\text{half-max}} = 2\sqrt{4^{1/3} - 1} \frac{h}{\gamma v}. \quad (4.99)$$

Problems

4.1 Compute the invariants $F^{\mu\nu} F_{\mu\nu}$ and $\varepsilon^{\mu\nu\rho\sigma} F_{\mu\nu} F_{\rho\sigma}$ in terms of the electric and magnetic fields.

4.2 Verify the vector identity in Eq. (4.31).

Fig. 4.3 As in Figure 4.2 for
E_y

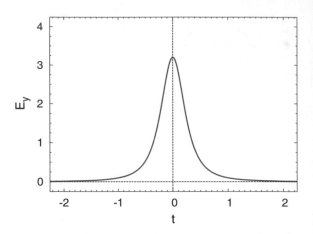

4.3 Consider a Cartesian coordinate system of an inertial reference frame in which we measure a constant uniform electric field $\mathbf{E} = (E, 0, 0)$. Calculate the electric field measured in the inertial reference frame moving with constant velocity $\mathbf{v} = (v, 0, 0)$ with respect to the former.

4.4 Let us consider a constant uniform electric field $\mathbf{E} = (E, 0, 0)$. Calculate the associated energy-momentum tensor.

Chapter 5
Riemannian Geometry

In the previous chapters, we studied non-gravitational phenomena in inertial reference frames, and often we limited our discussion to Cartesian coordinate systems. Now we want to include gravity, non-inertial reference frames, and general coordinate systems. The aim of this chapter is to introduce some mathematical tools necessary to achieve this goal. We will follow quite a heuristic approach. The term *Riemannian geometry* is used when we deal with a differentiable manifold equipped with a metric tensor (see Appendix C for the definition of the concept of differentiable manifold).

5.1 Motivations

As we will see better in the next chapter, gravity has quite a special property: for the same initial conditions, any test-particle[1] in an external gravitational field follows the same trajectory, regardless of its internal structure and composition. To be more explicit, we can consider the Newtonian case. Newton's Second Law reads $m_i \ddot{\mathbf{x}} = \mathbf{F}$, where m_i is the *inertial mass* of the particle. If $\mathbf{F}$ is the gravitational force on our particle generated by a point-like body with mass M, we have

$$\mathbf{F} = G_N \frac{M m_g}{r^2} \hat{\mathbf{r}}, \qquad (5.1)$$

where m_g is the *gravitational mass* of the particle (and we are assuming that $m_g \ll M$). In principle, m_i and m_g may be different, because the former has nothing to do with the gravitational force (it is well defined even in the absence of gravity!) and

---

[1] A test-particle must have a sufficiently small mass, size, etc. such that its mass does not significantly alter the background gravitational field, tidal forces can be ignored, etc.

© Springer Nature Singapore Pte Ltd. 2018
C. Bambi, *Introduction to General Relativity*, Undergraduate Lecture Notes in Physics, https://doi.org/10.1007/978-981-13-1090-4_5

the second is the "gravitational charge" of the particle. For instance, in the case of an electrostatic field, the force is given by the Coulomb force, which is proportional to the product of the electric charges of the two objects. The electric charge is completely independent of the inertial mass of a body. On the contrary, the ratio between the inertial and the gravitational masses, m_i/m_g, is a constant independent of the particle. This is an experimental result! We can thus choose units in which $m_i = m_g = m$, where m is just the mass of the particle. At this point, Newton's Second Law reads

$$\ddot{\mathbf{x}} = G_N \frac{M}{r^2} \hat{\mathbf{r}} , \tag{5.2}$$

and the solution is independent of m and the internal structure and composition of the particle: any test-particle follows the same trajectory for the same initial conditions.

The trajectory of a particle can be obtained by minimizing the path length between two events of the spacetime. We can thus think of writing an effective metric such that the equations of motion of the particle take into account the effect of the gravitational field. The example below can better illustrate this point.

In Newtonian mechanics, the Lagrangian of a particle in a gravitational field is $L = T - V$, where T is the particle kinetic energy, $V = m\Phi$ is the gravitational potential energy, and Φ is the gravitational potential; see Sect. 1.8. As seen in Chap. 3, in special relativity, for small velocities T is replaced by Eq. (3.11). The Lagrangian of a non-relativistic particle in a Newtonian gravitational field is thus

$$L = -mc^2 + \frac{1}{2}mv^2 - m\Phi . \tag{5.3}$$

Since

$$mc^2 \gg \frac{1}{2}mv^2 , \quad -m\Phi , \tag{5.4}$$

we can rewrite Eq. (5.3) as

$$L = -mc\sqrt{c^2 - v^2 + 2\Phi} , \tag{5.5}$$

and the corresponding action as

$$\begin{aligned} S &= -mc \int \sqrt{\left(1 + \frac{2\Phi}{c^2}\right)c^2 - \dot{x}^2 - \dot{y}^2 - \dot{z}^2} dt \\ &= -mc \int \sqrt{-g_{\mu\nu}\dot{x}^\mu \dot{x}^\nu} dt , \end{aligned} \tag{5.6}$$

where we have introduced the metric tensor $g_{\mu\nu}$ defined as

$$\|g_{\mu\nu}\| = \begin{pmatrix} -\left(1 + \frac{2\Phi}{c^2}\right) & 0 & 0 & 0 \\ 0 & 1 & 0 & 0 \\ 0 & 0 & 1 & 0 \\ 0 & 0 & 0 & 1 \end{pmatrix}. \tag{5.7}$$

If we apply the Least Action Principle to the action in Eq. (5.6), we obtain the geodesic equations for the metric $g_{\mu\nu}$. They are equivalent to the Euler–Lagrange equations for the Lagrangian in (5.3) by construction. So we can describe the gravitational field as a geometrical property of the spacetime.

With this simple example, we see how we can "absorb" the gravitational field into the metric tensor $g_{\mu\nu}$. The particle trajectories provided by the geodesic equations for the metric $g_{\mu\nu}$ are not straight lines, because $g_{\mu\nu}$ takes gravity into account. Note that $g_{\mu\nu}$ cannot be reduced to the Minkowski metric in the whole spacetime with a coordinate transformation and we say that the spacetime is *curved*. On the contrary, if we can recover the Minkowski metric $\eta_{\mu\nu}$ in the whole spacetime with a coordinate transformation, the spacetime is *flat*. In this second case, the reference frame in which the metric is not $\eta_{\mu\nu}$ either employs non-Cartesian coordinates or is a non-inertial reference frame (or both).

5.2 Covariant Derivative

The partial derivative of a scalar is a dual vector and it is easy to see that it transforms as a dual vector under a coordinate transformation

$$\frac{\partial \phi}{\partial x^\mu} \to \frac{\partial \phi}{\partial x'^\mu} = \frac{\partial x^\nu}{\partial x'^\mu} \frac{\partial \phi}{\partial x^\nu}. \tag{5.8}$$

The partial derivative of the components of a vector field is not a tensor field. Let V^μ be a vector and $x^\mu \to x'^\mu$ a coordinate transformation. We have

$$\frac{\partial V^\mu}{\partial x^\nu} \to \frac{\partial V'^\mu}{\partial x'^\nu} = \frac{\partial x^\rho}{\partial x'^\nu} \frac{\partial}{\partial x^\rho} \left(\frac{\partial x'^\mu}{\partial x^\sigma} V^\sigma \right)$$

$$= \frac{\partial x^\rho}{\partial x'^\nu} \frac{\partial x'^\mu}{\partial x^\sigma} \frac{\partial V^\sigma}{\partial x^\rho} + \frac{\partial x^\rho}{\partial x'^\nu} \frac{\partial^2 x'^\mu}{\partial x^\rho \partial x^\sigma} V^\sigma. \tag{5.9}$$

If the relation between the two coordinate systems is not linear, we have also the second term on the right hand side and we see that $\partial V^\mu / \partial x^\nu$ cannot be a vector. The reason is that dV^μ is the difference between two vectors at different points. With the terminology of Appendix C, vectors at different points belong to different tangent spaces. dV^μ transforms as

$$dV^\mu \rightarrow dV'^\mu = V'^\mu(x + dx) - V'^\mu(x)$$

$$= \left(\frac{\partial x'^\mu}{\partial x^\alpha}\right)_{x+dx} V^\alpha(x + dx) - \left(\frac{\partial x'^\mu}{\partial x^\alpha}\right)_x V^\alpha(x). \quad (5.10)$$

If $\partial x'^\mu / \partial x^\alpha$ in front of $V^\alpha(x + dx)$ were the same as that in front of $V^\alpha(x)$, then we would have

$$dV'^\mu = \frac{\partial x'^\mu}{\partial x^\alpha} \left[V^\alpha(x + dx) - V^\alpha(x)\right] = \frac{\partial x'^\mu}{\partial x^\alpha} dV^\alpha. \quad (5.11)$$

However, in general this is not the case: $\partial x'^\mu / \partial x^\alpha$ in front of $V^\alpha(x + dx)$ is evaluated at $x + dx$, that in front of $V^\alpha(x)$ is evaluated at x. In this section we want to introduce the concept of covariant derivative, which is the natural generalization of partial derivative in the case of arbitrary coordinates.

5.2.1   Definition

We know that dx^μ is a 4-vector and that the 4-velocity of a particle, $u^\mu = dx^\mu/d\tau$ is a 4-vector too, since $d\tau$ is a scalar. However, we know from Eq. (5.10) that du^μ is not a 4-vector.

In Sect. 1.7, we introduced the geodesic equations. Since $u^\mu = dx^\mu/d\tau$, we can rewrite the geodesic equations as

$$\frac{du^\mu}{d\tau} + \Gamma^\mu_{\nu\rho} u^\rho \frac{dx^\nu}{d\tau} = 0, \quad (5.12)$$

and also as

$$\frac{Du^\mu}{d\tau} = 0, \quad (5.13)$$

where we have defined Du^μ as

$$Du^\mu = du^\mu + \Gamma^\mu_{\nu\rho} u^\rho dx^\nu = \left(\frac{\partial u^\mu}{\partial x^\nu} + \Gamma^\mu_{\nu\rho} u^\rho\right) dx^\nu. \quad (5.14)$$

We will now show that Du^μ is the natural generalization of du^μ for general coordinate systems and that the partial derivative ∂_μ generalizes to the *covariant derivative* ∇_μ. In the case of a 4-vector like u^μ, we have $Du^\mu = (\nabla_\nu u^\mu)dx^\nu$, where ∇_ν is defined as

$$\nabla_\nu u^\mu = \frac{\partial u^\mu}{\partial x^\nu} + \Gamma^\mu_{\nu\rho} u^\rho. \quad (5.15)$$

First, we check that the components of $\nabla_\nu u^\mu$ transform as a tensor. The first term on the right hand side in Eq. (5.15) transforms as

$$
\frac{\partial u^\mu}{\partial x^\nu} \rightarrow \frac{\partial u'^\mu}{\partial x'^\nu} = \frac{\partial x^\alpha}{\partial x'^\nu} \frac{\partial}{\partial x^\alpha} \left(\frac{\partial x'^\mu}{\partial x^\beta} u^\beta \right)
$$

$$
= \frac{\partial x^\alpha}{\partial x'^\nu} \frac{\partial x'^\mu}{\partial x^\beta} \frac{\partial u^\beta}{\partial x^\alpha} + \frac{\partial x^\alpha}{\partial x'^\nu} \frac{\partial^2 x'^\mu}{\partial x^\alpha \partial x^\beta} u^\beta. \tag{5.16}
$$

The Christoffel symbols transform as

$$
\Gamma^\mu_{\nu\rho} \rightarrow \Gamma'^\mu_{\nu\rho} = \frac{1}{2} g'^{\mu\sigma} \left(\frac{\partial g'_{\sigma\rho}}{\partial x'^\nu} + \frac{\partial g'_{\nu\sigma}}{\partial x'^\rho} - \frac{\partial g'_{\nu\rho}}{\partial x'^\sigma} \right)
$$

$$
= \frac{1}{2} \frac{\partial x'^\mu}{\partial x^\alpha} \frac{\partial x'^\sigma}{\partial x^\beta} g^{\alpha\beta} \frac{\partial x^\gamma}{\partial x'^\nu} \frac{\partial}{\partial x^\gamma} \left(\frac{\partial x^\delta}{\partial x'^\sigma} \frac{\partial x^\varepsilon}{\partial x'^\rho} g_{\delta\varepsilon} \right)
$$

$$
+ \frac{1}{2} \frac{\partial x'^\mu}{\partial x^\alpha} \frac{\partial x'^\sigma}{\partial x^\beta} g^{\alpha\beta} \frac{\partial x^\gamma}{\partial x'^\rho} \frac{\partial}{\partial x^\gamma} \left(\frac{\partial x^\delta}{\partial x'^\nu} \frac{\partial x^\varepsilon}{\partial x'^\sigma} g_{\delta\varepsilon} \right)
$$

$$
- \frac{1}{2} \frac{\partial x'^\mu}{\partial x^\alpha} \frac{\partial x'^\sigma}{\partial x^\beta} g^{\alpha\beta} \frac{\partial x^\gamma}{\partial x'^\sigma} \frac{\partial}{\partial x^\gamma} \left(\frac{\partial x^\delta}{\partial x'^\nu} \frac{\partial x^\varepsilon}{\partial x'^\rho} g_{\delta\varepsilon} \right). \tag{5.17}
$$

Since the calculations become long, we consider the three terms on the right hand side in Eq. (5.17) separately. The first term is

$$
\frac{1}{2} \frac{\partial x'^\mu}{\partial x^\alpha} \frac{\partial x'^\sigma}{\partial x^\beta} g^{\alpha\beta} \frac{\partial x^\gamma}{\partial x'^\nu} \frac{\partial}{\partial x^\gamma} \left(\frac{\partial x^\delta}{\partial x'^\sigma} \frac{\partial x^\varepsilon}{\partial x'^\rho} g_{\delta\varepsilon} \right)
$$

$$
= \frac{1}{2} \frac{\partial x'^\mu}{\partial x^\alpha} \frac{\partial x'^\sigma}{\partial x^\beta} g^{\alpha\beta} \frac{\partial x^\gamma}{\partial x'^\nu} \frac{\partial^2 x^\delta}{\partial x'^\tau \partial x'^\sigma} \frac{\partial x'^\tau}{\partial x^\gamma} \frac{\partial x^\varepsilon}{\partial x'^\rho} g_{\delta\varepsilon}
$$

$$
+ \frac{1}{2} \frac{\partial x'^\mu}{\partial x^\alpha} \frac{\partial x'^\sigma}{\partial x^\beta} g^{\alpha\beta} \frac{\partial x^\gamma}{\partial x'^\nu} \frac{\partial x^\delta}{\partial x'^\sigma} \frac{\partial^2 x^\varepsilon}{\partial x'^\tau \partial x'^\rho} \frac{\partial x'^\tau}{\partial x^\gamma} g_{\delta\varepsilon}
$$

$$
+ \frac{1}{2} \frac{\partial x'^\mu}{\partial x^\alpha} \frac{\partial x'^\sigma}{\partial x^\beta} g^{\alpha\beta} \frac{\partial x^\gamma}{\partial x'^\nu} \frac{\partial x^\delta}{\partial x'^\sigma} \frac{\partial x^\varepsilon}{\partial x'^\rho} \frac{\partial g_{\delta\varepsilon}}{\partial x^\gamma}
$$

$$
= \frac{1}{2} \frac{\partial x'^\mu}{\partial x^\alpha} \frac{\partial x'^\sigma}{\partial x^\beta} g^{\alpha\beta} \frac{\partial^2 x^\delta}{\partial x'^\nu \partial x'^\sigma} \frac{\partial x^\varepsilon}{\partial x'^\rho} g_{\delta\varepsilon} + \frac{1}{2} \frac{\partial x'^\mu}{\partial x^\alpha} g^{\alpha\beta} \frac{\partial^2 x^\varepsilon}{\partial x'^\nu \partial x'^\rho} g_{\beta\varepsilon}
$$

$$
+ \frac{1}{2} \frac{\partial x'^\mu}{\partial x^\alpha} g^{\alpha\beta} \frac{\partial x^\gamma}{\partial x'^\nu} \frac{\partial x^\varepsilon}{\partial x'^\rho} \frac{\partial g_{\beta\varepsilon}}{\partial x^\gamma}. \tag{5.18}
$$

For the second term we have

$$\frac{1}{2}\frac{\partial x'^{\mu}}{\partial x^{\alpha}}\frac{\partial x'^{\sigma}}{\partial x^{\beta}}g^{\alpha\beta}\frac{\partial x^{\gamma}}{\partial x'^{\rho}}\frac{\partial}{\partial x^{\gamma}}\left(\frac{\partial x^{\delta}}{\partial x'^{\nu}}\frac{\partial x^{\varepsilon}}{\partial x'^{\sigma}}g_{\delta\varepsilon}\right)$$

$$=\frac{1}{2}\frac{\partial x'^{\mu}}{\partial x^{\alpha}}g^{\alpha\beta}\frac{\partial^{2}x^{\delta}}{\partial x'^{\nu}\partial x'^{\rho}}g_{\delta\beta}+\frac{1}{2}\frac{\partial x'^{\mu}}{\partial x^{\alpha}}\frac{\partial x'^{\sigma}}{\partial x^{\beta}}g^{\alpha\beta}\frac{\partial x^{\delta}}{\partial x'^{\nu}}\frac{\partial^{2}x^{\varepsilon}}{\partial x'^{\rho}\partial x'^{\sigma}}g_{\delta\varepsilon}$$

$$+\frac{1}{2}\frac{\partial x'^{\mu}}{\partial x^{\alpha}}g^{\alpha\beta}\frac{\partial x^{\gamma}}{\partial x'^{\rho}}\frac{\partial x^{\delta}}{\partial x'^{\nu}}\frac{\partial g_{\delta\beta}}{\partial x^{\gamma}}. \tag{5.19}$$

Lastly, the third term becomes

$$-\frac{1}{2}\frac{\partial x'^{\mu}}{\partial x^{\alpha}}\frac{\partial x'^{\sigma}}{\partial x^{\beta}}g^{\alpha\beta}\frac{\partial x^{\gamma}}{\partial x'^{\sigma}}\frac{\partial}{\partial x^{\gamma}}\left(\frac{\partial x^{\delta}}{\partial x'^{\nu}}\frac{\partial x^{\varepsilon}}{\partial x'^{\rho}}g_{\delta\varepsilon}\right)$$

$$=-\frac{1}{2}\frac{\partial x'^{\mu}}{\partial x^{\alpha}}\frac{\partial x'^{\sigma}}{\partial x^{\beta}}g^{\alpha\beta}\frac{\partial^{2}x^{\delta}}{\partial x'^{\nu}\partial x'^{\sigma}}\frac{\partial x^{\varepsilon}}{\partial x'^{\rho}}g_{\delta\varepsilon}-\frac{1}{2}\frac{\partial x'^{\mu}}{\partial x^{\alpha}}\frac{\partial x'^{\sigma}}{\partial x^{\beta}}g^{\alpha\beta}\frac{\partial x^{\delta}}{\partial x'^{\nu}}\frac{\partial^{2}x^{\varepsilon}}{\partial x'^{\rho}\partial x'^{\sigma}}g_{\delta\varepsilon}$$

$$-\frac{1}{2}\frac{\partial x'^{\mu}}{\partial x^{\alpha}}g^{\alpha\beta}\frac{\partial x^{\delta}}{\partial x'^{\nu}}\frac{\partial x^{\varepsilon}}{\partial x'^{\rho}}\frac{\partial g_{\delta\varepsilon}}{\partial x^{\beta}}. \tag{5.20}$$

If we combine the results in Eqs. (5.18)–(5.20), we find the transformation rule for the Christoffel symbols

$$\Gamma^{\mu}_{\nu\rho}\to\Gamma'^{\mu}_{\nu\rho}=\frac{\partial x'^{\mu}}{\partial x^{\alpha}}\frac{\partial x^{\delta}}{\partial x'^{\nu}}\frac{\partial x^{\varepsilon}}{\partial x'^{\rho}}\frac{1}{2}g^{\alpha\beta}\left(\frac{\partial g_{\beta\varepsilon}}{\partial x^{\delta}}+\frac{\partial g_{\delta\beta}}{\partial x^{\varepsilon}}-\frac{\partial g_{\delta\varepsilon}}{\partial x^{\beta}}\right)$$

$$+\frac{\partial x'^{\mu}}{\partial x^{\alpha}}\frac{\partial^{2}x^{\delta}}{\partial x'^{\nu}\partial x'^{\rho}}g^{\alpha\beta}g_{\delta\beta}$$

$$=\frac{\partial x'^{\mu}}{\partial x^{\alpha}}\frac{\partial x^{\delta}}{\partial x'^{\nu}}\frac{\partial x^{\varepsilon}}{\partial x'^{\rho}}\Gamma^{\alpha}_{\delta\varepsilon}+\frac{\partial x^{\alpha}}{\partial x'^{\nu}\partial x'^{\rho}}\frac{\partial x'^{\mu}}{\partial x^{\alpha}}. \tag{5.21}$$

We see here that the Christoffel symbols do not transform as the components of a tensor, so they cannot be the components of a tensor.

Let us now combine the results in Eqs. (5.16) and (5.21). We find

$$\frac{\partial u^{\mu}}{\partial x^{\nu}}+\Gamma^{\mu}_{\nu\rho}u^{\rho}\to\frac{\partial u'^{\mu}}{\partial x'^{\nu}}+\Gamma'^{\mu}_{\nu\rho}u'^{\rho}=\frac{\partial x^{\alpha}}{\partial x'^{\nu}}\frac{\partial x'^{\mu}}{\partial x^{\beta}}\frac{\partial u^{\beta}}{\partial x^{\alpha}}+\frac{\partial x^{\alpha}}{\partial x'^{\nu}}\frac{\partial^{2}x'^{\mu}}{\partial x^{\alpha}\partial x^{\beta}}u^{\beta}$$

$$+\frac{\partial x'^{\mu}}{\partial x^{\alpha}}\frac{\partial x^{\beta}}{\partial x'^{\nu}}\frac{\partial x^{\gamma}}{\partial x'^{\rho}}\Gamma^{\alpha}_{\beta\gamma}\frac{\partial x'^{\rho}}{\partial x^{\delta}}u^{\delta}$$

$$+\frac{\partial^{2}x^{\alpha}}{\partial x'^{\nu}\partial x'^{\rho}}\frac{\partial x'^{\mu}}{\partial x^{\alpha}}\frac{\partial x'^{\rho}}{\partial x^{\beta}}u^{\beta}$$

$$=\frac{\partial x^{\alpha}}{\partial x'^{\nu}}\frac{\partial x'^{\mu}}{\partial x^{\beta}}\frac{\partial u^{\beta}}{\partial x^{\alpha}}+\frac{\partial x^{\alpha}}{\partial x'^{\nu}}\frac{\partial^{2}x'^{\mu}}{\partial x^{\alpha}\partial x^{\beta}}u^{\beta}$$

$$+\frac{\partial x'^{\mu}}{\partial x^{\alpha}}\frac{\partial x^{\beta}}{\partial x'^{\nu}}\Gamma^{\alpha}_{\beta\gamma}u^{\gamma}$$

$$+\frac{\partial^{2}x^{\alpha}}{\partial x'^{\nu}\partial x'^{\rho}}\frac{\partial x'^{\mu}}{\partial x^{\alpha}}\frac{\partial x'^{\rho}}{\partial x^{\beta}}u^{\beta} \tag{5.22}$$

Note that

$$\frac{\partial x^\alpha}{\partial x'^\nu} \frac{\partial x'^\mu}{\partial x^\alpha} = \delta^\mu_\nu \,,$$

$$\frac{\partial}{\partial x^\beta} \left(\frac{\partial x^\alpha}{\partial x'^\nu} \frac{\partial x'^\mu}{\partial x^\alpha} \right) = \frac{\partial}{\partial x^\beta} \delta^\mu_\nu \,,$$

$$\frac{\partial^2 x^\alpha}{\partial x'^\nu \partial x'^\rho} \frac{\partial x'^\rho}{\partial x^\beta} \frac{\partial x'^\mu}{\partial x^\alpha} + \frac{\partial x^\alpha}{\partial x'^\nu} \frac{\partial^2 x'^\mu}{\partial x^\beta \partial x^\alpha} = 0 \,, \tag{5.23}$$

and therefore

$$\frac{\partial^2 x^\alpha}{\partial x'^\nu \partial x'^\rho} \frac{\partial x'^\rho}{\partial x^\beta} \frac{\partial x'^\mu}{\partial x^\alpha} = -\frac{\partial x^\alpha}{\partial x'^\nu} \frac{\partial^2 x'^\mu}{\partial x^\beta \partial x^\alpha} \,. \tag{5.24}$$

We use Eq. (5.24) in the last term on the right hand side in Eq. (5.22) and we see that the second and the last terms cancel each other. Equation (5.22) can thus be written as

$$\frac{\partial u'^\mu}{\partial x'^\nu} + \Gamma'^\mu_{\nu\rho} u'^\rho = \frac{\partial x^\alpha}{\partial x'^\nu} \frac{\partial x'^\mu}{\partial x^\beta} \left(\frac{\partial u^\beta}{\partial x^\alpha} + \Gamma^\beta_{\alpha\gamma} u^\gamma \right) \,, \tag{5.25}$$

and we see that $\nabla_\nu u^\mu$s transform as the components of a tensor of type (1, 1).

5.2.2 Parallel Transport

As it was pointed out at the beginning of this section, the partial derivative of a vector computes the difference of two vectors belonging to different points and for this reason the new object is not a tensor. The sum or the difference of two vectors is another vector if the two vectors belong to the same vector space, but this is not the case here. Intuitively, we should "transport" one of the two vectors to the point of the other vector and compute the difference there. This is what the covariant derivative indeed does and involves the concept of *parallel transport*.

Let us consider the example illustrated in Fig. 5.1. We have a 2-dimensional Euclidean space and we consider both Cartesian coordinates (x, y) and polar coordinates (r, θ). The vector **V** is at point $A = x_A = \{x^\mu\}$. Its components are (V^x, V^y) in Cartesian coordinates and (V^r, V^θ) in polar coordinates. If we think of "rigidly" transporting the vector **V** from point A to point $B = x_B = \{x^\mu + dx^\mu\}$, as shown in Fig. 5.1, the Cartesian coordinates do not change

$$(V^x, V^y) \to (V^x, V^y) \,. \tag{5.26}$$

However, the polar coordinates change. This operation is called parallel transport and in what follows we want to show that the components of the parallel transported

Fig. 5.1 The vector **V** at x^μ is parallel transported to the point $x^\mu + dx^\mu$. While the vector **V** does not change under parallel transport, its components change in general

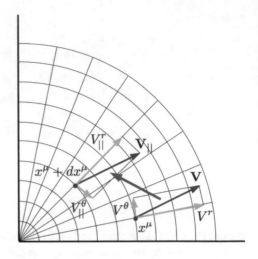

vector are given by

$$V^\mu_{A\to B} = V^\mu_A - \Gamma^\mu_{\nu\rho}(x_A)V^\nu_A dx^\rho \ . \tag{5.27}$$

First, the relations between Cartesian and polar coordinates are

$$x = r\cos\theta \ , \quad y = r\sin\theta \ , \tag{5.28}$$

$$r = \sqrt{x^2 + y^2} \ , \quad \theta = \arctan\frac{y}{x} \ . \tag{5.29}$$

If the vector **V** has Cartesian coordinates (V^x, V^y), its polar coordinates (V^r, V^θ) are

$$V^r = \frac{\partial r}{\partial x}V^x + \frac{\partial r}{\partial y}V^y = \cos\theta V^x + \sin\theta V^y \ ,$$

$$V^\theta = \frac{\partial\theta}{\partial x}V^x + \frac{\partial\theta}{\partial y}V^y = -\frac{\sin\theta}{r}V^x + \frac{\cos\theta}{r}V^y \ . \tag{5.30}$$

If we parallel transport the vector **V** from the point (r, θ) to the point $(r + dr, \theta + d\theta)$, we have the vector $\mathbf{V}_\parallel$ in Fig. 5.1. The radial coordinate $V^r_\parallel$ is

$$V^r_\parallel = \cos(\theta + d\theta)V^x + \sin(\theta + d\theta)V^y$$
$$= (\cos\theta - \sin\theta d\theta)V^x + (\sin\theta + \cos\theta d\theta)V^y \ . \tag{5.31}$$

where we have neglected $O(d\theta^2)$ terms

$$\cos(\theta + d\theta) = \cos\theta\cos d\theta - \sin\theta\sin d\theta$$
$$= \cos\theta - \sin\theta d\theta + O(d\theta^2),$$
$$\sin(\theta + d\theta) = \sin\theta\cos d\theta + \cos\theta\sin d\theta$$
$$= \sin\theta + \cos\theta d\theta + O(d\theta^2). \tag{5.32}$$

For the polar coordinate $V_{\|}^{\theta}$, we have

$$V_{\|}^{\theta} = -\frac{\sin(\theta + d\theta)}{(r + dr)}V^x + \frac{\cos(\theta + d\theta)}{(r + dr)}V^y. \tag{5.33}$$

Since

$$\frac{\sin(\theta + d\theta)}{(r + dr)} = \frac{\sin\theta}{r} + \frac{\cos\theta}{r}d\theta - \frac{\sin\theta}{r^2}dr + O(dr^2, drd\theta, d\theta^2),$$
$$\frac{\cos(\theta + d\theta)}{(r + dr)} = \frac{\cos\theta}{r} - \frac{\sin\theta}{r}d\theta - \frac{\cos\theta}{r^2}dr + O(dr^2, drd\theta, d\theta^2), \tag{5.34}$$

Equation (5.33) becomes

$$V_{\|}^{\theta} = \left(-\frac{\sin\theta}{r} - \frac{\cos\theta}{r}d\theta + \frac{\sin\theta}{r^2}dr\right)V^x$$
$$+ \left(\frac{\cos\theta}{r} - \frac{\sin\theta}{r}d\theta - \frac{\cos\theta}{r^2}dr\right)V^y. \tag{5.35}$$

In polar coordinates, the line element reads

$$dl^2 = dr^2 + r^2 d\theta^2. \tag{5.36}$$

As we have seen in Sect. 1.7, the Christoffel symbols can be more quickly calculated from the comparison of the Euler–Lagrange equations for a free particle with the geodesic equations. The Lagrangian to employ is

$$L = \frac{1}{2}\left(\dot{r}^2 + r^2\dot{\theta}^2\right). \tag{5.37}$$

The Euler–Lagrange equation for the Lagrangian coordinate r is

$$\frac{d\dot{r}}{dt} - r\dot{\theta}^2 = 0,$$
$$\ddot{r} - r\dot{\theta}^2 = 0. \tag{5.38}$$

For the Lagrangian coordinate θ, we have

$$\frac{d}{dt}\left(r^2\dot\theta\right) = 0\,,$$

$$\ddot\theta + \frac{2}{r}\dot r\dot\theta = 0\,. \tag{5.39}$$

If we compare Eqs. (5.38) and (5.39) with the geodesic equations, we see that the non-vanishing Christoffel symbols are

$$\Gamma^r_{\theta\theta} = -r\,, \quad \Gamma^\theta_{r\theta} = \Gamma^\theta_{\theta r} = \frac{1}{r}\,. \tag{5.40}$$

Now we want to see that, if we parallel transport the vector **V** from point A to point B, the coordinates of the vector at B are given by

$$V^\mu_{A\to B} = V^\mu_A - \Gamma^\mu_{\nu\rho}(x_A)V^\nu_A dx^\rho\,. \tag{5.41}$$

For the radial coordinate we have

$$\begin{aligned}
V^r_{A\to B} &= V^r_A + rV^\theta_A d\theta \\
&= \cos\theta V^x + \sin\theta V^y + r\left(-\frac{\sin\theta}{r}V^x + \frac{\cos\theta}{r}V^y\right)d\theta \\
&= (\cos\theta - \sin\theta d\theta)\,V^x + (\sin\theta + \cos\theta d\theta)\,V^y\,.
\end{aligned} \tag{5.42}$$

For the polar coordinate we have

$$\begin{aligned}
V^\theta_{A\to B} &= V^\theta_A - \frac{1}{r}V^r_A d\theta - \frac{1}{r}V^\theta_A dr \\
&= -\frac{\sin\theta}{r}V^x + \frac{\cos\theta}{r}V^y - \frac{1}{r}\left(\cos\theta V^x + \sin\theta V^y\right)d\theta \\
&\quad - \frac{1}{r}\left(-\frac{\sin\theta}{r}V^x + \frac{\cos\theta}{r}V^y\right)dr \\
&= \left(-\frac{\sin\theta}{r} - \frac{\cos\theta}{r}d\theta + \frac{\sin\theta}{r^2}dr\right)V^x \\
&\quad + \left(\frac{\cos\theta}{r} - \frac{\sin\theta}{r}d\theta - \frac{\cos\theta}{r^2}dr\right)V^y\,.
\end{aligned} \tag{5.43}$$

We thus see that we recover the results in Eqs. (5.31) and (5.35).

When we compute the covariant derivative of a vector V^μ we are calculating

$$\begin{aligned}
\nabla_\nu V^\mu &= \frac{\partial V^\mu}{\partial x^\nu} + \Gamma^\mu_{\nu\rho}V^\rho \\
&= \lim_{dx^\nu\to 0}\frac{V^\mu(x+dx) - V^\mu(x) + \Gamma^\mu_{\nu\rho}V^\rho dx^\nu}{dx^\nu}
\end{aligned}$$

$$= \lim_{dx^\nu \to 0} \frac{V^\mu_{x+dx \to x}(x) - V^\mu(x)}{dx^\nu}, \tag{5.44}$$

that is, we calculate the difference between the vector $V^\mu(x + dx)$ parallel transported to x and the vector $V^\mu(x)$. Note that now the sign in front of the Christoffel
symbols is plus because we are transporting a vector from $x + dx$ to x, while in
Eq. (5.41) we have the opposite case.

5.2.3 Properties of the Covariant Derivative

From the discussion above, it is clear that the covariant derivative of a scalar reduces
to the ordinary partial derivative

$$\nabla_\mu \phi = \frac{\partial \phi}{\partial x^\mu}. \tag{5.45}$$

Indeed a scalar is just a number and the parallel transport is trivial: $\phi_{A \to B} = \phi_A$.
    The covariant derivative of a dual vector is given by

$$\nabla_\mu V_\nu = \frac{\partial V_\nu}{\partial x^\mu} - \Gamma^\rho_{\mu\nu} V_\rho. \tag{5.46}$$

Indeed, if we consider any vector V^μ and dual vector W_μ, $V^\mu W_\mu$ is a scalar. Enforcing
Leibniz's rule for the covariant derivative, we have

$$\nabla_\mu (V^\nu W_\nu) = \left(\nabla_\mu V^\nu\right) W_\nu + V^\nu \left(\nabla_\mu W_\nu\right)$$
$$= \frac{\partial V^\nu}{\partial x^\mu} W_\nu + \Gamma^\nu_{\mu\rho} V^\rho W_\nu + V^\nu \left(\nabla_\mu W_\nu\right). \tag{5.47}$$

Since ∇_μ becomes ∂_μ for a scalar function

$$\nabla_\mu (V^\nu W_\nu) = \frac{\partial V^\nu}{\partial x^\mu} W_\nu + V^\nu \frac{\partial W_\nu}{\partial x^\mu}, \tag{5.48}$$

and therefore, equating Eqs. (5.47) and (5.48), we must have

$$\nabla_\mu W_\nu = \frac{\partial W_\nu}{\partial x^\mu} - \Gamma^\rho_{\mu\nu} W_\rho. \tag{5.49}$$

    The generalization to tensors of any type is straightforward and we have

$$\nabla_\lambda T^{\mu_1 \mu_2 \dots \mu_r}_{\nu_1 \nu_2 \dots \nu_s} = \frac{\partial}{\partial x^\lambda} T^{\mu_1 \mu_2 \dots \mu_r}_{\nu_1 \nu_2 \dots \nu_s}$$

$$\underbrace{+\Gamma^{\mu_1}_{\lambda\sigma}T^{\sigma\mu_2\dots\mu_r}_{\nu_1\nu_2\dots\nu_s} + \Gamma^{\mu_2}_{\lambda\sigma}T^{\mu_1\sigma\dots\mu_r}_{\nu_1\nu_2\dots\nu_s} + \dots + \Gamma^{\mu_r}_{\lambda\sigma}T^{\mu_1\mu_2\dots\sigma}_{\nu_1\nu_2\dots\nu_s}}_{r \text{ terms}}$$

$$\underbrace{-\Gamma^{\sigma}_{\lambda\nu_1}T^{\mu_1\mu_2\dots\mu_r}_{\sigma\nu_2\dots\nu_s} - \Gamma^{\sigma}_{\lambda\nu_2}T^{\mu_1\mu_2\dots\mu_r}_{\nu_1\sigma\dots\nu_s} - \dots - \Gamma^{\sigma}_{\lambda\nu_s}T^{\mu_1\mu_2\dots\mu_r}_{\nu_1\nu_2\dots\sigma}}_{s \text{ terms}} \quad (5.50)$$

It is worth noting that the covariant derivative of the metric tensor vanishes. If we calculate $\nabla_\mu g_{\nu\rho}$, we find

$$\begin{aligned}
\nabla_\mu g_{\nu\rho} &= \frac{\partial g_{\nu\rho}}{\partial x^\mu} - \Gamma^\kappa_{\mu\nu}g_{\kappa\rho} - \Gamma^\kappa_{\mu\rho}g_{\nu\kappa} \\
&= \frac{\partial g_{\nu\rho}}{\partial x^\mu} - \frac{1}{2}g^{\kappa\lambda}\left(\frac{\partial g_{\lambda\nu}}{\partial x^\mu} + \frac{\partial g_{\mu\lambda}}{\partial x^\nu} - \frac{\partial g_{\mu\nu}}{\partial x^\lambda}\right)g_{\kappa\rho} \\
&\quad - \frac{1}{2}g^{\kappa\lambda}\left(\frac{\partial g_{\lambda\rho}}{\partial x^\mu} + \frac{\partial g_{\mu\lambda}}{\partial x^\rho} - \frac{\partial g_{\mu\rho}}{\partial x^\lambda}\right)g_{\nu\kappa}
\end{aligned} \quad (5.51)$$

Since $g^{\kappa\lambda}g_{\kappa\rho} = \delta^\lambda_\rho$ and $g^{\kappa\lambda}g_{\nu\kappa} = \delta^\lambda_\nu$, we obtain

$$\begin{aligned}
\nabla_\mu g_{\nu\rho} &= \frac{\partial g_{\nu\rho}}{\partial x^\mu} - \frac{1}{2}\frac{\partial g_{\rho\nu}}{\partial x^\mu} - \frac{1}{2}\frac{\partial g_{\mu\rho}}{\partial x^\nu} + \frac{1}{2}\frac{\partial g_{\mu\nu}}{\partial x^\rho} \\
&\quad - \frac{1}{2}\frac{\partial g_{\nu\rho}}{\partial x^\mu} - \frac{1}{2}\frac{\partial g_{\mu\nu}}{\partial x^\rho} + \frac{1}{2}\frac{\partial g_{\mu\rho}}{\partial x^\nu} = 0.
\end{aligned} \quad (5.52)$$

5.3   Useful Expressions

In this section we will derive a number of useful expressions and identities involving the metric tensor, the Christoffel symbols, and the covariant derivative.

We indicate with g the determinant of the metric tensor $g_{\mu\nu}$ and with $\tilde{g}_{\mu\nu}$ the cofactor (μ, ν). The determinant g is defined as

$$g = \sum_\nu g_{\mu\nu}\tilde{g}_{\mu\nu} \quad (\text{no summation over } \mu). \quad (5.53)$$

The cofactor $\tilde{g}_{\mu\nu}$ can be written in terms of the determinant g and of the inverse of the metric tensor as

$$\tilde{g}_{\mu\nu} = gg^{\mu\nu}. \quad (5.54)$$

Indeed, if we have an invertible square matrix A, its inverse is given by

$$A^{-1} = \frac{1}{\det(A)}C^T, \quad (5.55)$$

where C is the cofactor matrix and C^T is the transpose of C (the proof of this formula can be found in a textbook on linear algebra). If we apply Eq. (5.55) to the metric tensor, we recover Eq. (5.54). If we plug Eq. (5.54) into (5.53), we find $g = g$, which confirms that Eq. (5.54) is correct.

With the help of Eq. (5.54) we can write

$$\frac{\partial g}{\partial g_{\mu\nu}} = \tilde{g}_{\mu\nu} = g g^{\mu\nu}. \tag{5.56}$$

as well as

$$\frac{\partial g}{\partial x^\sigma} = \frac{\partial g}{\partial g_{\mu\nu}} \frac{\partial g_{\mu\nu}}{\partial x^\sigma} = g g^{\mu\nu} \frac{\partial g_{\mu\nu}}{\partial x^\sigma}. \tag{5.57}$$

This last expression will be used later.

Let us write the formula for Christoffel symbols

$$\Gamma^\kappa_{\nu\sigma} = \frac{1}{2} g^{\kappa\lambda} \left(\frac{\partial g_{\lambda\sigma}}{\partial x^\nu} + \frac{\partial g_{\nu\lambda}}{\partial x^\sigma} - \frac{\partial g_{\nu\sigma}}{\partial x^\lambda} \right). \tag{5.58}$$

We multiply both sides in Eq. (5.58) by $g_{\kappa\mu}$ and we get

$$g_{\kappa\mu} \Gamma^\kappa_{\nu\sigma} = \frac{1}{2} \left(\frac{\partial g_{\mu\sigma}}{\partial x^\nu} + \frac{\partial g_{\nu\mu}}{\partial x^\sigma} - \frac{\partial g_{\nu\sigma}}{\partial x^\mu} \right). \tag{5.59}$$

We rewrite Eq. (5.59) exchanging the indices μ and ν

$$g_{\kappa\nu} \Gamma^\kappa_{\mu\sigma} = \frac{1}{2} \left(\frac{\partial g_{\nu\sigma}}{\partial x^\mu} + \frac{\partial g_{\mu\nu}}{\partial x^\sigma} - \frac{\partial g_{\mu\sigma}}{\partial x^\nu} \right). \tag{5.60}$$

We sum Eq. (5.59) with (5.60) and we obtain

$$\frac{\partial g_{\mu\nu}}{\partial x^\sigma} = g_{\kappa\mu} \Gamma^\kappa_{\nu\sigma} + g_{\kappa\nu} \Gamma^\kappa_{\mu\sigma}. \tag{5.61}$$

Now we can multiply both sides in Eq. (5.61) by $g g^{\mu\nu}$ and employ Eq. (5.57)

$$\frac{\partial g}{\partial x^\sigma} = g g^{\mu\nu} \left(g_{\kappa\mu} \Gamma^\kappa_{\nu\sigma} + g_{\kappa\nu} \Gamma^\kappa_{\mu\sigma} \right) = g \left(\Gamma^\nu_{\sigma\nu} + \Gamma^\mu_{\sigma\mu} \right) = 2 g \Gamma^\mu_{\sigma\mu}. \tag{5.62}$$

We rewrite Eq. (5.62) as

$$\frac{\partial}{\partial x^\sigma} \ln \sqrt{-g} = \frac{1}{2g} \frac{\partial g}{\partial x^\sigma} = \Gamma^\mu_{\sigma\mu}. \tag{5.63}$$

With the help of Eq. (5.63), the covariant divergence of a generic vector A^μ can be written as

$$
\nabla_\mu A^\mu = \frac{\partial A^\mu}{\partial x^\mu} + \Gamma^\mu_{\sigma\mu} A^\sigma = \frac{\partial A^\mu}{\partial x^\mu} + A^\sigma \frac{\partial}{\partial x^\sigma} \ln \sqrt{-g}
$$
$$
= \frac{1}{\sqrt{-g}} \frac{\partial}{\partial x^\mu} \left(A^\mu \sqrt{-g} \right) . \tag{5.64}
$$

In the case of a generic tensor $A^{\mu\nu}$ of type $(2, 0)$, we can write

$$
\nabla_\mu A^{\mu\nu} = \frac{\partial A^{\mu\nu}}{\partial x^\mu} + \Gamma^\mu_{\sigma\mu} A^{\sigma\nu} + \Gamma^\nu_{\sigma\mu} A^{\mu\sigma}
$$
$$
= \frac{1}{\sqrt{-g}} \frac{\partial}{\partial x^\mu} \left(A^{\mu\nu} \sqrt{-g} \right) + \Gamma^\nu_{\sigma\mu} A^{\mu\sigma} . \tag{5.65}
$$

Note that, if $A^{\mu\nu}$ is antisymmetric, namely $A^{\mu\nu} = -A^{\nu\mu}$, $\Gamma^\nu_{\sigma\mu} A^{\mu\sigma} = 0$, and Eq. (5.65) simplifies to

$$
\nabla_\mu A^{\mu\nu} = \frac{1}{\sqrt{-g}} \frac{\partial}{\partial x^\mu} \left(A^{\mu\nu} \sqrt{-g} \right) . \tag{5.66}
$$

5.4 Riemann Tensor

5.4.1 Definition

We know that if the first partial derivatives are differentiable then the partial derivatives commute, i.e. $\partial_\mu \partial_\nu = \partial_\nu \partial_\mu$; see e.g. [1]. In general, covariant derivatives do not. We can introduce the *Riemann tensor* $R^\lambda_{\rho\nu\mu}$ as the tensor of type $(1, 3)$ defined as

$$
\nabla_\mu \nabla_\nu A_\rho - \nabla_\nu \nabla_\mu A_\rho = R^\lambda_{\rho\nu\mu} A_\lambda , \tag{5.67}
$$

where A_μ is a generic dual vector. $R^\lambda_{\rho\nu\mu}$ is a tensor because $\nabla_\mu \nabla_\nu A_\rho$ and $\nabla_\nu \nabla_\mu A_\rho$ are tensors.

In order to find the explicit expression of the Riemann tensor, first we calculate $\nabla_\mu \nabla_\nu A_\rho$

$$
\nabla_\mu \nabla_\nu A_\rho = \nabla_\mu \left(\frac{\partial A_\rho}{\partial x^\nu} - \Gamma^\lambda_{\nu\rho} A_\lambda \right)
$$
$$
= \frac{\partial}{\partial x^\mu} \left(\frac{\partial A_\rho}{\partial x^\nu} - \Gamma^\lambda_{\nu\rho} A_\lambda \right) - \Gamma^\kappa_{\mu\rho} \left(\frac{\partial A_\kappa}{\partial x^\nu} - \Gamma^\lambda_{\kappa\nu} A_\lambda \right) - \Gamma^\kappa_{\mu\nu} \left(\frac{\partial A_\rho}{\partial x^\kappa} - \Gamma^\lambda_{\kappa\rho} A_\lambda \right)
$$
$$
= \frac{\partial^2 A_\rho}{\partial x^\mu \partial x^\nu} - \frac{\partial \Gamma^\lambda_{\nu\rho}}{\partial x^\mu} A_\lambda - \Gamma^\lambda_{\nu\rho} \frac{\partial A_\lambda}{\partial x^\mu} - \Gamma^\kappa_{\mu\rho} \frac{\partial A_\kappa}{\partial x^\nu} + \Gamma^\kappa_{\mu\rho} \Gamma^\lambda_{\kappa\nu} A_\lambda
$$

$$- \Gamma^\kappa_{\mu\nu} \frac{\partial A_\rho}{\partial x^\kappa} + \Gamma^\kappa_{\mu\nu} \Gamma^\lambda_{\kappa\rho} A_\lambda \,. \tag{5.68}$$

The expression for $\nabla_\nu \nabla_\mu A_\rho$ is

$$\nabla_\nu \nabla_\mu A_\rho = \frac{\partial^2 A_\rho}{\partial x^\nu \partial x^\mu} - \frac{\partial \Gamma^\lambda_{\mu\rho}}{\partial x^\nu} A_\lambda - \Gamma^\lambda_{\mu\rho} \frac{\partial A_\lambda}{\partial x^\nu} - \Gamma^\kappa_{\nu\rho} \frac{\partial A_\kappa}{\partial x^\mu} + \Gamma^\kappa_{\nu\rho} \Gamma^\lambda_{\kappa\mu} A_\lambda$$
$$- \Gamma^\kappa_{\nu\mu} \frac{\partial A_\rho}{\partial x^\kappa} + \Gamma^\kappa_{\nu\mu} \Gamma^\lambda_{\kappa\rho} A_\lambda \,. \tag{5.69}$$

We combine Eqs. (5.68) and (5.69) and we find

$$\nabla_\mu \nabla_\nu A_\rho - \nabla_\nu \nabla_\mu A_\rho = \left(\frac{\partial \Gamma^\lambda_{\mu\rho}}{\partial x^\nu} - \frac{\partial \Gamma^\lambda_{\nu\rho}}{\partial x^\mu} + \Gamma^\kappa_{\mu\rho} \Gamma^\lambda_{\kappa\nu} - \Gamma^\kappa_{\nu\rho} \Gamma^\lambda_{\kappa\mu} \right) A_\lambda \,. \tag{5.70}$$

We can now write the Riemann tensor $R^\lambda{}_{\rho\nu\mu}$ in terms of the Christoffel symbols as follows

$$R^\mu{}_{\nu\rho\sigma} = \frac{\partial \Gamma^\mu_{\nu\sigma}}{\partial x^\rho} - \frac{\partial \Gamma^\mu_{\nu\rho}}{\partial x^\sigma} + \Gamma^\lambda_{\nu\sigma} \Gamma^\mu_{\rho\lambda} - \Gamma^\lambda_{\nu\rho} \Gamma^\mu_{\sigma\lambda} \,. \tag{5.71}$$

It is also useful to have the explicit expression of $R_{\mu\nu\rho\sigma}$. From Eq. (5.71), we lower the upper index with the metric tensor

$$R_{\mu\nu\rho\sigma} = g_{\mu\lambda} R^\lambda{}_{\nu\rho\sigma}$$
$$= g_{\mu\lambda} \left(\frac{\partial \Gamma^\lambda_{\nu\sigma}}{\partial x^\rho} - \frac{\partial \Gamma^\lambda_{\nu\rho}}{\partial x^\sigma} + \Gamma^\kappa_{\nu\sigma} \Gamma^\lambda_{\rho\kappa} - \Gamma^\kappa_{\nu\rho} \Gamma^\lambda_{\sigma\kappa} \right) , \tag{5.72}$$

The first term on the right hand side in Eq. (5.72) can be written as

$$g_{\mu\lambda} \frac{\partial \Gamma^\lambda_{\nu\sigma}}{\partial x^\rho} = g_{\mu\lambda} \frac{\partial}{\partial x^\rho} \left[\frac{1}{2} g^{\lambda\kappa} \left(\frac{\partial g_{\kappa\sigma}}{\partial x^\nu} + \frac{\partial g_{\nu\kappa}}{\partial x^\sigma} - \frac{\partial g_{\nu\sigma}}{\partial x^\kappa} \right) \right]$$
$$= \frac{1}{2} g_{\mu\lambda} \frac{\partial g^{\lambda\kappa}}{\partial x^\rho} \left(\frac{\partial g_{\kappa\sigma}}{\partial x^\nu} + \frac{\partial g_{\nu\kappa}}{\partial x^\sigma} - \frac{\partial g_{\nu\sigma}}{\partial x^\kappa} \right)$$
$$+ \frac{1}{2} g^\kappa_\mu \left(\frac{\partial^2 g_{\kappa\sigma}}{\partial x^\rho \partial x^\nu} + \frac{\partial^2 g_{\nu\kappa}}{\partial x^\rho \partial x^\sigma} - \frac{\partial^2 g_{\nu\sigma}}{\partial x^\rho \partial x^\kappa} \right)$$
$$= -\frac{1}{2} \frac{\partial g_{\mu\lambda}}{\partial x^\rho} g^{\lambda\kappa} \left(\frac{\partial g_{\kappa\sigma}}{\partial x^\nu} + \frac{\partial g_{\nu\kappa}}{\partial x^\sigma} - \frac{\partial g_{\nu\sigma}}{\partial x^\kappa} \right)$$
$$+ \frac{1}{2} \left(\frac{\partial^2 g_{\mu\sigma}}{\partial x^\rho \partial x^\nu} + \frac{\partial^2 g_{\nu\mu}}{\partial x^\rho \partial x^\sigma} - \frac{\partial^2 g_{\nu\sigma}}{\partial x^\rho \partial x^\mu} \right)$$
$$= -\Gamma^\lambda_{\nu\sigma} \frac{\partial g_{\mu\lambda}}{\partial x^\rho} + \frac{1}{2} \left(\frac{\partial^2 g_{\mu\sigma}}{\partial x^\rho \partial x^\nu} + \frac{\partial^2 g_{\nu\mu}}{\partial x^\rho \partial x^\sigma} - \frac{\partial^2 g_{\nu\sigma}}{\partial x^\rho \partial x^\mu} \right) . \tag{5.73}$$

We use Eq. (5.61) to rewrite $\partial g_{\mu\lambda}/\partial x^\rho$ and Eq. (5.73) becomes

$$g_{\mu\lambda} \frac{\partial \Gamma^\lambda_{\nu\sigma}}{\partial x^\rho} = \frac{1}{2} \left(\frac{\partial^2 g_{\mu\sigma}}{\partial x^\rho \partial x^\nu} + \frac{\partial^2 g_{\nu\mu}}{\partial x^\rho \partial x^\sigma} - \frac{\partial^2 g_{\nu\sigma}}{\partial x^\mu \partial x^\rho} \right)$$
$$- g_{\mu\kappa} \Gamma^\lambda_{\nu\sigma} \Gamma^\kappa_{\lambda\rho} - g_{\kappa\lambda} \Gamma^\lambda_{\nu\sigma} \Gamma^\kappa_{\mu\rho} . \tag{5.74}$$

In the same way, we can rewrite the second term on the right hand side of Eq. (5.72)

$$g_{\mu\lambda} \frac{\partial \Gamma^\lambda_{\nu\rho}}{\partial x^\sigma} = \frac{1}{2} \left(\frac{\partial^2 g_{\mu\rho}}{\partial x^\sigma \partial x^\nu} + \frac{\partial^2 g_{\nu\mu}}{\partial x^\sigma \partial x^\rho} - \frac{\partial^2 g_{\nu\rho}}{\partial x^\mu \partial x^\sigma} \right)$$
$$- g_{\mu\kappa} \Gamma^\lambda_{\nu\rho} \Gamma^\kappa_{\lambda\sigma} - g_{\kappa\lambda} \Gamma^\lambda_{\nu\rho} \Gamma^\kappa_{\mu\sigma} . \tag{5.75}$$

With Eqs. (5.74) and (5.75) we can rewrite $R_{\mu\nu\rho\sigma}$ as

$$R_{\mu\nu\rho\sigma} = \frac{1}{2} \left(\frac{\partial^2 g_{\mu\sigma}}{\partial x^\nu \partial x^\rho} + \frac{\partial^2 g_{\nu\rho}}{\partial x^\mu \partial x^\sigma} - \frac{\partial^2 g_{\mu\rho}}{\partial x^\nu \partial x^\sigma} - \frac{\partial^2 g_{\nu\sigma}}{\partial x^\mu \partial x^\rho} \right)$$
$$+ g_{\kappa\lambda} \left(\Gamma^\lambda_{\nu\rho} \Gamma^\kappa_{\mu\sigma} - \Gamma^\lambda_{\nu\sigma} \Gamma^\kappa_{\mu\rho} \right) . \tag{5.76}$$

The Riemann tensor $R_{\mu\nu\rho\sigma}$ is antisymmetric in the first and second indices as well as in the third and forth indices, while it is symmetric if we exchange the first and second indices respectively with the third and fourth indices:

$$R_{\mu\nu\rho\sigma} = -R_{\nu\mu\rho\sigma} = -R_{\mu\nu\sigma\rho} = R_{\rho\sigma\mu\nu} . \tag{5.77}$$

Note that, if $R^\mu_{\nu\rho\sigma} = 0$ in a certain coordinate system, it vanishes in any coordinate system. This follows from the transformation rule of tensors

$$R^\mu_{\nu\rho\sigma} \rightarrow R'^\mu_{\nu\rho\sigma} = \frac{\partial x'^\mu}{\partial x^\alpha} \frac{\partial x^\beta}{\partial x'^\nu} \frac{\partial x^\gamma}{\partial x'^\rho} \frac{\partial x^\delta}{\partial x'^\sigma} R^\alpha_{\beta\gamma\delta} . \tag{5.78}$$

In particular, since in flat spacetime in Cartesian coordinates the Riemann tensor vanishes, it vanishes in any coordinate system even if the Christoffel symbols may not vanish. So in flat spacetime all the components of the Riemann tensor are identically zero. Note that such a statement is not true for the Christoffel symbols, because they transform with the rule in Eq. (5.21), where the last term may be non-zero under a certain coordinate transformation.

5.4.2 Geometrical Interpretation

Here we want to show that the result of parallel transport of a vector depends on the path. With reference to Fig. 5.2, we have the vector **V** at point $A = x_A = \{x^\mu\}$. The vector has components

$$V_A^\mu = V^\mu . \tag{5.79}$$

Let us now parallel transport the vector to point $B = x_B = \{x^\mu + p^\mu\}$, where p^μ is an infinitesimal displacement. After parallel transport, the components of the vector are

$$V_{A \to B}^\mu = V^\mu - \Gamma_{\nu\rho}^\mu(x_A)V^\nu p^\rho . \tag{5.80}$$

Lastly, we parallel transport the vector to point $D = x_D = \{x^\mu + p^\mu + q^\mu\}$, where q^μ is an infinitesimal displacement too. At point D the components of the vector are

$$
\begin{aligned}
V_{A \to B \to D}^\mu &= V_{A \to B}^\mu - \Gamma_{\nu\rho}^\mu(x_B)V_{A \to B}^\nu q^\rho \\
&= V^\mu - \Gamma_{\nu\rho}^\mu(x_A)V^\nu p^\rho - \left[\Gamma_{\nu\rho}^\mu(x_A) + \frac{\partial\Gamma_{\nu\rho}^\mu}{\partial x^\sigma}(x_A)p^\sigma\right]\left[V^\nu - \Gamma_{\tau\upsilon}^\nu(x_A)V^\tau p^\upsilon\right]q^\rho \\
&= V^\mu - \Gamma_{\nu\rho}^\mu(x_A)V^\nu p^\rho - \Gamma_{\nu\rho}^\mu(x_A)V^\nu q^\rho - \frac{\partial\Gamma_{\nu\rho}^\mu}{\partial x^\sigma}(x_A)p^\sigma V^\nu q^\rho \\
&\quad + \Gamma_{\nu\rho}^\mu(x_A)\Gamma_{\tau\upsilon}^\nu(x_A)V^\tau p^\upsilon q^\rho ,
\end{aligned}
\tag{5.81}
$$

where we have neglected terms of order higher than second in the infinitesimal displacements p^μ and q^μ.

Let us now do the same changing path. We start from the vector $\mathbf{V}$ at point A and we parallel transport it to point $C = x_C = \{x^\mu + q^\mu\}$, as shown in Fig. 5.2. The result is the vector $\mathbf{V}_{A \to C}$. We continue and we parallel transport the vector to point D, where the vector components are

$$
\begin{aligned}
V_{A \to C \to D}^\mu &= V^\mu - \Gamma_{\nu\rho}^\mu(x_A)V^\nu q^\rho - \Gamma_{\nu\rho}^\mu(x_A)V^\nu p^\rho - \frac{\partial\Gamma_{\nu\rho}^\mu}{\partial x^\sigma}(x_A)q^\sigma V^\nu p^\rho \\
&\quad + \Gamma_{\nu\rho}^\mu(x_A)\Gamma_{\tau\upsilon}^\nu(x_A)V^\tau q^\upsilon p^\rho .
\end{aligned}
\tag{5.82}
$$

If we compare Eqs. (5.81) and (5.82), we find that

$$
\begin{aligned}
V_{A \to B \to D}^\mu - V_{A \to C \to D}^\mu &= \left(\frac{\partial\Gamma_{\tau\rho}^\mu}{\partial x^\nu} - \frac{\partial\Gamma_{\tau\upsilon}^\mu}{\partial x^\rho} + \Gamma_{\nu\upsilon}^\mu\Gamma_{\tau\rho}^\nu - \Gamma_{\nu\rho}^\mu\Gamma_{\tau\upsilon}^\nu\right)_{x=x_A} V^\tau q^\upsilon p^\rho \\
&= R_{\tau\upsilon\rho}^\mu V^\tau q^\upsilon p^\rho .
\end{aligned}
\tag{5.83}
$$

The difference in the parallel transport between the two paths is regulated by the Riemann tensor. In flat spacetime, the Riemann tensor vanishes, and, indeed, if we parallel transport a vector from one point to another the result is independent of the choice of the path.

Fig. 5.2 If the vector $\mathbf{V}_A$ at $A = \{x^\mu\}$ is parallel transported to point $B = \{x^\mu + p^\mu\}$ and then to point $D = \{x^\mu + p^\mu + q^\mu\}$, we obtain the vector $\mathbf{V}_{A \to B \to D}$. If $\mathbf{V}_A$ is parallel transported to point $C = \{x^\mu + q^\mu\}$ and then to point $D = \{x^\mu + p^\mu + q^\mu\}$, we obtain the vector $\mathbf{V}_{A \to C \to D}$. In general, $\mathbf{V}_{A \to B \to D}$ and $\mathbf{V}_{A \to C \to D}$ are not the same vector

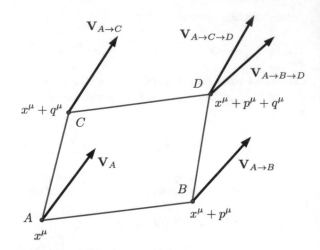

5.4.3   Ricci Tensor and Scalar Curvature

From the Riemann tensor, we can define the Ricci tensor and the scalar curvature after contracting its indices. The *Ricci tensor* is a tensor of second order defined as

$$R_{\mu\nu} = R^\lambda{}_{\mu\lambda\nu} = \frac{\partial \Gamma^\lambda_{\mu\nu}}{\partial x^\lambda} - \frac{\partial \Gamma^\lambda_{\mu\lambda}}{\partial x^\nu} + \Gamma^\kappa_{\mu\nu}\Gamma^\lambda_{\kappa\lambda} - \Gamma^\kappa_{\mu\lambda}\Gamma^\lambda_{\nu\kappa} . \tag{5.84}$$

The Ricci tensor is symmetric

$$R_{\mu\nu} = R_{\nu\mu} . \tag{5.85}$$

Contracting the indices of the Ricci tensor we obtain the *scalar curvature*

$$R = R^\mu{}_\mu = g^{\mu\nu} R_{\mu\nu} . \tag{5.86}$$

With the Ricci tensor and the scalar curvature we can define the *Einstein tensor* as

$$G_{\mu\nu} = R_{\mu\nu} - \frac{1}{2} g_{\mu\nu} R . \tag{5.87}$$

Since both $R_{\mu\nu}$ and $g_{\mu\nu}$ are symmetric tensors of second order, the Einstein tensor is a symmetric tensor of second order as well.

5.4.4 Bianchi Identities

The Bianchi Identities are two important identities involving the Riemann tensor. The *First Bianchi Identity* reads

$$R^{\mu}{}_{\nu\rho\sigma} + R^{\mu}{}_{\rho\sigma\nu} + R^{\mu}{}_{\sigma\nu\rho} = 0, \qquad (5.88)$$

and it can be easily verified by using the explicit expression of the Riemann tensor in Eq. (5.71). Indeed we have

$$
\begin{aligned}
R^{\mu}{}_{\nu\rho\sigma} + R^{\mu}{}_{\rho\sigma\nu} + R^{\mu}{}_{\sigma\nu\rho} &= \frac{\partial \Gamma^{\mu}_{\nu\sigma}}{\partial x^{\rho}} - \frac{\partial \Gamma^{\mu}_{\nu\rho}}{\partial x^{\sigma}} + \Gamma^{\lambda}_{\nu\sigma}\Gamma^{\mu}_{\rho\lambda} - \Gamma^{\lambda}_{\nu\rho}\Gamma^{\mu}_{\sigma\lambda} \\
&\quad + \frac{\partial \Gamma^{\mu}_{\rho\nu}}{\partial x^{\sigma}} - \frac{\partial \Gamma^{\mu}_{\rho\sigma}}{\partial x^{\nu}} + \Gamma^{\lambda}_{\rho\nu}\Gamma^{\mu}_{\sigma\lambda} - \Gamma^{\lambda}_{\rho\sigma}\Gamma^{\mu}_{\nu\lambda} \\
&\quad + \frac{\partial \Gamma^{\mu}_{\sigma\rho}}{\partial x^{\nu}} - \frac{\partial \Gamma^{\mu}_{\sigma\nu}}{\partial x^{\rho}} + \Gamma^{\lambda}_{\sigma\rho}\Gamma^{\mu}_{\nu\lambda} - \Gamma^{\lambda}_{\sigma\nu}\Gamma^{\mu}_{\rho\lambda} \\
&= 0. \qquad (5.89)
\end{aligned}
$$

The *Second Bianchi Identity* reads

$$\nabla_{\mu} R^{\kappa}{}_{\lambda\nu\rho} + \nabla_{\nu} R^{\kappa}{}_{\lambda\rho\mu} + \nabla_{\rho} R^{\kappa}{}_{\lambda\mu\nu} = 0. \qquad (5.90)$$

The first term on the left hand side in Eq. (5.90) can be written as

$$\nabla_{\mu} R^{\kappa}{}_{\lambda\nu\rho} = \nabla_{\mu}\left(\frac{\partial \Gamma^{\kappa}_{\lambda\rho}}{\partial x^{\nu}} - \frac{\partial \Gamma^{\kappa}_{\lambda\nu}}{\partial x^{\rho}} + \Gamma^{\sigma}_{\lambda\rho}\Gamma^{\kappa}_{\nu\sigma} - \Gamma^{\sigma}_{\lambda\nu}\Gamma^{\kappa}_{\rho\sigma} \right). \qquad (5.91)$$

If we choose a coordinate system in which the Christoffel symbols vanish at a certain point (this is always possible, see Sect. 6.4.2), at that point Eq. (5.91) becomes

$$\nabla_{\mu} R^{\kappa}{}_{\lambda\nu\rho} = \frac{\partial^{2} \Gamma^{\kappa}_{\lambda\rho}}{\partial x^{\mu}\partial x^{\nu}} - \frac{\partial^{2} \Gamma^{\kappa}_{\lambda\nu}}{\partial x^{\mu}\partial x^{\rho}}, \qquad (5.92)$$

because in the case of vanishing Christoffel symbols the covariant derivative reduces to the partial one. The second and third terms on the left hand side in Eq. (5.90) read, respectively,

$$\nabla_{\nu} R^{\kappa}{}_{\lambda\rho\mu} = \frac{\partial^{2} \Gamma^{\kappa}_{\lambda\mu}}{\partial x^{\nu}\partial x^{\rho}} - \frac{\partial \Gamma^{\kappa}_{\lambda\rho}}{\partial x^{\nu}\partial x^{\mu}}, \qquad (5.93)$$

$$\nabla_{\rho} R^{\kappa}{}_{\lambda\mu\nu} = \frac{\partial \Gamma^{\kappa}_{\lambda\nu}}{\partial x^{\rho}\partial x^{\mu}} - \frac{\partial \Gamma^{\kappa}_{\lambda\mu}}{\partial x^{\rho}\partial x^{\nu}}. \qquad (5.94)$$

The Second Bianchi Identity can thus be written as

$$\nabla_\mu R^\kappa{}_{\lambda\nu\rho} + \nabla_\nu R^\kappa{}_{\lambda\rho\mu} + \nabla_\rho R^\kappa{}_{\lambda\mu\nu} = \frac{\partial^2 \Gamma^\kappa_{\lambda\rho}}{\partial x^\mu \partial x^\nu} - \frac{\partial^2 \Gamma^\kappa_{\lambda\nu}}{\partial x^\mu \partial x^\rho}$$

$$+ \frac{\partial^2 \Gamma^\kappa_{\lambda\mu}}{\partial x^\nu \partial x^\rho} - \frac{\partial \Gamma^\kappa_{\lambda\rho}}{\partial x^\nu \partial x^\mu}$$

$$+ \frac{\partial \Gamma^\kappa_{\lambda\nu}}{\partial x^\rho \partial x^\mu} - \frac{\partial \Gamma^\kappa_{\lambda\mu}}{\partial x^\rho \partial x^\nu}$$

$$= 0 . \tag{5.95}$$

Since the left hand side is a tensor, if all its components vanish in a certain coordinate system they vanish in any coordinate system, and this concludes the proof of the identity.

From the Second Bianchi Identity we find that the covariant divergence of the Einstein tensor vanishes. As we will see in Sect. 7.1, this is of fundamental importance in Einstein's gravity. If we multiply the Second Bianchi Identity in Eq. (5.90) by g^ν_κ and we sum over the indices κ and ν, we find

$$g^\nu_\kappa \left(\nabla_\mu R^\kappa{}_{\lambda\nu\rho} + \nabla_\nu R^\kappa{}_{\lambda\rho\mu} + \nabla_\rho R^\kappa{}_{\lambda\mu\nu} \right) = 0 ,$$
$$\nabla_\mu \left(g^\nu_\kappa R^\kappa{}_{\lambda\nu\rho} \right) + g^\nu_\kappa \nabla_\nu R^\kappa{}_{\lambda\rho\mu} - \nabla_\rho \left(g^\nu_\kappa R^\kappa{}_{\lambda\nu\mu} \right) = 0 ,$$
$$\nabla_\mu R_{\lambda\rho} + \nabla_\kappa R^\kappa{}_{\lambda\rho\mu} - \nabla_\rho R_{\lambda\mu} = 0 . \tag{5.96}$$

We multiply by $g^{\lambda\rho}$ and we sum over the indices λ and ρ

$$g^{\lambda\rho} \left(\nabla_\mu R_{\lambda\rho} + \nabla_\kappa R^\kappa{}_{\lambda\rho\mu} - \nabla_\rho R_{\lambda\mu} \right) = 0 ,$$
$$\nabla_\mu \left(g^{\lambda\rho} R_{\lambda\rho} \right) - \nabla_\kappa \left(g^{\lambda\rho} g^{\kappa\sigma} R_{\lambda\sigma\rho\mu} \right) - \nabla_\rho \left(g^{\lambda\rho} R_{\lambda\mu} \right) = 0 ,$$
$$\nabla_\mu R - \nabla_\kappa R^\kappa_\mu - \nabla_\rho R^\rho_\mu = 0 ,$$
$$\nabla_\kappa \left(g^\kappa_\mu R - 2 R^\kappa_\mu \right) = 0 . \tag{5.97}$$

This is equivalent to

$$\nabla_\mu G^{\mu\nu} = 0 . \tag{5.98}$$

Problems

5.1 Write the components of the following tensors:

$$\nabla_\mu A_{\alpha\beta} , \quad \nabla_\mu A^{\alpha\beta} , \quad \nabla_\mu A^\alpha{}_\beta , \quad \nabla_\mu A_\alpha{}^\beta , \tag{5.99}$$

5.2 Write the non-vanishing components of the Riemann tensor, the Ricci tensor, and the scalar curvature for the Minkowski spacetime in spherical coordinates.

5.3 Check that the Ricci tensor is symmetric.

Reference

1. M.H. Protter, C.B. Morrey, *A First Course in Real Analysis* (Springer, New York, 1991)

Chapter 6
General Relativity

The theory of special relativity discussed in Chaps. 2–4 is based on the Einstein Principle of Relativity and requires flat spacetime and inertial reference frames. The aim of this chapter is to discuss the extension of such a theoretical framework in order to include gravity and non-inertial reference frames.

From simple considerations, it is clear that Newtonian gravity needs a profound revision. As was already pointed out in Sect. 2.1, Newton's Law of Universal Gravitation is inconsistent with the postulate that there exists a maximum velocity for the propagation of interactions. However, this is not all. In Newton's Law of Universal Gravitation, there is the distance between the two bodies, but distances depend on the reference frame. Moreover, we learned that in special relativity we can transform mass into energy and vice versa. Thus, we have to expect that massless particles also feel and generate gravitational fields, and we need a framework to include them. Lastly, if gravity couples to energy, it should couple to the gravitational energy itself. We should thus expect that the theory is non-linear, which is not the case in Newtonian gravity, where the gravitational field of a multi-body system is simply the sum of the gravitational fields of the single bodies.

6.1 General Covariance

The theory of *general relativity* is based on the *General Principle of Relativity*.

General Principle of Relativity. The laws of physics are the same in all reference frames.

© Springer Nature Singapore Pte Ltd. 2018
C. Bambi, *Introduction to General Relativity*, Undergraduate Lecture Notes
in Physics, https://doi.org/10.1007/978-981-13-1090-4_6

An alternative formulation of the idea that the laws of physics are independent of the choice of reference frame is the *Principle of General Covariance*.

Principle of General Covariance. The form of the laws of physics is invariant under arbitrary differentiable coordinate transformations.

Both the General Principle of Relativity and the Principle of General Covariance are principles: they cannot be proved by theoretical arguments but only tested by experiments. If the latter confirm the validity of these principles, then we can take them as our postulates and formulate physical theories that are consistent with these assumptions.

A *general covariant transformation* is a transformation between two arbitrary reference frames (i.e. not necessarily inertial).[1] An equation is *manifestly covariant* if it is written in terms of tensors only. Indeed we know how tensors change under a coordinate transformation and therefore, if an equation is written in a manifestly covariant form, it is easy to write it in any coordinate system.

Note that in this textbook we distinguish the theory of general relativity from Einstein's gravity. As we have already pointed out, here we call the theory of general relativity the theoretical framework based on the General Principle of Relativity or, equivalently, on the Principle of General Covariance. With the term *Einstein's gravity* we refer instead to the general covariant theory of gravity based on the Einstein–Hilbert action (which will be discussed in the next chapter). Note, however, that different authors/textbooks can use a different terminology and call general relativity the theory of gravity based on the Einstein–Hilbert action.

With the theory of general relativity we can treat non-inertial reference frames and phenomena in gravitational fields. As we have seen at the beginning of the previous chapter, the fact that the motion of a body in a gravitational field is independent of its internal structure and composition permits us to absorb the gravitational field into the metric tensor of the spacetime. The same is true in the case of a non-inertial reference frame. If we are in an inertial reference frame, all free particles move at a constant speed along a straight line. If we consider a non-inertial reference frame, free particles may not move at a constant speed along a straight line any longer, but still they move in the same way independent of their internal structure and composition. We can thus think of absorbing the effects related to the non-inertial reference frame into the spacetime metric.

An example can clarify this point. We are in an elevator with two bodies, as shown in Fig. 6.1. In case A (left picture), the elevator is not moving, but it is in an external gravitational field with acceleration g. The two bodies feel the acceleration g. In case B (central picture), there is no gravitational field, but the elevator has acceleration $a = -g$ as shown in Fig. 6.1. This time the two bodies in the elevator feel an acceleration $-a = g$. If we are inside the elevator, we cannot distinguish

---

[1]Note that in special relativity we talked about "covariance" or "manifestly Lorentz-invariance". "General covariance" is their extension to arbitrary reference frames.

Fig. 6.1 The elevator experiment. See the text for the details

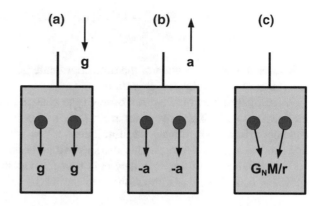

cases A and B. There is no experiment that can do so. This suggests that we should be able to find a common framework to describe physical phenomena in gravitational fields and in non-inertial reference frames.

However, such an analogy holds only locally. If we can perform our experiment for a "sufficiently" long time, where here "sufficiently" depends on the accuracy/precision of our measurements, we can distinguish the cases of a gravitational field and of a non-inertial reference frame. This is illustrated by case C (right picture in Fig. 6.1). The gravitational field generated by a massive body is not perfectly homogeneous. If the gravitational field is generated by a body of finite size, the spacetime should indeed tend to the Minkowski one sufficiently far from the object. Mathematically, this is related to the fact that it is always possible to change reference frame from a non-inertial to an inertial one and recover the physics of special relativity in all spacetime, but it is never possible to remove the gravitational field in *all* spacetime simply by changing coordinate system. Otherwise, the gravitational field would not be real! We can at most consider a coordinate transformation to move to a locally inertial frame (see Sect. 6.4).

An example similar to that introduced at the beginning of the previous chapter can better explain this point. Let us consider a flat spacetime. In the inertial reference frame with the Cartesian coordinates (ct, x, y, z) the line element reads

$$ds^2 = -c^2 dt^2 + dx^2 + dy^2 + dz^2 . \tag{6.1}$$

Then we consider a rotating Cartesian coordinate system (ct', x', y', z') related to (ct, x, y, z) by

$$
\begin{aligned}
x' &= x \cos \Omega t + y \sin \Omega t , \\
y' &= -x \sin \Omega t + y \cos \Omega t , \\
z' &= z ,
\end{aligned}
\tag{6.2}
$$

where Ω is the angular velocity of the rotating system. The line element becomes

$$ds^2 = -\left(c^2 - \Omega^2 x'^2 - \Omega^2 y'^2\right) dt^2 - 2\Omega y'\, dt dx' + 2\Omega x'\, dt dy'$$
$$+dx'^2 + dy'^2 + dz'^2\,. \tag{6.3}$$

Whatever the transformation of the time coordinate is, we see that the line element is not that of the Minkowski spacetime.

We already know how tensors of any type change under a coordinate transformation $x^\mu \rightarrow x'^\mu$. This was given in Eq. (1.30). The difference is that now we can consider arbitrary coordinate transformations $x'^\mu = x'^\mu(x)$, even those to move to non-inertial reference frames. This also shows that the spacetime coordinates cannot be the components of a vector; they only transform as the components of a vector for linear coordinate transformations.

6.2 Einstein Equivalence Principle

The observational fact that the ratio between the inertial and the gravitational masses, m_i/m_g, is a constant independent of the body is encoded in the so-called Weak Equivalence Principle.

> **Weak Equivalence Principle**. The trajectory of a freely-falling test-particle is independent of its internal structure and composition.

Here "freely-falling" means that the particle is in a gravitational field, but there are no other forces acting on it. "Test-particle" means that the particle is too small to be affected by tidal gravitational forces and to alter the gravitational field with its presence (i.e. the so-called "back-reaction" is negligible).

The Einstein Equivalence Principle is the fundamental pillar of the theory of general relativity.

> **Einstein Equivalence Principle**.
> 1. The Weak Equivalence Principle holds.
> 2. The outcome of any local non-gravitational experiment is independent of the velocity of the freely-falling reference frame in which it is performed (*Local Lorentz Invariance*).
> 3. The outcome of any local non-gravitational experiment is independent of where and when it is performed (*Local Position Invariance*).

The Local Lorentz Invariance and the Local Position Invariance replace the Einstein Principle of Relativity in the case of non-inertial reference frames. The Local Lorentz Invariance implies that – locally – we can always find a quasi-inertial reference frame. How this can be implemented will be shown in the next sections of this chapter. The Local Position Invariance requires that the non-gravitational laws of physics are the same at all points of the spacetime; for example, this forbids variation of fundamental constants.

The theories of gravity that satisfy the Einstein Equivalence Principle are the so-called *metric theories of gravity*, which are defined as follows:

Metric theories of gravity.
1. The spacetime is equipped with a symmetric metric.
2. The trajectories of freely-falling test-particles are the geodesics of the metric.
3. In local freely-falling reference frames, the non-gravitational laws of physics are the same as in special relativity.

6.3 Connection to the Newtonian Potential

We have already seen in Sect. 5.1 how to connect the gravitational field of Newtonian gravity, Φ, with the spacetime metric $g_{\mu\nu}$. In this section, we propose an alternative derivation.

We employ Cartesian coordinates and require the following conditions in order to recover Newton's gravity: (i) the gravitational field is "weak", so we can write $g_{\mu\nu}$ as the Minkowski metric plus a small correction, (ii) the gravitational field is stationary, i.e. independent of time, and (iii) the motion of particles is non-relativistic, i.e. the particle speed is much smaller than the speed of light. These three conditions can be written as

$$(i) \quad g_{\mu\nu} = \eta_{\mu\nu} + h_{\mu\nu} \quad |h_{\mu\nu}| \ll 1 , \tag{6.4}$$

$$(ii) \quad \frac{\partial g_{\mu\nu}}{\partial t} = 0 , \tag{6.5}$$

$$(iii) \quad \frac{dx^i}{dt} \ll c . \tag{6.6}$$

From the condition in (6.6), the geodesic equations become

$$\ddot{x}^\mu + \Gamma^\mu_{tt} c^2 \dot{t}^2 = 0 , \tag{6.7}$$

because $\dot{x}^i = (dx^i/dt)\dot{t} \ll c\dot{t}$. Note that here the dot ˙ indicates the derivative with respect to the proper time τ. In the Christoffel symbols Γ^μ_{tt}s, we only consider the linear terms in h and we ignore those of higher order

Table 6.1 Mass, radius, and value of $|2\Phi/c^2|$ at their radius for the Sun, Earth, and a proton. As we can see, $|2\Phi/c^2| \ll 1$

Object	Mass	Radius	$|2\Phi/c^2|$
Sun	$1.99 \cdot 10^{33}$ g	$6.96 \cdot 10^5$ km	$4.24 \cdot 10^{-6}$
Earth	$5.97 \cdot 10^{27}$ g	$6.38 \cdot 10^3$ km	$1.39 \cdot 10^{-9}$
Proton	$1.67 \cdot 10^{-24}$ g	0.8 fm	$3 \cdot 10^{-39}$

$$\Gamma_{tt}^t = \frac{1}{2} g^{tv} \left(\frac{1}{c} \frac{\partial g_{vt}}{\partial t} + \frac{1}{c} \frac{\partial g_{tv}}{\partial t} - \frac{\partial g_{tt}}{\partial x^v} \right) = O\left(h^2\right) ,$$

$$\Gamma_{tt}^i = \frac{1}{2} g^{iv} \left(\frac{1}{c} \frac{\partial g_{vt}}{\partial t} + \frac{1}{c} \frac{\partial g_{tv}}{\partial t} - \frac{\partial g_{tt}}{\partial x^v} \right) = -\frac{1}{2} \eta^{ij} \frac{\partial h_{tt}}{\partial x^j} + O\left(h^2\right) . \quad (6.8)$$

The geodesic equations reduce to

$$\ddot{t} = 0, \quad (6.9)$$

$$\ddot{x}^i = \frac{1}{2} \frac{\partial h_{tt}}{\partial x^i} c^2 \dot{t}^2 . \quad (6.10)$$

$\dot{x}^i = (dx^i/dt)\dot{t}$ and, from Eq. (6.9), $\ddot{x}^i = (d^2x^i/dt^2)\dot{t}^2$. Equation (6.10) becomes

$$\frac{d^2x^i}{dt^2} = \frac{c^2}{2} \frac{\partial h_{tt}}{\partial x^i} . \quad (6.11)$$

In Newtonian gravity in Cartesian coordinates, Newton's Second Law reads

$$\frac{d^2x^i}{dt^2} = -\frac{\partial \Phi}{\partial x^i} . \quad (6.12)$$

From the comparison of Eqs. (6.11) and (6.12), we see that

$$h_{tt} = -\frac{2\Phi}{c^2} + C , \quad (6.13)$$

where C is a constant that should be zero because at large radii $h_{tt} \to 0$ in order to recover the Minkowski metric. We thus obtain again the result

$$g_{tt} = -\left(1 + \frac{2\Phi}{c^2} \right) . \quad (6.14)$$

Table 6.1 shows the value of $|2\Phi/c^2|$ at their surface for the Sun, the Earth, and a proton. This quantity is always very small and a posteriori justifies our assumption (6.4).

6.4 Locally Inertial Frames

6.4.1 Locally Minkowski Reference Frames

Since the metric is a symmetric tensor, in an n-dimensional spacetime $g_{\mu\nu}$ has $n(n + 1)/2$ distinct components. If $n = 4$, there are 10 distinct components. In a curved spacetime, it is not possible to reduce $g_{\mu\nu}$ to the Minkowski metric $\eta_{\mu\nu}$ over the whole spacetime with a coordinate transformation. If it were possible, the spacetime would be flat and not curved by definition. From the mathematical point of view, the fact that such a transformation is not possible in general is because one should solve six differential equations for the off-diagonal components of the metric tensor

$$g_{\mu\nu} \rightarrow g'_{\mu\nu} = \frac{\partial x^\alpha}{\partial x'^\mu} \frac{\partial x^\beta}{\partial x'^\nu} g_{\alpha\beta} = 0 \qquad \mu \neq \nu , \tag{6.15}$$

for the four functions $x'^\mu = x'^\mu(x)$.

On the contrary, it is always possible to find a coordinate transformation that reduces the metric tensor to $\eta_{\mu\nu}$ at a point of the spacetime. In such a case we have to make diagonal a symmetric matrix with constant coefficients (the metric tensor in a certain reference frame and evaluated at a certain point of the spacetime) and then rescale the coordinates to reduce to ± 1 the diagonal elements. Formally, we perform the following coordinate transformation

$$dx^\mu \rightarrow d\hat{x}^{(\alpha)} = E_\mu^{(\alpha)} dx^\mu , \tag{6.16}$$

such that the new metric tensor is given by the Minkowski metric $\mathrm{diag}(-1, 1, 1, 1)$

$$g_{\mu\nu} \rightarrow \eta_{(\alpha)(\beta)} = E_{(\alpha)}^\mu E_{(\beta)}^\nu g_{\mu\nu} . \tag{6.17}$$

A similar coordinate system is called *locally Minkowski reference frame.*

$E_{(\alpha)}^\mu$s are the inverse of $E_\mu^{(\alpha)}$s and we can write

$$E_\mu^{(\alpha)} E_{(\alpha)}^\nu = \delta_\mu^\nu , \quad E_\mu^{(\alpha)} E_{(\beta)}^\mu = \delta_{(\beta)}^{(\alpha)} . \tag{6.18}$$

The coefficients $E_{(\alpha)}^\mu$s are called the *vierbeins* if the spacetime has dimension $n = 4$ and *vielbeins* for n arbitrary.

If a vector (dual vector) has components V^μ (V_μ) in the coordinate system $\{x^\mu\}$, the components of the vector and of the dual vector in the locally Minkowski reference frame are

$$V^{(\alpha)} = E_\mu^{(\alpha)} V^\mu , \quad V_{(\alpha)} = E_{(\alpha)}^\mu V_\mu . \tag{6.19}$$

From the definition of vierbeins, it is straightforward to see that

$$V^\mu = E_{(\alpha)}^\mu V^{(\alpha)} , \quad V_\mu = E_\mu^{(\alpha)} V_{(\alpha)} . \tag{6.20}$$

6.4.2 Locally Inertial Reference Frames

A *locally inertial reference frame* represents the best local approximation to a Minkowski spacetime that we can always find in a generic curved spacetime. In the case of a locally Minkowski spacetime, the metric at a point is given by the Minkowski metric, but we have not yet locally removed the gravitational field.

Let us consider an arbitrary coordinate system $\{x^\mu\}$. We expand the metric tensor around the origin

$$g_{\mu\nu}(x) = g_{\mu\nu}(0) + \left.\frac{\partial g_{\mu\nu}}{\partial x^\rho}\right|_0 x^\rho + \frac{1}{2}\left.\frac{\partial^2 g_{\mu\nu}}{\partial x^\rho \partial x^\sigma}\right|_0 x^\rho x^\sigma + \cdots . \tag{6.21}$$

We consider the following coordinate transformation $x^\mu \to x'^\mu$

$$x^\mu \to x'^\mu = x^\mu + \frac{1}{2}\Gamma^\mu_{\rho\sigma}(0)x^\rho x^\sigma + \cdots , \tag{6.22}$$

with inverse

$$x^\mu = x'^\mu - \frac{1}{2}\Gamma^\mu_{\rho\sigma}(0)x'^\rho x'^\sigma + \cdots . \tag{6.23}$$

Since

$$\frac{\partial x^\alpha}{\partial x'^\mu} = \delta^\alpha_\mu - \Gamma^\alpha_{\mu\nu}(0)x'^\nu + \cdots , \tag{6.24}$$

the metric tensor in the new coordinate system $\{x'^\mu\}$ is

$$\begin{aligned}
g'_{\mu\nu} &= g_{\alpha\beta}\frac{\partial x^\alpha}{\partial x'^\mu}\frac{\partial x^\beta}{\partial x'^\nu} \\
&= g_{\alpha\beta}\left(\delta^\alpha_\mu - \Gamma^\alpha_{\mu\rho}(0)x'^\rho + \cdots\right)\left(\delta^\beta_\nu - \Gamma^\beta_{\nu\sigma}(0)x'^\sigma + \cdots\right) \\
&= g_{\mu\nu} - g_{\mu\beta}\Gamma^\beta_{\nu\sigma}(0)x'^\sigma - g_{\alpha\nu}\Gamma^\alpha_{\mu\rho}(0)x'^\rho + \cdots .
\end{aligned} \tag{6.25}$$

For simplicity, in what follows we omit ... at the end of the expressions to indicate that we are only interested in the leading order terms.

The partial derivative of the metric tensor evaluated at the origin is

$$\left.\frac{\partial g'_{\mu\nu}}{\partial x'^\rho}\right|_0 = \left.\frac{\partial g_{\mu\nu}}{\partial x^\rho}\right|_0 - g_{\mu\beta}(0)\Gamma^\beta_{\nu\rho}(0) - g_{\alpha\nu}(0)\Gamma^\alpha_{\mu\rho}(0) . \tag{6.26}$$

The second term on the right hand side in Eq. (6.26) is

$$\Gamma^{\beta}_{\nu\rho} = \frac{1}{2} g^{\beta\gamma} \left(\frac{\partial g_{\gamma\rho}}{\partial x^{\nu}} + \frac{\partial g_{\nu\gamma}}{\partial x^{\rho}} - \frac{\partial g_{\nu\rho}}{\partial x^{\gamma}} \right),$$

$$g_{\mu\beta} \Gamma^{\beta}_{\nu\rho} = \frac{1}{2} \left(\frac{\partial g_{\mu\rho}}{\partial x^{\nu}} + \frac{\partial g_{\nu\mu}}{\partial x^{\rho}} - \frac{\partial g_{\nu\rho}}{\partial x^{\mu}} \right). \tag{6.27}$$

The third term on the right hand side in Eq. (6.26) is

$$g_{\alpha\nu} \Gamma^{\alpha}_{\mu\rho} = \frac{1}{2} \left(\frac{\partial g_{\nu\rho}}{\partial x^{\mu}} + \frac{\partial g_{\mu\nu}}{\partial x^{\rho}} - \frac{\partial g_{\mu\rho}}{\partial x^{\nu}} \right). \tag{6.28}$$

We sum the expressions in Eqs. (6.27) and (6.28)

$$g_{\mu\beta} \Gamma^{\beta}_{\nu\rho} + g_{\alpha\nu} \Gamma^{\alpha}_{\mu\rho} = \frac{\partial g_{\mu\nu}}{\partial x^{\rho}}, \tag{6.29}$$

and we thus see that the expression in Eq. (6.26) vanishes. The metric tensor $g'_{\mu\nu}$ around the origin is

$$g'_{\mu\nu}(x') = g'_{\mu\nu}(0) + \frac{1}{2} \frac{\partial^2 g'_{\mu\nu}}{\partial x'^{\rho} \partial x'^{\sigma}} \bigg|_{0} x'^{\rho} x'^{\sigma}, \tag{6.30}$$

because all its first partial derivatives vanish. The geodesic equations around the origin are

$$\frac{d^2 x'^{\mu}}{d\lambda^2} = 0. \tag{6.31}$$

The motion of a test-particle around the origin in this reference frame is the same as in special relativity in Cartesian coordinates and we can thus say that we have locally removed the gravitational field.

6.5 Measurements of Time Intervals

In general relativity, the choice of the coordinate system is arbitrary and we can move from one reference frame to another one with a coordinate transformation. In general, the value of the coordinates of a certain reference frame has no physical meaning; the coordinate system is just a tool to describe the points of the spacetime. When we want to compare theoretical predictions with observations, we need to consider the reference frame associated with the observer performing the experiment and compute the theoretical predictions of the physical phenomenon under consideration there.

In Sect. 2.4, we discussed the relations between time intervals measured by clocks in different reference frames. Here we want to extend that discussion in the presence of gravitational fields.

Let us consider an observer in a gravitational field described by the metric tensor $g_{\mu\nu}$. The observer is equipped with a locally Minkowski reference frame. His/her proper time is the time measured by a clock at the origin of his/her locally Minkowski reference frame and is given by

$$- c^2 d\tau^2 = ds^2 = g_{\mu\nu} dx^\mu dx^\nu \,, \tag{6.32}$$

where ds is the line element of his/her trajectory (which is an invariant) and $\{x^\mu\}$ is the coordinate system associated to the metric $g_{\mu\nu}$.

A special case is represented by an observer at rest in the coordinate system $\{x^\mu\}$; that is, the observer's spatial coordinates are constant in the time coordinate: $dx^i = 0$ in Eq. (6.32). In such a situation we have the following relation between the proper time of the observer, τ, and the temporal coordinate of the coordinate system, t,

$$d\tau^2 = -g_{tt} dt^2 \,. \tag{6.33}$$

In the case of a weak gravitational field, g_{tt} is given by Eq. (6.14), and we find

$$d\tau^2 = \left(1 + \frac{2\Phi}{c^2}\right) dt^2 \,. \tag{6.34}$$

If the gravitational field is generated by a spherically symmetric source of mass M, $\Phi = -G_{\rm N} M/r$ and

$$d\tau = \sqrt{1 - \frac{2G_{\rm N} M}{c^2 r}} \, dt < dt \,. \tag{6.35}$$

In this specific example, the temporal coordinate t coincides with the proper time τ for $r \to \infty$, which means that the reference frame corresponds to the coordinate system of an observer at infinity. In general, $\Delta\tau < \Delta t$, and the clock of the observer in the gravitational field is slower than the clock of an observer at infinity.

6.6 Example: GPS Satellites

The Global Position System (GPS) is the most famous global navigation satellite system. It consists of a constellation of satellites orbiting at an altitude of about 20,200 km from the ground and with an orbital speed of about 14,000 km/h. With a typical GPS receiver, one can quickly determine his/her position on Earth with an accuracy of 5–10 m.

GPS satellites carry very stable atomic clocks and continually broadcast a signal that includes their current time and position. From the signal of at least four satellites, a GPS receiver can determine its position on Earth. Even if the gravitational field of Earth is weak and the speed of the satellites is relatively low in comparison to the speed of light, special and general relativistic effects are important and cannot be ignored.

A GPS receiver on Earth is a quasi-inertial observer, while the GPS satellites are not and move with a speed of about 14,000 km/h with respect to the ground, so we have

$$\frac{v}{c} = 1.3 \cdot 10^{-5}, \quad \gamma \approx 1 + \frac{1}{2}\frac{v^2}{c^2} = 1 + 8.4 \cdot 10^{-11}. \tag{6.36}$$

If $\Delta\tau$ is the measurement of a certain time interval by the GPS receiver and $\Delta\tau'$ is the measurement of the same time interval by the GPS satellites, we have that $\Delta\tau = \gamma \Delta\tau'$. After 24 h, the clocks of the GPS satellites would have a delay of 7 μs due to the orbital motion of the satellites.

Assuming that the Earth is a spherically symmetric body of mass M, the Newtonian gravitational potential is

$$\Phi = -\frac{G_N M}{r}, \tag{6.37}$$

where r is the distance from the center of Earth. The GPS receiver is on the Earth surface, so $r = 6,400$ km. For the GPS satellites, the distance from the center of Earth is $r = 26,600$ km. From Eq. (6.35), we can write the relation between the proper time interval of the GPS receiver $\Delta\tau$ and the proper time interval of the GPS satellites $\Delta\tau'$

$$\frac{\Delta\tau}{\sqrt{1 + 2\Phi_{\text{rec}}/c^2}} = \frac{\Delta\tau'}{\sqrt{1 + 2\Phi_{\text{sat}}/c^2}}, \tag{6.38}$$

where Φ_{rec} and Φ_{sat} are, respectively, the Newtonian gravitational potential at the position of the GPS receiver on the surface of Earth and at the positions of the GPS satellites at an altitude of about 20,200 km from the ground. We have

$$\Delta\tau = \sqrt{\frac{1 + 2\Phi_{\text{rec}}/c^2}{1 + 2\Phi_{\text{sat}}/c^2}} \Delta\tau' \approx \left(1 + \frac{\Phi_{\text{rec}}}{c^2} - \frac{\Phi_{\text{sat}}}{c^2}\right)\Delta\tau'. \tag{6.39}$$

$\Phi_{\text{rec}}/c^2 = -6.9 \cdot 10^{-10}$, $\Phi_{\text{sat}}/c^2 = -1.7 \cdot 10^{-10}$, and, after 24 h, the clock of the GPS receiver would have a delay of 45 μs with respect to the clocks of the GPS satellites due to the Earth's gravitational field.

If we combine the effect of the orbital motion of the GPS satellites with the effect of the Earth's gravitational field, we find that, after 24 h, the difference between the time of the clock of the GPS receiver and that of the clocks of the GPS satellites is

$$\delta t = \Delta\tau - \Delta\tau' \approx \left(\frac{\Phi_{\text{rec}}}{c^2} - \frac{\Phi_{\text{sat}}}{c^2} + \frac{1}{2}\frac{v^2}{c^2}\right)\Delta\tau'$$

$$= -45 \, \mu s + 7 \, \mu s = -38 \, \mu s. \tag{6.40}$$

Since the communication between GPS satellites and GPS receiver is through electromagnetic signals moving at the speed of light, an error of 38µs is equivalent to an error of $c\delta t \approx 10$ km in space position, which would make GPS navigation systems in cars and smartphones completely useless.

6.7 Non-gravitational Phenomena in Curved Spacetimes

Now we want to figure out how to write the laws of physics in the presence of a gravitational field. In other words, we want to find a "recipe" to apply the Principle of General Covariance: we know the mathematical expression describing a certain non-gravitational phenomenon in the Minkowski spacetime in Cartesian coordinates and we want to write the same physical law in the presence of a gravitational field in arbitrary coordinates. The case of non-inertial reference frames in flat spacetime will be automatically included.

As it will be more clear from the examples below, there are some ambiguities when we want to translate the laws of physics from flat to curved spacetimes. This means that purely theoretical arguments are not enough to do it and experiments have to confirm if the new equations are right or not.

As the first step to solve this problem, we can start considering how to write a physical law valid in flat spacetime and Cartesian coordinates for the case of flat spacetime and non-Cartesian coordinates. This has been partially discussed in the previous chapters. We start from manifestly Lorentz-invariant expressions and we make the following substitutions:

1. we replace the Minkowski metric in Cartesian coordinates with the metric in the new coordinates: $\eta_{\mu\nu} \to g_{\mu\nu}$;
2. partial derivatives become covariant derivatives: $\partial_\mu \to \nabla_\mu$;
3. if we have an integral over the whole spacetime (e.g. an action), we have to integrate over the correct volume element $d^4\Omega$: $d^4x \to |J|d^4x$, where J is the Jacobian. We will see below that $J = \sqrt{-g}$, where g is the determinant of the metric tensor.

The new equations are written in terms of tensors and therefore they are manifestly covariant. We could apply the same recipe to write the physical laws in curved spacetimes. For the time being, experiments confirm that we obtain the right equations.

For the point 3 above, we know that, if we move from an inertial reference frame with the Minkowski metric $\eta_{\mu\nu}$ and the Cartesian coordinates $\{x^\mu\}$ to another reference frame with the coordinates $\{x'^\mu\}$, the new metric is

$$g'_{\mu\nu} = \frac{\partial x^\alpha}{\partial x'^\mu} \frac{\partial x^\beta}{\partial x'^\nu} \eta_{\alpha\beta}, \tag{6.41}$$

and therefore the determinant is

$$\det \left| g'_{\mu\nu} \right| = \det \left| \frac{\partial x^\alpha}{\partial x'^\mu} \frac{\partial x^\beta}{\partial x'^\nu} \eta_{\alpha\beta} \right| = \det \left| \frac{\partial x^\alpha}{\partial x'^\mu} \right| \det \left| \frac{\partial x^\beta}{\partial x'^\nu} \right| \det \left| \eta_{\alpha\beta} \right| . \quad (6.42)$$

Note that[2]

$$\det \left| g'_{\mu\nu} \right| = -g , \quad \det \left| \frac{\partial x^\alpha}{\partial x'^\mu} \right| = J , \quad \det \left| \eta_{\alpha\beta} \right| = 1 , \quad (6.43)$$

where J is the Jacobian of the coordinate transformation $x'^\mu \to x^\mu$. From Eq. (6.42), we see that $J = \sqrt{-g}$, and therefore we find that we need the following substitution when we have an integral and we move to an arbitrary reference frame

$$\int d^4x \to \int \sqrt{-g}\, d^4x . \quad (6.44)$$

Let us consider the electromagnetic field. The fundamental variable is the 4-potential A_μ. In the Faraday tensor $F_{\mu\nu}$, we replace the partial derivatives with the covariant ones, but there is no difference

$$F_{\mu\nu} = \nabla_\mu A_\nu - \nabla_\nu A_\mu = \partial_\mu A_\nu - \Gamma^\sigma_{\mu\nu} A_\sigma - \partial_\nu A_\mu + \Gamma^\sigma_{\nu\mu} A_\sigma$$
$$= \partial_\mu A_\nu - \partial_\nu A_\mu . \quad (6.45)$$

The Maxwell equations in the manifestly Lorentz-invariant form, in flat spacetime, and in Cartesian coordinates are given in Eqs. (4.40) and (4.48). We replace the partial derivatives with the covariant ones and we have

$$\nabla_\mu F_{\nu\rho} + \nabla_\nu F_{\rho\mu} + \nabla_\rho F_{\mu\nu} = 0 , \quad (6.46)$$

$$\nabla_\mu F^{\mu\nu} = -\frac{4\pi}{c} J^\nu . \quad (6.47)$$

Let us now talk about possible ambiguities of our recipe. First, we have to replace partial derivatives with covariant ones. However, partial derivatives commute while covariant derivatives do not. This may be a problem, but in practical examples it is not because it is possible to find a way to figure out the right order.

Second, in general we may expect that tensors that vanish in flat spacetime may affect the physical phenomenon when gravity is turned on. For example, the Lagrangian density of a real scalar field in Minkowski spacetime in Cartesian coordinates is

$$\mathcal{L} = -\frac{\hbar}{2} \eta^{\mu\nu} \left(\partial_\mu \phi \right) \left(\partial_\nu \phi \right) - \frac{1}{2} \frac{m^2 c^2}{\hbar} \phi^2 , \quad (6.48)$$

---

[2]We indicate with g the determinant of the metric $g'_{\mu\nu}$. Since $g < 0$ in a (3+1)-dimensional spacetime, the absolute value of the determinant of $g'_{\mu\nu}$ is $\det \left| g'_{\mu\nu} \right| = -g$.

and the action is

$$S = \frac{1}{c} \int \mathscr{L} d^4x .$$

(6.49)

If we apply our recipe, the Lagrangian density describing our real scalar field in curved spacetime should be (we remind that, for scalars, ∇_ν reduces to ∂_ν)

$$\mathscr{L} = -\frac{\hbar}{2} g^{\mu\nu} \left(\partial_\mu \phi\right) \left(\partial_\nu \phi\right) - \frac{1}{2} \frac{m^2 c^2}{\hbar} \phi^2 .$$

(6.50)

The action is

$$S = \frac{1}{c} \int \mathscr{L} \sqrt{-g} \, d^4x .$$

(6.51)

We say that such a scalar field is *minimally coupled* because we have applied our recipe that holds for both non-inertial reference frames and gravitational fields. However, we cannot exclude that the Lagrangian density in curved spacetime is

$$\mathscr{L} = -\frac{\hbar}{2} g^{\mu\nu} \left(\partial_\mu \phi\right) \left(\partial_\nu \phi\right) - \frac{1}{2} \frac{m^2 c^2}{\hbar} \phi^2 + \xi R \phi^2 ,$$

(6.52)

where ξ is a coupling constant. In this case, we say that the scalar field is *non-minimally coupled* because we have a term coupling the scalar field with the gravitational field. The Lagrangian density in Eq. (6.52) reduces to Eq. (6.48) when gravity is turned off. There are no theoretical reasons (such as the violation of some fundamental principle) to rule out the expression in Eq. (6.52), which is instead more general than the Lagrangian density in (6.50) and therefore more "natural" – in theoretical physics, it is common to follow the principle according to which whatever is not forbidden is allowed. For the time being, there are no experiments capable of testing the presence of the term $\xi R \phi^2$, in the sense that it is only possible to constrain the parameter ξ to be below some huge unnatural value.

Up to now, there are no experiments that require that some physical law requires a non-minimal coupling when we move from flat to curved spacetimes. However, we have to consider that the effects of gravitational fields via non-minimal coupling are extremely weak in comparison to non-gravitational interactions. This is because the Planck mass $M_{\text{Pl}} = \sqrt{\hbar c / G_{\text{N}}} = 1.2 \cdot 10^{19} \, \text{GeV}$ is huge in comparison with the energy scales in particle physics and there are no environments in the Universe today where gravity is "strong" in terms of the Planck mass.

Problems

6.1 In the Schwarzschild metric, the line element reads

$$ds^2 = -f(r)c^2 dt^2 + \frac{dr^2}{f(r)} + r^2 d\theta^2 + r^2 \sin^2 \theta d\phi^2 , \qquad (6.53)$$

where

$$f(r) = 1 - \frac{r_{\text{Sch}}}{r} , \qquad (6.54)$$

and $r_{\text{Sch}} = 2G_N M/c^2$. The Schwarzschild metric describes the spacetime around a static and spherically symmetric object of mass M. Write the vierbeins to move to a locally Minkowski reference frame.

6.2 In the Kerr metric, the line element reads

$$ds^2 = -\left(1 - \frac{r_{\text{Sch}} r}{\Sigma}\right) c^2 dt^2 + \frac{\Sigma}{\Delta} dr^2 + \Sigma d\theta^2$$
$$+ \left(r^2 + a^2 + \frac{r_{\text{Sch}} r a^2 \sin^2 \theta}{\Sigma}\right) \sin^2 \theta d\phi^2 - \frac{2 r_{\text{Sch}} r a \sin^2 \theta}{\Sigma} c dt d\phi , \quad (6.55)$$

where

$$\Sigma = r^2 + a^2 \cos^2 \theta , \quad \Delta = r^2 - r_{\text{Sch}} r + a^2 , \qquad (6.56)$$

$r_{\text{Sch}} = 2G_N M/c^2$, and $a = J/(Mc)$. The Kerr metric describes the spacetime around a stationary and axisymmetric black hole of mass M and spin angular momentum J. Write the relation between the proper time of an observer with constant spatial coordinates and the time coordinate t.

6.3 In Sect. 4.3, we met the equation of the conservation of the electric current in flat spacetime in Cartesian coordinates: $\partial_\mu J^\mu = 0$. Rewrite this equation in: (i) a general reference frame, and (ii) flat spacetime in spherical coordinates.

6.4 The field equation for the scalar field ϕ that we obtain from the Lagrangian density (6.48) in Cartesian coordinates is the Klein–Gordon equation

$$\left(\partial_\mu \partial^\mu - \frac{m^2 c^2}{\hbar^2}\right) \phi = 0 . \qquad (6.57)$$

Write the Klein–Gordon equation in curved spacetime, first assuming minimal coupling and then non-minimal coupling.

6.5 Write the Lagrangian in (6.52) and the associated Klein–Gordon equation in the case of a metric with signature $(+ - --)$.

6.6 The energy-momentum tensor of a perfect fluid in flat spacetime and Cartesian coordinates is given in Eq. (3.104). Write its expression for a general reference frame.

Chapter 7
Einstein's Gravity

In Chap. 6, we discussed non-gravitational phenomena in curved spacetimes for a generic metric theory of gravity. General covariance is the basic principle: once we have the metric of the spacetime, we can describe non-gravitational phenomena. A different issue is the calculation of the metric of the spacetime. The Principle of General Covariance is not enough to determine the metric. We need the field equations of a specific gravity theory. With Einstein's gravity, we refer to the gravity theory described by the Einstein–Hilbert action or, equivalently, by the Einstein equations.

7.1 Einstein Equations

As in the case of the construction of the Lagrangian of a physical system, we do not have any direct way to infer the field equations of the gravity theory we are looking for. Thus, we have to start by listing some "reasonable" requirements that our theory and its field equations should satisfy and then test their predictions with observations.

1. The gravitational field should be completely described by the metric tensor of the spacetime. As we saw in Chaps. 5 and 6, the spacetime metric is potentially capable of describing non-gravitational phenomena in a gravitational field if the Einstein Equivalence Principle holds. While additional degrees of freedom cannot be excluded, the requirement that the gravitational field is only described by the metric tensor is the *minimal* scenario and thus the first one to explore.
2. The field equations must be tensor equations; that is, they should be written in manifestly general covariant form in order to be explicitly independent of the choice of the coordinate system.
3. The field equations should be partial differential equations at most of second order, in analogy with the field equations of the known physical systems. As in point 1, this is the minimal scenario.

© Springer Nature Singapore Pte Ltd. 2018, corrected publication 2020
C. Bambi, *Introduction to General Relativity*, Undergraduate Lecture Notes
in Physics, https://doi.org/10.1007/978-981-13-1090-4_7

4. The field equations must have the correct Newtonian limit and therefore we must recover the Poisson equation $\Delta\Phi = 4\pi G_{\mathrm{N}}\rho$, where ρ is the mass density.
5. Since in Newton's gravity the source of the gravitational field is the mass density, now the source must be somehow related to the energy density. Since we want a tensor equation, the best candidate seems to be the matter energy-momentum tensor $T^{\mu\nu}$.
6. In the absence of matter, we must recover the Minkowski spacetime.

From conditions 2 and 5, the field equations can be written as

$$G^{\mu\nu} = \kappa T^{\mu\nu}\,, \tag{7.1}$$

where $G^{\mu\nu}$ is the tensor to find and κ is a proportionality constant somehow related to G_{N}. Since the matter energy-momentum tensor is covariantly conserved and symmetric

$$\nabla_\mu T^{\mu\nu} = 0\,, \quad T^{\mu\nu} = T^{\nu\mu}\,, \tag{7.2}$$

we need that

$$\nabla_\mu G^{\mu\nu} = 0\,, \quad G^{\mu\nu} = G^{\nu\mu}\,. \tag{7.3}$$

Conditions 1, 3, and 6 are compatible with the following choice

$$G_{\mu\nu} = R_{\mu\nu} - \frac{1}{2}g_{\mu\nu}R\,. \tag{7.4}$$

In the next section, we will show that this choice also meets condition 4 of the correct Newtonian limit.

If we relax conditions 4 and 6, we can also write

$$G_{\mu\nu} = R_{\mu\nu} - \frac{1}{2}g_{\mu\nu}R + \Lambda g_{\mu\nu}\,, \tag{7.5}$$

where Λ is called the *cosmological constant*. If its value is sufficiently small, the choice (7.5) can also be consistent with observations. For the moment, we assume that $\Lambda = 0$, but a non-vanishing value of the cosmological constant can be relevant in cosmological models (see Chap. 11).

The field equations in Einstein's gravity are the *Einstein equations* and read

$$R_{\mu\nu} - \frac{1}{2}g_{\mu\nu}R = \kappa T_{\mu\nu}\,, \tag{7.6}$$

where κ is Einstein's constant of gravitation (we will find its relation to Newton's constant of gravitation G_{N} in the next section). If we want to consider the possibility of a non-vanishing cosmological constant, we have[1]

$$R_{\mu\nu} - \frac{1}{2}g_{\mu\nu}R + \Lambda g_{\mu\nu} = \kappa T_{\mu\nu}. \tag{7.8}$$

In this section, we have arrived at these equations by imposing conditions 1–6. However, there is not a recipe to obtain the right field equations. Every theory has its own field equations. In Einstein's gravity, the field equations are the Einstein equations in (7.6) or in (7.8). Its predictions agree well with current observational data. However, there are also significant efforts to find alternative gravity theories with different field equations. The latter should be able to explain experimental data in order to be considered viable candidates as alternative theories. Once we find an observation that cannot be explained by one of these theories (Einstein's gravity included), the theory is ruled out.

From the Einstein equations, we can write the scalar curvature in terms of the matter content as follows

$$g^{\mu\nu}\left(R_{\mu\nu} - \frac{1}{2}g_{\mu\nu}R\right) = \kappa g^{\mu\nu}T_{\mu\nu},$$
$$R - 2R = \kappa T,$$
$$R = -\kappa T, \tag{7.9}$$

where $T = T^{\mu}_{\mu}$ is the trace of the matter energy-momentum tensor. In Einstein's gravity, the scalar curvature thus vanishes either in vacuum or for $T = 0$ (e.g. the energy-momentum tensor of an electromagnetic field, see Sect. 4.5). In other theories of gravity, this may not be true. The Einstein equations can be rewritten as

$$R_{\mu\nu} = \kappa \left(T_{\mu\nu} - \frac{1}{2}g_{\mu\nu}T\right). \tag{7.10}$$

Note that $R_{\mu\nu} = 0$ does not imply no gravitational field, but just no matter at that point of the spacetime. For example, if we consider a distribution of matter of finite extension, $R_{\mu\nu} \neq 0$ in the region with matter and $R_{\mu\nu} = 0$ in the exterior region.

In four dimensions, the Einstein equations are a system of 10 differential equations to determine the 10 components of the metric tensor $g_{\mu\nu}$ ($G_{\mu\nu}$ and $g_{\mu\nu}$ have 16 components, but the tensors are symmetric). Even if we fix the initial conditions, there is not a unique solution because it is always possible to perform a coordinate transformation: even if the solution looks different after a coordinate transformation,

---

[1]If we use the convention of a metric with signature $(+ - - -)$, Eq. (7.8) reads

$$R_{\mu\nu} - \frac{1}{2}g_{\mu\nu}R - \Lambda g_{\mu\nu} = \kappa T_{\mu\nu}, \tag{7.7}$$

i.e. the sign in front of Λ is $-$ instead of $+$ (employing the common convention of positive/negative Λ).

it is physically equivalent. This, in particular, means that we do not have 10 physical degrees of freedom.

Lastly, note that the Einstein equations relate the geometry of the spacetime (on the left hand side) to the matter content (on the right hand side). If we know the matter content, we can determine the spacetime metric. In principle, we can obtain any kind of spacetime for a proper choice of the matter energy-momentum tensor. For example, even unphysical spacetimes with closed time-like curves (i.e. trajectories in which massive particles can go backwards in time, as in a time machine) are possible for an unphysical matter energy-momentum tensor. In other words, the Einstein equations can make clear predictions only when we clearly specify the matter content. If this is not the case, every metric is allowed.

7.2 Newtonian Limit

In order to be consistent with observations, the Einstein equations must be able to recover the correct Newtonian limit. This should also provide the relation between Einstein's constant of gravitation κ appearing in (7.6) and Newton's constant of gravitation G_N appearing in Newton's Law of Universal Gravitation.

As in Sect. 6.3, we impose that the gravitational field is weak and stationary; that is,

$$g_{\mu\nu} = \eta_{\mu\nu} + h_{\mu\nu} \quad |h_{\mu\nu}| \ll 1, \tag{7.11}$$

$$\frac{\partial g_{\mu\nu}}{\partial t} = 0. \tag{7.12}$$

Let us also choose a coordinate system in which all the components of the matter energy-momentum tensor vanish with the exception of the tt component, which describes the energy density and in the Newtonian limit reduces to the mass density ρ multiplied by the square of the speed of light c^2,

$$T_{tt} = \rho c^2. \tag{7.13}$$

This assumption is justified by the fact that in Newton's gravity the source of the gravitational field is only the mass. With our choice, the trace of the matter energy-momentum tensor is $T = -\rho c^2$ and the tt component of Eq. (7.10) turns out to be

$$R_{tt} = \kappa \left(\rho c^2 + \frac{1}{2}\eta_{tt}\rho c^2 \right) = \frac{\kappa c^2}{2}\rho. \tag{7.14}$$

Neglecting terms of second order in $h_{\mu\nu}$ and employing Eq. (6.8), we have

$$R_{tt} = \frac{\partial \Gamma^i_{tt}}{\partial x^i} + O(h^2) = -\frac{1}{2}\frac{\partial}{\partial x^i}\left(\eta^{ij}\frac{\partial h_{tt}}{\partial x^j}\right) + O(h^2)$$

$$= -\frac{1}{2}\Delta h_{tt} + O(h^2), \tag{7.15}$$

where $\Delta = \partial^2_x + \partial^2_y + \partial^2_z$ is the Laplacian. As seen in Sect. 6.3, $h_{tt} = -2\Phi/c^2$, and therefore

$$R_{tt} = \frac{\Delta \Phi}{c^2}. \tag{7.16}$$

After replacing R_{tt} with $\Delta\Phi/c^2$ in Eq. (7.14), we have

$$\Delta\Phi = \frac{\kappa c^4}{2}\rho. \tag{7.17}$$

The formal solution is

$$\Phi(\mathbf{x}) = -\frac{\kappa c^4}{8\pi}\int \frac{\rho(\tilde{\mathbf{x}})}{|\mathbf{x} - \tilde{\mathbf{x}}|}d^3\tilde{\mathbf{x}}. \tag{7.18}$$

If we compare Eq. (7.17) with the Poisson equation $\Delta\Phi = 4\pi G_N\rho$ that holds in the Newtonian theory, we find the relation between κ and G_N

$$\kappa = \frac{8\pi G_N}{c^4}. \tag{7.19}$$

Note that in the presence of a non-vanishing cosmological constant Eq. (7.17) would be

$$\frac{\Delta\Phi}{c^2} = \frac{\kappa c^2}{2}\rho - \Lambda, \tag{7.20}$$

and we cannot recover the Poisson equation.

7.3 Einstein–Hilbert Action

It is sometimes convenient to have the action of a certain theory and be able to derive the equations governing the dynamics of the system by employing the Least Action Principle. While it is not guaranteed that such an action exists, for the known physical systems we have one. Einstein's gravity is not an exception. In this section we want thus to discuss the action that, when we impose the Least Action Principle, provides the Einstein equations.

The total action will be the sum of the action of the gravitational sector, say S_g, and of the action of the matter sector, say S_m,

$$S = S_g + S_m \, . \tag{7.21}$$

The natural candidates for the Lagrangian coordinates of the action of the gravitational field are the metric coefficients and their first derivatives, namely $g_{\mu\nu}$ and $\partial_\rho g_{\mu\nu}$. The matter sector will have its own Lagrangian coordinates. A coupling constant will connect the gravity and the matter sectors and establish the strength of the interaction. The Einstein equations should be obtained by considering the following variation

$$g_{\mu\nu} \to g'_{\mu\nu} = g_{\mu\nu} + \delta g_{\mu\nu} \, , \tag{7.22}$$

with $\delta g_{\mu\nu} = 0$ at the boundary of the integration region:

$$\delta S_g + \delta S_m = 0 \quad \Rightarrow \quad G^{\mu\nu} - \kappa T^{\mu\nu} = 0 \, . \tag{7.23}$$

The field equations of the matter sector in the gravitational field should instead be obtained by considering variations with respect to the Lagrangian coordinates of the matter sector

The Einstein equations can be obtained by applying the Least Action Principle to the *Einstein–Hilbert action*, which reads

$$S_{EH} = \frac{1}{2\kappa c} \int R\sqrt{-g}\, d^4x \, , \tag{7.24}$$

or, with Newton's constant of gravitation G_N instead of κ,

$$S_{EH} = \frac{c^3}{16\pi G_N} \int R\sqrt{-g}\, d^4x \, . \tag{7.25}$$

S_{EH} describes the action of the gravitational sector, S_g in Eq. (7.21). The Einstein equations require also the action for the matter sector, S_m. In what follows, we will check that, through the Least Action Principle, the Einstein–Hilbert action provides the left hand side part in the Einstein equations.

Let us calculate the effect of a variation of the metric coefficients of the form in (7.22) on $g_{\rho\sigma}g^{\sigma\nu}$. We find

$$0 = \delta g^\nu_\rho = \delta \left(g_{\rho\sigma} g^{\sigma\nu} \right) = \left(\delta g_{\rho\sigma} \right) g^{\sigma\nu} + g_{\rho\sigma} \left(\delta g^{\sigma\nu} \right) , \tag{7.26}$$

and therefore

$$g_{\rho\sigma} \delta g^{\sigma\nu} = -g^{\sigma\nu} \delta g_{\rho\sigma} \, . \tag{7.27}$$

$\delta g^{\mu\nu}$ is thus

$$\delta g^{\mu\nu} = g^{\mu\rho} g_{\rho\sigma} \delta g^{\sigma\nu} = -g^{\mu\rho} g^{\nu\sigma} \delta g_{\rho\sigma} \, . \qquad (7.28)$$

Let us now consider the effect of a variation of the metric coefficients on $\sqrt{-g}$. Employing Eq. (5.56), we have

$$\delta\sqrt{-g} = \frac{\partial\sqrt{-g}}{\partial g_{\mu\nu}} \delta g_{\mu\nu} = -\frac{1}{2} \frac{1}{\sqrt{-g}} \frac{\partial g}{\partial g_{\mu\nu}} \delta g_{\mu\nu} = -\frac{1}{2} \frac{1}{\sqrt{-g}} g g^{\mu\nu} \delta g_{\mu\nu}$$

$$= \frac{1}{2} \sqrt{-g} g^{\mu\nu} \delta g_{\mu\nu} \, , \qquad (7.29)$$

Lastly, we need to calculate the effect of a variation of the metric coefficients on the Ricci tensor

$$\delta R_{\mu\nu} = \delta \left(\partial_\rho \Gamma^\rho_{\mu\nu} - \partial_\nu \Gamma^\rho_{\mu\rho} + \Gamma^\sigma_{\mu\nu} \Gamma^\rho_{\sigma\rho} - \Gamma^\sigma_{\mu\rho} \Gamma^\rho_{\nu\sigma} \right)$$

$$= \partial_\rho \left(\delta\Gamma^\rho_{\mu\nu} \right) - \partial_\nu \left(\delta\Gamma^\rho_{\mu\rho} \right) + \left(\delta\Gamma^\sigma_{\mu\nu} \right) \Gamma^\rho_{\sigma\rho} + \Gamma^\sigma_{\mu\nu} \left(\delta\Gamma^\rho_{\sigma\rho} \right)$$

$$- \left(\delta\Gamma^\sigma_{\mu\rho} \right) \Gamma^\rho_{\nu\sigma} - \Gamma^\sigma_{\mu\rho} \left(\delta\Gamma^\rho_{\nu\sigma} \right) \, . \qquad (7.30)$$

If we add and subtract the quantity $\Gamma^\sigma_{\rho\nu} \left(\delta\Gamma^\rho_{\mu\sigma} \right)$ to the previous expression, we have

$$\delta R_{\mu\nu} = \delta R_{\mu\nu} - \Gamma^\sigma_{\nu\rho} \left(\delta\Gamma^\rho_{\mu\sigma} \right) + \Gamma^\sigma_{\rho\nu} \left(\delta\Gamma^\rho_{\mu\sigma} \right)$$

$$= \left[\partial_\rho \left(\delta\Gamma^\rho_{\mu\nu} \right) + \Gamma^\rho_{\sigma\rho} \left(\delta\Gamma^\sigma_{\mu\nu} \right) - \Gamma^\sigma_{\mu\rho} \left(\delta\Gamma^\rho_{\nu\sigma} \right) - \Gamma^\sigma_{\nu\rho} \left(\delta\Gamma^\rho_{\mu\sigma} \right) \right]$$

$$- \left[\partial_\nu \left(\delta\Gamma^\rho_{\mu\rho} \right) + \Gamma^\rho_{\nu\sigma} \left(\delta\Gamma^\sigma_{\mu\rho} \right) - \Gamma^\sigma_{\mu\nu} \left(\delta\Gamma^\rho_{\sigma\rho} \right) - \Gamma^\sigma_{\rho\nu} \left(\delta\Gamma^\rho_{\mu\sigma} \right) \right] \, . \qquad (7.31)$$

Note that the quantity

$$\delta\Gamma^\rho_{\mu\nu} = \Gamma'^\rho_{\mu\nu} - \Gamma^\rho_{\mu\nu} \qquad (7.32)$$

is a tensor of type $(1, 2)$. Indeed we know that the Christoffel symbols transform with the rule in Eq. (5.21). Under a coordinate transformation $x^\mu \to \tilde{x}^\mu$, $\delta\Gamma^\rho_{\mu\nu}$ transforms as[2]

$$\delta\Gamma^\rho_{\mu\nu} \to \delta\tilde{\Gamma}^\rho_{\mu\nu} = \frac{\partial\tilde{x}^\rho}{\partial x^\alpha} \frac{\partial x^\beta}{\partial\tilde{x}^\mu} \frac{\partial x^\gamma}{\partial\tilde{x}^\nu} \left(\Gamma'^\alpha_{\beta\gamma} - \Gamma^\alpha_{\beta\gamma} \right) \, . \qquad (7.33)$$

Moreover, $\delta\Gamma^\rho_{\mu\nu}$ is made of objects evaluated at the same point (objects belonging to the same tangent space with the terminology of Appendix C), and therefore it is a tensor. The covariant derivative of a tensor of type $(1, 2)$ is

---

[2]Note that $\Gamma^\rho_{\mu\nu}$s and $\Gamma'^\rho_{\mu\nu}$s are the Christoffel symbols associated, respectively, to the metric tensors $g_{\mu\nu}$ and $g_{\mu\nu} + \delta g_{\mu\nu}$, both in the coordinates x^μ.

$$\nabla_\mu A^\nu{}_{\rho\sigma} = \partial_\mu A^\nu{}_{\rho\sigma} + \Gamma^\nu_{\lambda\mu} A^\lambda{}_{\rho\sigma} - \Gamma^\lambda_{\rho\mu} A^\nu{}_{\lambda\sigma} - \Gamma^\lambda_{\sigma\mu} A^\nu{}_{\rho\lambda} , \tag{7.34}$$

and Eq. (7.31) reduces to

$$\delta R_{\mu\nu} = \nabla_\rho \left(\delta \Gamma^\rho_{\mu\nu} \right) - \nabla_\nu \left(\delta \Gamma^\rho_{\mu\rho} \right) . \tag{7.35}$$

Equation (7.35) is called the *Palatini Identity*.

Employing Eqs. (7.28), (7.29), and (7.35), we can write the effect of the variation (7.22) on $\sqrt{-g}R$

$$
\begin{aligned}
\delta \left(\sqrt{-g} R \right) &= \left(\delta \sqrt{-g} \right) R + \sqrt{-g} \left(\delta g^{\rho\sigma} \right) R_{\rho\sigma} + \sqrt{-g} g^{\mu\nu} \left(\delta R_{\mu\nu} \right) \\
&= \frac{1}{2} \sqrt{-g} g^{\mu\nu} \left(\delta g_{\mu\nu} \right) R - \sqrt{-g} g^{\rho\mu} g^{\sigma\nu} \left(\delta g_{\mu\nu} \right) R_{\rho\sigma} \\
&\quad + \sqrt{-g} g^{\mu\nu} \left[\nabla_\rho \left(\delta \Gamma^\rho_{\mu\nu} \right) - \nabla_\nu \left(\delta \Gamma^\rho_{\mu\rho} \right) \right] \\
&= \left(\frac{1}{2} g^{\mu\nu} R - R^{\mu\nu} \right) \sqrt{-g} \left(\delta g_{\mu\nu} \right) \\
&\quad + \sqrt{-g} \left\{ \nabla_\rho \left[g^{\mu\nu} \left(\delta \Gamma^\rho_{\mu\nu} \right) \right] - \nabla_\nu \left[g^{\mu\nu} \left(\delta \Gamma^\rho_{\mu\rho} \right) \right] \right\} \\
&= \left(\frac{1}{2} g^{\mu\nu} R - R^{\mu\nu} \right) \sqrt{-g} \left(\delta g_{\mu\nu} \right) \\
&\quad + \sqrt{-g} \left\{ \nabla_\rho \left[g^{\mu\nu} \left(\delta \Gamma^\rho_{\mu\nu} \right) \right] - \nabla_\rho \left[g^{\mu\rho} \left(\delta \Gamma^\nu_{\mu\nu} \right) \right] \right\} .
\end{aligned}
\tag{7.36}
$$

If we define

$$H^\rho = g^{\mu\nu} \left(\delta \Gamma^\rho_{\mu\nu} \right) - g^{\mu\rho} \left(\delta \Gamma^\nu_{\mu\nu} \right) , \tag{7.37}$$

we can rewrite the variation of the Einstein–Hilbert action as

$$
\begin{aligned}
\delta S_{\text{EH}} &= \frac{1}{2\kappa c} \int \left(\frac{1}{2} g^{\mu\nu} R - R^{\mu\nu} \right) \sqrt{-g} \left(\delta g_{\mu\nu} \right) d^4 x \\
&\quad + \frac{1}{2\kappa c} \int \nabla_\rho H^\rho \sqrt{-g} \, d^4 x .
\end{aligned}
\tag{7.38}
$$

Note that the last term in this expression is a divergence [see Eq. (5.64)]

$$\sqrt{-g} \nabla_\rho H^\rho = \sqrt{-g} \frac{1}{\sqrt{-g}} \frac{\partial}{\partial x^\rho} \left(\sqrt{-g} H^\rho \right) = \frac{\partial}{\partial x^\rho} \left(\sqrt{-g} H^\rho \right) , \tag{7.39}$$

and therefore we can apply Gauss's theorem to reduce the second integral on the right hand side of Eq. (7.38) into a surface integral. The Least Action Principle requires that the variations of the Lagrangian coordinates vanish at the boundary of the integration region, and therefore any surface integral vanishes too. At this point we have the following expression

$$\delta S = \frac{1}{2\kappa c} \int \left(\frac{1}{2} g^{\mu\nu} R - R^{\mu\nu} \right) \sqrt{-g} \left(\delta g_{\mu\nu} \right) d^4x + \delta S_{\rm m} , \qquad (7.40)$$

In the next section, we will show that with the contribution from $\delta S_{\rm m}$ we recover the Einstein equations.

7.4 Matter Energy-Momentum Tensor

7.4.1 Definition

Let us now consider the effect of a variation of the metric coefficients on the action of the matter sector

$$S_{\rm m} = \frac{1}{c} \int \mathcal{L}_{\rm m} \sqrt{-g} \, d^4x . \qquad (7.41)$$

At this point we *define* as the matter energy-momentum tensor appearing in the Einstein equations the tensor $T^{\mu\nu}$ given by

$$\delta S_{\rm m} = \frac{1}{2c} \int T^{\mu\nu} \sqrt{-g} \left(\delta g_{\mu\nu} \right) d^4x . \qquad (7.42)$$

Such a tensor is symmetric by construction, since $g_{\mu\nu}$ is symmetric. We can also check a posteriori that the definition (7.42) of $T^{\mu\nu}$ provides the same matter energy-momentum tensor that we obtain starting from that of special relativity (Sect. 3.9) and proceeding as discussed in Sect. 6.7.

Eventually the variation of the total action (gravity and matter sectors) gives

$$\delta S = \frac{1}{2\kappa c} \int \left(\frac{1}{2} g^{\mu\nu} R - R^{\mu\nu} + \kappa T^{\mu\nu} \right) \sqrt{-g} \left(\delta g_{\mu\nu} \right) d^4x . \qquad (7.43)$$

$\delta S = 0$ for any choice of $\delta g_{\mu\nu}$ only if the Einstein equations hold.

7.4.2 Examples

As a first example, let us consider the action of the electromagnetic field in curved spacetime

$$S = -\frac{1}{16\pi c} \int F_{\mu\nu} F^{\mu\nu} \sqrt{-g} \, d^4x \qquad (7.44)$$

When we consider a variation of $g_{\mu\nu}$, we have[3]

$$
\begin{aligned}
\delta \left(F_{\mu\nu} F_{\rho\sigma} g^{\mu\rho} g^{\nu\sigma} \sqrt{-g} \right) &= F_{\mu\nu} F_{\rho\sigma} \left(\delta g^{\mu\rho} \right) g^{\nu\sigma} \sqrt{-g} + F_{\mu\nu} F_{\rho\sigma} g^{\mu\rho} \left(\delta g^{\nu\sigma} \right) \sqrt{-g} \\
&\quad + F_{\mu\nu} F_{\rho\sigma} g^{\mu\rho} g^{\nu\sigma} \left(\delta \sqrt{-g} \right) \\
&= -F_{\mu\nu} F_{\rho\sigma} g^{\mu\alpha} g^{\rho\beta} \left(\delta g_{\alpha\beta} \right) g^{\nu\sigma} \sqrt{-g} \\
&\quad - F_{\mu\nu} F_{\rho\sigma} g^{\mu\rho} g^{\nu\alpha} g^{\sigma\beta} \left(\delta g_{\alpha\beta} \right) \sqrt{-g} \\
&\quad + \frac{1}{2} F_{\mu\nu} F_{\rho\sigma} g^{\mu\rho} g^{\nu\sigma} \sqrt{-g} g^{\alpha\beta} \left(\delta g_{\alpha\beta} \right) .
\end{aligned}
\tag{7.47}
$$

Changing the indices, we can rewrite the last expression as

$$
\delta \left(F_{\mu\nu} F^{\mu\nu} \sqrt{-g} \right) = \left(-F^{\mu\rho} F^{\nu}{}_{\rho} - F^{\rho\mu} F_{\rho}{}^{\nu} + \frac{1}{2} F_{\rho\sigma} F^{\rho\sigma} g^{\mu\nu} \right) \sqrt{-g} \left(\delta g_{\mu\nu} \right) .
\tag{7.48}
$$

The variation of the action is thus

$$
\delta S = \frac{1}{2c} \int \left(\frac{1}{4\pi} F^{\mu\rho} F^{\nu}{}_{\rho} - \frac{1}{16\pi} F_{\rho\sigma} F^{\rho\sigma} g^{\mu\nu} \right) \sqrt{-g} \left(\delta g_{\mu\nu} \right) d^4 x ,
\tag{7.49}
$$

and the energy-momentum tensor of the electromagnetic field is

$$
T^{\mu\nu} = \frac{1}{4\pi} F^{\mu\rho} F^{\nu}{}_{\rho} - \frac{1}{16\pi} F_{\rho\sigma} F^{\rho\sigma} g^{\mu\nu} ,
\tag{7.50}
$$

in agreement with the expression found in Sect. 4.5 when the spacetime metric is $\eta_{\mu\nu}$.

The action for a free point-like particle is [see Eq. (3.22)]

$$
S = \frac{1}{2} \int m g_{\mu\nu} \dot{x}^{\mu} \dot{x}^{\nu} \, d\tau .
\tag{7.51}
$$

The definition of the energy-momentum tensor requires an action as in Eq. (7.41). We thus have to rewrite the mass of the particle m as a mass density ρ integrated over the 4-volume of the spacetime. Since the particle is point-like, the mass density

---

[3] Remember that the fundamental variables of the electromagnetic sector are A_μ and $\partial_\nu A_\mu$, while A^μ and $\partial^\nu A^\mu$ have the metric tensor inside. For this reason, in Eq. (7.47) we consider the variation of

$$
F_{\mu\nu} F_{\rho\sigma} g^{\mu\rho} g^{\nu\sigma} \sqrt{-g}
\tag{7.45}
$$

and not of

$$
F^{\mu\nu} F^{\rho\sigma} g_{\mu\rho} g_{\nu\sigma} \sqrt{-g} .
\tag{7.46}
$$

The variation of Eq. (7.46) would provide a different (and wrong) result.

is

$$\rho = m\delta^4 \left[x^\sigma - \tilde{x}^\sigma(\tau) \right]$$

$$= \frac{m}{\sqrt{-g}} \delta \left[x^0 - \tilde{x}^0(\tau) \right] \delta \left[x^1 - \tilde{x}^1(\tau) \right] \delta \left[x^2 - \tilde{x}^2(\tau) \right] \delta \left[x^3 - \tilde{x}^3(\tau) \right]$$

$$= \frac{m}{\sqrt{-g}} \prod_\sigma \delta \left[x^\sigma - \tilde{x}^\sigma(\tau) \right] . \tag{7.52}$$

where $\{\tilde{x}^\mu(\tau)\}$ are the coordinates of the particle trajectory. The action of the free point-like particle becomes

$$S = \frac{1}{2c} \int m\delta^4 \left[x^\sigma - \tilde{x}^\sigma(\tau) \right] g_{\mu\nu} \dot{x}^\mu \dot{x}^\nu \sqrt{-g} \, d^4x \, d\tau$$

$$= \frac{1}{2c} \int m \left[\prod_\sigma \delta \left[x^\sigma - \tilde{x}^\sigma(\tau) \right] \right] g_{\mu\nu} \dot{x}^\mu \dot{x}^\nu \, d^4x \, d\tau . \tag{7.53}$$

When we consider a variation of the metric tensor, we find

$$\delta S = \frac{1}{2c} \int m \left[\prod_\sigma \delta \left[x^\sigma - \tilde{x}^\sigma(\tau) \right] \right] \dot{x}^\mu \dot{x}^\nu \left(\delta g_{\mu\nu} \right) d^4x \, d\tau$$

$$= \frac{1}{2c} \int \left\{ \int \frac{m}{\sqrt{-g}} \left[\prod_\sigma \delta \left[x^\sigma - \tilde{x}^\sigma(\tau) \right] \right] \dot{x}^\mu \dot{x}^\nu \, d\tau \right\} \sqrt{-g} \left(\delta g_{\mu\nu} \right) d^4x ,$$

$$\tag{7.54}$$

and therefore the energy-momentum tensor of the free point-like particle is

$$T^{\mu\nu} = \int \frac{m}{\sqrt{-g}} \left[\prod_\sigma \delta \left[x^\sigma - \tilde{x}^\sigma(\tau) \right] \right] \dot{x}^\mu \dot{x}^\nu \, d\tau . \tag{7.55}$$

Let us consider the special case of an inertial reference frame in the Minkowski spacetime. For Cartesian coordinates $\sqrt{-g} = 1$ and we can write $d\tau$ as

$$d\tau = \frac{d\tau}{dt} dt = \frac{dt}{\gamma} , \tag{7.56}$$

where γ is the Lorentz factor of the particle. Integrating over dt, Eq. (7.55) becomes

$$T^{\mu\nu} = m\delta^3 \left(\mathbf{x} - \tilde{\mathbf{x}}(\tau(t)) \right) \frac{\dot{x}^\mu \dot{x}^\nu}{\gamma} , \tag{7.57}$$

and we recover the result of Eq. (3.101).

The action for a real scalar field in curved spacetime is

$$S = -\frac{\hbar}{2c} \int \left[g^{\mu\nu} \left(\partial_\mu \phi \right) \left(\partial_\nu \phi \right) + \frac{m^2 c^2}{\hbar^2} \phi^2 \right] \sqrt{-g} \, d^4 x \,. \tag{7.58}$$

When we consider a variation of the metric coefficients, we find

$$\begin{aligned}
\delta S &= -\frac{\hbar}{2c} \int \left\{ \left(\delta g^{\mu\nu} \right) \left(\partial_\mu \phi \right) \left(\partial_\nu \phi \right) \sqrt{-g} \right. \\
&\quad \left. + \left[g^{\mu\nu} \left(\partial_\mu \phi \right) \left(\partial_\nu \phi \right) + \frac{m^2 c^2}{\hbar^2} \phi^2 \right] \left(\delta \sqrt{-g} \right) \right\} d^4 x \\
&= \frac{1}{2c} \int \hbar \left\{ g^{\mu\rho} g^{\nu\sigma} \left(\partial_\mu \phi \right) \left(\partial_\nu \phi \right) \right. \\
&\quad \left. - \frac{1}{2} g^{\rho\sigma} \left[g^{\mu\nu} \left(\partial_\mu \phi \right) \left(\partial_\nu \phi \right) + \frac{m^2 c^2}{\hbar^2} \phi^2 \right] \right\} \sqrt{-g} \left(\delta g_{\rho\sigma} \right) d^4 x \,. \tag{7.59}
\end{aligned}$$

The resulting energy-momentum tensor is

$$T^{\mu\nu} = \hbar \left(\partial^\mu \phi \right) \left(\partial^\nu \phi \right) - \frac{\hbar}{2} g^{\mu\nu} \left[g^{\rho\sigma} \left(\partial_\rho \phi \right) \left(\partial_\sigma \phi \right) + \frac{m^2 c^2}{\hbar^2} \phi^2 \right] \,. \tag{7.60}$$

7.4.3 Covariant Conservation of the Matter Energy-Momentum Tensor

As discussed in Sect. 3.9, in the Minkowski spacetime in Cartesian coordinates, the equation $\partial_\mu T^{\mu\nu} = 0$ is associated with the conservation of the 4-momentum of the system. In curved spacetime, the conservation equation $\partial_\mu T^{\mu\nu} = 0$ becomes

$$\nabla_\mu T^{\mu\nu} = 0 \,, \tag{7.61}$$

which also follows from the Einstein equations. Note, however, that Eq. (7.61) does not imply the conservation of the 4-momentum. With the formula in Eqs. (5.65), (7.61) can be written as

$$\frac{1}{\sqrt{-g}} \frac{\partial}{\partial x^\mu} \left(T^{\mu\nu} \sqrt{-g} \right) + \Gamma^\nu_{\sigma\mu} T^{\mu\sigma} = 0 \,. \tag{7.62}$$

This is not a conservation law because it cannot be written as a partial derivative and therefore we cannot apply the Gauss theorem and proceed as in Sect. 3.9. The physical reason is that, in the presence of a gravitational field, we do not have the conservation of the matter 4-momentum, but the conservation of the 4-momentum of the whole system, including both that of matter and that of the gravitational field.

7.5 Pseudo-Tensor of Landau–Lifshitz

As discussed in the previous section, the covariant conservation of the matter energy-momentum tensor, $\nabla_\mu T^{\mu\nu} = 0$, is not a conservation equation. Since it is often necessary to evaluate energy balances of physical processes, it would be useful to find a non-covariant formulation of the theory in such a way that we can write something like

$$\partial_\mu \mathscr{T}^{\mu\nu} = 0, \tag{7.63}$$

where $\mathscr{T}^{\mu\nu}$ is a quantity connected to the 4-momentum of the whole system and associated with conserved physical quantities. This issue has been studied since the advent of general relativity and solved in different ways. In this section, we will follow the approach proposed by Landau and Lifshitz [1].

Let us consider a locally inertial frame (LIF) at the point x_0 (see Sect. 6.4.2 for the definition of locally inertial frame). Here all the first derivatives of the metric vanish, the Christoffel symbols vanish as well, and Eq. (7.61) becomes

$$\partial_\mu T^{\mu\nu}_{\text{LIF}} = 0. \tag{7.64}$$

The matter energy-momentum tensor $T^{\mu\nu}_{\text{LIF}}$ can be obtained from the Einstein equations

$$T^{\mu\nu}_{\text{LIF}} = \frac{c^4}{8\pi G_{\text{N}}} \left(R^{\mu\nu} - \frac{1}{2} g^{\mu\nu} R \right)_{\text{LIF}}. \tag{7.65}$$

Let us now evaluate the terms on the right hand side of this equation at x_0. $R^{\mu\nu}_{\text{LIF}}$ is given by (remember that at x_0 all the Christoffel symbols vanish)

$$
\begin{aligned}
R^{\mu\nu}_{\text{LIF}} &= g^{\mu\rho} g^{\nu\sigma} \left(\frac{\partial \Gamma^\lambda_{\rho\sigma}}{\partial x^\lambda} - \frac{\partial \Gamma^\lambda_{\rho\lambda}}{\partial x^\sigma} \right) \\
&= g^{\mu\rho} g^{\nu\sigma} \frac{\partial}{\partial x^\lambda} \left[\frac{1}{2} g^{\lambda\kappa} \left(\frac{\partial g_{\kappa\sigma}}{\partial x^\rho} + \frac{\partial g_{\rho\kappa}}{\partial x^\sigma} - \frac{\partial g_{\rho\sigma}}{\partial x^\kappa} \right) \right] \\
&\quad - g^{\mu\rho} g^{\nu\sigma} \frac{\partial}{\partial x^\sigma} \left(\frac{1}{2g} \frac{\partial g}{\partial x^\rho} \right),
\end{aligned} \tag{7.66}
$$

where in the last passage we used Eq. (5.63). Since the first derivatives of the metric tensor vanish, we can write

$$R_{\text{LIF}}^{\mu\nu} = \frac{1}{2}\frac{\partial}{\partial x^\lambda}\left[g^{\mu\rho}g^{\nu\sigma}g^{\lambda\kappa}\left(\frac{\partial g_{\kappa\sigma}}{\partial x^\rho} + \frac{\partial g_{\rho\kappa}}{\partial x^\sigma} - \frac{\partial g_{\rho\sigma}}{\partial x^\kappa}\right)\right]$$

$$-\frac{\partial}{\partial x^\sigma}\left(g^{\mu\rho}g^{\nu\sigma}\frac{1}{2g}\frac{\partial g}{\partial x^\rho}\right)$$

$$= -\frac{1}{2}\frac{\partial}{\partial x^\lambda}\left(g^{\mu\rho}g^{\nu\sigma}\frac{\partial g^{\lambda\kappa}}{\partial x^\rho}g_{\kappa\sigma} + g^{\mu\rho}g^{\nu\sigma}\frac{\partial g^{\lambda\kappa}}{\partial x^\sigma}g_{\rho\kappa} - \frac{\partial g^{\mu\rho}}{\partial x^\kappa}g^{\nu\sigma}g^{\lambda\kappa}g_{\rho\sigma}\right)$$

$$-\frac{\partial}{\partial x^\sigma}\left[\frac{1}{2g}\frac{\partial}{\partial x^\rho}(gg^{\mu\rho}g^{\nu\sigma})\right] + \frac{1}{2}\frac{\partial^2}{\partial x^\sigma \partial x^\rho}(g^{\mu\rho}g^{\nu\sigma})$$

$$= -\frac{1}{2}\frac{\partial}{\partial x^\lambda}\left[g^{\mu\rho}\frac{\partial g^{\lambda\nu}}{\partial x^\rho} + g^{\nu\sigma}\frac{\partial g^{\lambda\mu}}{\partial x^\sigma} - \frac{\partial g^{\mu\nu}}{\partial x^\kappa}g^{\lambda\kappa}\right]$$

$$-\frac{\partial}{\partial x^\sigma}\left[\frac{1}{2g}\frac{\partial}{\partial x^\rho}(gg^{\mu\rho}g^{\nu\sigma})\right] + \frac{1}{2}\frac{\partial^2 g^{\mu\rho}}{\partial x^\sigma \partial x^\rho}g^{\nu\sigma} + \frac{1}{2}\frac{\partial^2 g^{\nu\sigma}}{\partial x^\sigma \partial x^\rho}g^{\mu\rho}$$

$$= -\frac{1}{2}g^{\mu\rho}\frac{\partial^2 g^{\lambda\nu}}{\partial x^\lambda \partial x^\rho} - \frac{1}{2}g^{\nu\sigma}\frac{\partial^2 g^{\lambda\mu}}{\partial x^\lambda \partial x^\sigma} + \frac{1}{2}\frac{\partial^2 g^{\mu\nu}}{\partial x^\lambda \partial x^\kappa}g^{\lambda\kappa}$$

$$-\frac{\partial}{\partial x^\sigma}\left[\frac{1}{2g}\frac{\partial}{\partial x^\rho}(gg^{\mu\rho}g^{\nu\sigma})\right] + \frac{1}{2}\frac{\partial^2 g^{\mu\rho}}{\partial x^\sigma \partial x^\rho}g^{\nu\sigma} + \frac{1}{2}\frac{\partial^2 g^{\nu\sigma}}{\partial x^\sigma \partial x^\rho}g^{\mu\rho}$$

$$= \frac{1}{2}\frac{\partial^2 g^{\mu\nu}}{\partial x^\lambda \partial x^\kappa}g^{\lambda\kappa} - \frac{1}{2}\frac{\partial}{\partial x^\sigma}\left[\frac{1}{g}\frac{\partial}{\partial x^\rho}(gg^{\mu\rho}g^{\nu\sigma})\right]. \tag{7.67}$$

We can proceed in a similar way to calculate $g^{\mu\nu}R$ in the locally inertial frame. We have

$$(g^{\mu\nu}R)_{\text{LIF}} = g^{\mu\nu}g^{\rho\sigma}R_{\rho\sigma}$$

$$= g^{\mu\nu}g^{\rho\sigma}\left(\frac{\partial \Gamma_{\rho\sigma}^\lambda}{\partial x^\lambda} - \frac{\partial \Gamma_{\rho\lambda}^\lambda}{\partial x^\sigma}\right)$$

$$= g^{\mu\nu}g^{\rho\sigma}\frac{\partial}{\partial x^\lambda}\left[\frac{1}{2}g^{\lambda\kappa}\left(\frac{\partial g_{\kappa\sigma}}{\partial x^\rho} + \frac{\partial g_{\rho\kappa}}{\partial x^\sigma} - \frac{\partial g_{\rho\sigma}}{\partial x^\kappa}\right)\right]$$

$$-g^{\mu\nu}g^{\rho\sigma}\frac{\partial}{\partial x^\sigma}\left(\frac{1}{2g}\frac{\partial g}{\partial x^\rho}\right)$$

$$= \frac{1}{2}\frac{\partial}{\partial x^\lambda}\left[g^{\mu\nu}g^{\rho\sigma}g^{\lambda\kappa}\left(\frac{\partial g_{\kappa\sigma}}{\partial x^\rho} + \frac{\partial g_{\rho\kappa}}{\partial x^\sigma} - \frac{\partial g_{\rho\sigma}}{\partial x^\kappa}\right)\right]$$

$$-\frac{\partial}{\partial x^\sigma}\left(g^{\mu\nu}g^{\rho\sigma}\frac{1}{2g}\frac{\partial g}{\partial x^\rho}\right)$$

$$= \frac{1}{2}\frac{\partial}{\partial x^\lambda}\left(g^{\mu\nu}g^{\rho\sigma}g^{\lambda\kappa}\frac{\partial g_{\kappa\sigma}}{\partial x^\rho} + g^{\mu\nu}g^{\rho\sigma}g^{\lambda\kappa}\frac{\partial g_{\rho\kappa}}{\partial x^\sigma} - g^{\mu\nu}g^{\rho\sigma}g^{\lambda\kappa}\frac{\partial g_{\rho\sigma}}{\partial x^\kappa}\right)$$

$$-\frac{1}{2}\frac{\partial}{\partial x^\sigma}\left[\frac{1}{g}\frac{\partial}{\partial x^\rho}(gg^{\mu\nu}g^{\rho\sigma})\right] + \frac{1}{2}\frac{\partial^2}{\partial x^\sigma \partial x^\rho}(g^{\mu\nu}g^{\rho\sigma}). \tag{7.68}$$

Using the formula in Eq. (5.57), we write

$$
(g^{\mu\nu}R)_{\text{LIF}} = -\frac{1}{2}\frac{\partial}{\partial x^\lambda}\left(g^{\mu\nu}g^{\rho\sigma}\frac{\partial g^{\lambda\kappa}}{\partial x^\rho}g_{\kappa\sigma} + g^{\mu\nu}g^{\rho\sigma}\frac{\partial g^{\lambda\kappa}}{\partial x^\sigma}g_{\rho\kappa} + g^{\mu\nu}g^{\lambda\kappa}\frac{1}{g}\frac{\partial g}{\partial x^\kappa}\right)
$$

$$
-\frac{1}{2}\frac{\partial}{\partial x^\sigma}\left[\frac{1}{g}\frac{\partial}{\partial x^\rho}(gg^{\mu\nu}g^{\rho\sigma})\right] + \frac{1}{2}\frac{\partial^2 g^{\mu\nu}}{\partial x^\sigma \partial x^\rho}g^{\rho\sigma} + \frac{1}{2}g^{\mu\nu}\frac{\partial^2 g^{\rho\sigma}}{\partial x^\sigma \partial x^\rho}
$$

$$
= -\frac{1}{2}\frac{\partial}{\partial x^\lambda}\left(g^{\mu\nu}\frac{\partial g^{\lambda\rho}}{\partial x^\rho} + g^{\mu\nu}\frac{\partial g^{\lambda\sigma}}{\partial x^\sigma} + g^{\mu\nu}g^{\lambda\kappa}\frac{1}{g}\frac{\partial g}{\partial x^\kappa}\right)
$$

$$
-\frac{1}{2}\frac{\partial}{\partial x^\sigma}\left[\frac{1}{g}\frac{\partial}{\partial x^\rho}(gg^{\mu\nu}g^{\rho\sigma})\right] + \frac{1}{2}\frac{\partial^2 g^{\mu\nu}}{\partial x^\sigma \partial x^\rho}g^{\rho\sigma} + \frac{1}{2}g^{\mu\nu}\frac{\partial^2 g^{\rho\sigma}}{\partial x^\sigma \partial x^\rho}
$$

$$
= -g^{\mu\nu}\frac{\partial^2 g^{\lambda\rho}}{\partial x^\lambda \partial x^\rho} - \frac{1}{2}\frac{\partial}{\partial x^\lambda}\left[\frac{1}{g}\frac{\partial}{\partial x^\kappa}(gg^{\mu\nu}g^{\lambda\kappa})\right] + \frac{1}{2}\frac{\partial^2}{\partial x^\lambda \partial x^\kappa}(g^{\mu\nu}g^{\lambda\kappa})
$$

$$
-\frac{1}{2}\frac{\partial}{\partial x^\sigma}\left[\frac{1}{g}\frac{\partial}{\partial x^\rho}(gg^{\mu\nu}g^{\rho\sigma})\right] + \frac{1}{2}\frac{\partial^2 g^{\mu\nu}}{\partial x^\sigma \partial x^\rho}g^{\rho\sigma} + \frac{1}{2}g^{\mu\nu}\frac{\partial^2 g^{\rho\sigma}}{\partial x^\sigma \partial x^\rho}
$$

$$
= -\frac{\partial}{\partial x^\sigma}\left[\frac{1}{g}\frac{\partial}{\partial x^\rho}(gg^{\mu\nu}g^{\rho\sigma})\right] + \frac{\partial^2 g^{\mu\nu}}{\partial x^\sigma \partial x^\rho}g^{\rho\sigma}. \tag{7.69}
$$

The matter energy-momentum tensor in the locally inertial reference frame turns out to be

$$
T_{\text{LIF}}^{\mu\nu} = \frac{c^4}{8\pi G_{\text{N}}}\left(R^{\mu\nu} - \frac{1}{2}g^{\mu\nu}R\right)_{\text{LIF}}
$$

$$
= \frac{c^4}{8\pi G_{\text{N}}}\left\{\frac{1}{2}\frac{\partial^2 g^{\mu\nu}}{\partial x^\lambda \partial x^\kappa}g^{\lambda\kappa} - \frac{1}{2}\frac{\partial}{\partial x^\sigma}\left[\frac{1}{g}\frac{\partial}{\partial x^\rho}(gg^{\mu\rho}g^{\nu\sigma})\right]\right.
$$

$$
\left. + \frac{1}{2}\frac{\partial}{\partial x^\sigma}\left[\frac{1}{g}\frac{\partial}{\partial x^\rho}(gg^{\mu\nu}g^{\rho\sigma})\right] - \frac{1}{2}\frac{\partial^2 g^{\mu\nu}}{\partial x^\sigma \partial x^\rho}g^{\rho\sigma}\right\}
$$

$$
= \frac{\partial}{\partial x^\sigma}\left\{\frac{c^4}{16\pi G_{\text{N}}}\frac{1}{(-g)}\frac{\partial}{\partial x^\rho}\left[(-g)(g^{\mu\nu}g^{\rho\sigma} - g^{\mu\rho}g^{\nu\sigma})\right]\right\}
$$

$$
= \frac{1}{(-g)}\frac{\partial}{\partial x^\sigma}\left\{\frac{c^4}{16\pi G_{\text{N}}}\frac{\partial}{\partial x^\rho}\left[(-g)(g^{\mu\nu}g^{\rho\sigma} - g^{\mu\rho}g^{\nu\sigma})\right]\right\}, \tag{7.70}
$$

which we can rewrite as

$$
(-g)T_{\text{LIF}}^{\mu\nu} = \frac{\partial}{\partial x^\sigma}\tau^{\mu\nu\sigma}, \tag{7.71}
$$

where we have introduced

$$
\tau^{\mu\nu\sigma} = \frac{c^4}{16\pi G_{\text{N}}}\frac{\partial}{\partial x^\rho}\left[(-g)(g^{\mu\nu}g^{\rho\sigma} - g^{\mu\rho}g^{\nu\sigma})\right]. \tag{7.72}
$$

Equation (7.71) has been obtained in a locally inertial reference frame. In a generic reference frame, the equality between the left and right hand sides does not hold. We will have another piece that we call $(-g)\, t^{\mu\nu}$

$$(-g)\,(T^{\mu\nu} + t^{\mu\nu}) = \frac{\partial}{\partial x^\sigma} \tau^{\mu\nu\sigma}\,. \tag{7.73}$$

By construction, $\partial_\mu \partial_\sigma \tau^{\mu\nu\sigma} = 0$, because $\tau^{\mu\nu\sigma}$ is antisymmetric in the indices μ and σ, and therefore

$$\frac{\partial}{\partial x^\mu}\left[(-g)\,(T^{\mu\nu} + t^{\mu\nu})\right] = 0\,. \tag{7.74}$$

$t^{\mu\nu}$ is called the *pseudo-tensor of Landau–Lifshitz*. From the Einstein equations in a generic reference frame, we can obtain the expression of $t^{\mu\nu}$

$$
\begin{aligned}
t^{\mu\nu} = \frac{c^4}{16\pi\,G_N}\Bigg[& \left(2\Gamma^\rho_{\kappa\lambda}\Gamma^\sigma_{\rho\sigma} - \Gamma^\rho_{\kappa\sigma}\Gamma^\sigma_{\lambda\rho} - \Gamma^\rho_{\kappa\rho}\Gamma^\sigma_{\lambda\sigma}\right)\left(g^{\mu\kappa}g^{\nu\lambda} - g^{\mu\nu}g^{\kappa\lambda}\right) \\
& + g^{\mu\kappa}g^{\lambda\rho}\left(\Gamma^\nu_{\kappa\sigma}\Gamma^\sigma_{\lambda\rho} + \Gamma^\nu_{\lambda\rho}\Gamma^\sigma_{\kappa\sigma} - \Gamma^\nu_{\rho\sigma}\Gamma^\sigma_{\kappa\lambda} - \Gamma^\nu_{\kappa\lambda}\Gamma^\sigma_{\rho\sigma}\right) \\
& + g^{\nu\kappa}g^{\lambda\rho}\left(\Gamma^\mu_{\kappa\sigma}\Gamma^\sigma_{\lambda\rho} + \Gamma^\mu_{\lambda\rho}\Gamma^\sigma_{\kappa\sigma} - \Gamma^\mu_{\sigma\rho}\Gamma^\sigma_{\kappa\lambda} - \Gamma^\mu_{\kappa\lambda}\Gamma^\sigma_{\sigma\rho}\right) \\
& + g^{\kappa\lambda}g^{\rho\sigma}\left(\Gamma^\mu_{\kappa\rho}\Gamma^\nu_{\lambda\sigma} - \Gamma^\mu_{\kappa\lambda}\Gamma^\nu_{\rho\sigma}\right)\Bigg]\,.
\end{aligned}
\tag{7.75}
$$

$t^{\mu\nu}$ is not a tensor being a combination of Christoffel symbols that are not tensors. However, it transforms as a tensor under a linear coordinate transformation (hence the name "pseudo-tensor").

Problems

7.1 Let us consider the following action of a scalar field ϕ non-minimally coupled to gravity

$$S = -\frac{1}{c}\int\left[\frac{\hbar}{2}g^{\mu\nu}\,(\partial_\mu\phi)\,(\partial_\nu\phi) + \frac{1}{2}\frac{m^2c^2}{\hbar}\phi^2 - \xi\,R\phi^2\right]\sqrt{-g}\,d^4x\,. \tag{7.76}$$

Evaluate the energy-momentum tensor of the scalar field.

7.2 Consider the action in Eq. (7.76). Write the equations of motion.

7.3 Rewrite the Einstein–Hilbert action in Eq. (7.24) in order to obtain the Einstein equations in (7.8) with a cosmological constant Λ.

Reference

1. L.D. Landau, E.M. Lifshitz, *The Classical Theory of Fields* (Butterworth-Heinemann, Oxford, 1980)

Chapter 8
Schwarzschild Spacetime

The Einstein equations (7.6) relate the geometry of the spacetime, encoded in the Einstein tensor $G^{\mu\nu}$, to the matter content, described by the matter energy-momentum tensor $T^{\mu\nu}$. If we know the matter content, in principle we can solve the Einstein equations and find the spacetime metric $g_{\mu\nu}$ in some coordinate system. However, in general it is highly non-trivial to solve the Einstein equations, because they are second order non-linear partial differential equations for ten components of the metric tensor $g_{\mu\nu}$. Analytic solutions of the Einstein equations can be found when the spacetime has some special symmetries.

The *Schwarzschild metric* is a relevant example of an exact solution of the Einstein equations with important physical applications. It is the only spherically symmetric vacuum solution of the Einstein equations and usually it can well approximate the gravitational field of slowly-rotating astrophysical objects like stars and planets.

8.1 Spherically Symmetric Spacetimes

First, we want to find the most general form for the line element of a spherically symmetric spacetime. Note that at this stage we are not assuming the Einstein equations, which means that the same form of the line element holds in any theory of gravity in which the spacetime geometry is described by the metric tensor of the spacetime. To achieve our goal, we choose a particular coordinate system in which the metric tensor $g_{\mu\nu}$ clearly shows the symmetries of the spacetime.

As our starting point, we employ *isotropic coordinates* (ct, x, y, z), we choose the origin of the coordinate system of the 3-space $x = y = z = 0$ at the center of symmetry, and we require that the line element of the 3-space, dl, only depends on the time t and on the distance from the origin. dl^2 should thus have the following form

© Springer Nature Singapore Pte Ltd. 2018, corrected publication 2020
C. Bambi, *Introduction to General Relativity*, Undergraduate Lecture Notes
in Physics, https://doi.org/10.1007/978-981-13-1090-4_8

$$dl^2 = g\left(t, \sqrt{x^2 + y^2 + z^2}\right)\left(dx^2 + dy^2 + dz^2\right), \tag{8.1}$$

where g is an unknown function of t and $\sqrt{x^2 + y^2 + z^2}$. Note that, in general, $\sqrt{x^2 + y^2 + z^2}$ is not the proper distance from the origin. However, points with the same value of $\sqrt{x^2 + y^2 + z^2}$ have the same proper distance from the origin, which is enough for us because we have not yet specified the function g.

We move to spherical-like coordinates (t, r, θ, ϕ) with the following coordinate transformation

$$
\begin{aligned}
x &= r \sin\theta \cos\phi, \\
y &= r \sin\theta \sin\phi, \\
x &= r \cos\theta.
\end{aligned}
\tag{8.2}
$$

The line element of the 3-space is now

$$dl^2 = g(t, r)\left(dr^2 + r^2 d\theta^2 + r^2 \sin^2\theta d\phi^2\right). \tag{8.3}$$

Let us now construct the form of the 4-metric $g_{\mu\nu}$. For $dt \neq 0$, ds^2 should be separately invariant under the following coordinate transformations

$$\theta \to \tilde{\theta} = -\theta, \quad \phi \to \tilde{\phi} = -\phi, \tag{8.4}$$

which implies $g_{t\theta} = g_{t\phi} = 0$. The line element of the 4-dimensional spacetime can thus be written as

$$ds^2 = -f(t, r)c^2 dt^2 + g(t, r)dr^2 + h(t, r)dtdr + g(t, r)r^2\left(d\theta^2 + \sin^2\theta d\phi^2\right), \quad (8.5)$$

where f and h are unknown functions of t and r only.

The expression in Eq. (8.5) can be further simplified. We can still consider a coordinate transformation

$$t \to \tilde{t} = \tilde{t}(t, r), \quad r \to \tilde{r} = \tilde{r}(t, r), \tag{8.6}$$

such that

$$\tilde{r}^2 = g(t, r)r^2, \quad g_{\tilde{t}\tilde{r}} = 0. \tag{8.7}$$

Note that the coordinate $\tilde{r}$ has a clear geometrical meaning. It corresponds to the value of the radial coordinate defining the 2-dimensional spherical surface of area $4\pi\tilde{r}^2$. Note also that, in general, $\tilde{r}$ does not describe the real distance from the center $\tilde{r} = 0$ (see Sect. 8.3 later). Eventually, the line element of the spacetime can be written as

$$ds^2 = -f(t, r)c^2 dt^2 + g(t, r)dr^2 + r^2\left(d\theta^2 + \sin^2\theta d\phi^2\right). \tag{8.8}$$

Since we are interested in gravitational fields generated by a matter distribution with a finite extension, the spacetime must be asymptotically flat, namely we must recover the Minkowski metric at large radii. The boundary conditions are thus

$$\lim_{r \to \infty} f(t, r) = \lim_{r \to \infty} g(t, r) = 1 \,. \tag{8.9}$$

As we have already stressed at the beginning of this section, the expression in (8.8) is the most general form for the line element of a spherically symmetric spacetime. If we assume to be in Einstein's gravity, we can solve the Einstein equations with the ansatz in (8.8) to find the explicit form of the functions f and g in Einstein's gravity for a certain matter distribution.

8.2 Birkhoff's Theorem

Birkhoff's theorem is an important uniqueness theorem valid in Einstein's gravity.

Birkhoff's Theorem. The only spherically symmetric solution of the vacuum Einstein equations is the *Schwarzschild metric*.

Let us first prove the theorem and then discuss its implications. We have to solve the Einstein equations with the ansatz (8.8) for the metric tensor and with $T^{\mu\nu} = 0$ on the right hand side. Since we are in vacuum, the scalar curvature vanishes, $R = 0$, and the Einstein equations reduce to $R^{\mu\nu} = 0$, see Eq. (7.10). The strategy is to calculate the Christoffel symbols and then the Ricci tensor from the formula

$$R_{\mu\nu} = \frac{\partial \Gamma^{\lambda}_{\mu\nu}}{\partial x^{\lambda}} - \frac{\partial \Gamma^{\lambda}_{\mu\lambda}}{\partial x^{\nu}} + \Gamma^{\lambda}_{\mu\nu} \Gamma^{\rho}_{\lambda\rho} - \Gamma^{\lambda}_{\mu\rho} \Gamma^{\rho}_{\nu\lambda} \,. \tag{8.10}$$

The calculations may be somewhat long and boring, but they can be a good exercise to better understand the formalism of general relativity. The fastest way to calculate the Christoffel symbols is to write the geodesic equations from the Euler–Lagrange equations of the Lagrangian[1]

$$L = -\frac{f}{2} \left(\frac{dt}{d\lambda} \right)^2 + \frac{g}{2} \left(\frac{dr}{d\lambda} \right)^2 + \frac{r^2}{2} \left(\frac{d\theta}{d\lambda} \right)^2 + \frac{1}{2} r^2 \sin^2 \theta \left(\frac{d\phi}{d\lambda} \right)^2 \,, \tag{8.11}$$

---

[1]In these calculations we ignore the dimensional difference between t and the space coordinates and we do not write the speed of light c to simplify the equations. This is equivalent to employing units in which $c = 1$, which is a convention widely used among the gravity and particle physics communities.

where λ is the particle proper time (for time-like geodesics) or an affine parameter (for null geodesics). For $x^\mu = t$, the Euler–Lagrange equation is

$$\frac{d}{d\lambda}\left(-f\frac{dt}{d\lambda}\right) + \frac{\dot{f}}{2}\left(\frac{dt}{d\lambda}\right)^2 - \frac{\dot{g}}{2}\left(\frac{dr}{d\lambda}\right)^2 = 0,$$

$$f\frac{d^2t}{d\lambda^2} + \dot{f}\left(\frac{dt}{d\lambda}\right)^2 + f'\frac{dt}{d\lambda}\frac{dr}{d\lambda} - \frac{\dot{f}}{2}\left(\frac{dt}{d\lambda}\right)^2 + \frac{\dot{g}}{2}\left(\frac{dr}{d\lambda}\right)^2 = 0,$$

$$\frac{d^2t}{d\lambda^2} + \frac{\dot{f}}{2f}\left(\frac{dt}{d\lambda}\right)^2 + \frac{f'}{f}\frac{dt}{d\lambda}\frac{dr}{d\lambda} + \frac{\dot{g}}{2f}\left(\frac{dr}{d\lambda}\right)^2 = 0, \qquad (8.12)$$

where here and in what follows we use the dot $\dot{}$ to indicate the derivative with respect to the time t, i.e. $\dot{} = \partial_t$, and the prime $'$ to indicate the derivative with respect to the radial coordinate r, namely $' = \partial_r$. If we compare the geodesic equations with the last expression in (8.12), we see that the only non-vanishing Christoffel symbols of the type $\Gamma^t_{\mu\nu}$ are

$$\Gamma^t_{tt} = \frac{\dot{f}}{2f}, \quad \Gamma^t_{tr} = \Gamma^t_{rt} = \frac{f'}{2f}, \quad \Gamma^t_{rr} = \frac{\dot{g}}{2f}. \qquad (8.13)$$

For $x^\mu = r$, the Euler–Lagrange equation is

$$\frac{d}{d\lambda}\left(g\frac{dr}{d\lambda}\right) + \frac{f'}{2}\left(\frac{dt}{d\lambda}\right)^2 - \frac{g'}{2}\left(\frac{dr}{d\lambda}\right)^2 - r\left(\frac{d\theta}{d\lambda}\right)^2 - r\sin^2\theta\left(\frac{d\phi}{d\lambda}\right)^2 = 0,$$

$$g\frac{d^2r}{d\lambda^2} + \dot{g}\frac{dt}{d\lambda}\frac{dr}{d\lambda} + g'\left(\frac{dr}{d\lambda}\right)^2 + \frac{f'}{2}\left(\frac{dt}{d\lambda}\right)^2 - \frac{g'}{2}\left(\frac{dr}{d\lambda}\right)^2$$

$$-r\left(\frac{d\theta}{d\lambda}\right)^2 - r\sin^2\theta\left(\frac{d\phi}{d\lambda}\right)^2 = 0,$$

$$\frac{d^2r}{d\lambda^2} + \frac{f'}{2g}\left(\frac{dt}{d\lambda}\right)^2 + \frac{\dot{g}}{g}\frac{dt}{d\lambda}\frac{dr}{d\lambda} + \frac{g'}{2g}\left(\frac{dr}{d\lambda}\right)^2$$

$$-\frac{r}{g}\left(\frac{d\theta}{d\lambda}\right)^2 - \frac{r\sin^2\theta}{g}\left(\frac{d\phi}{d\lambda}\right)^2 = 0, \qquad (8.14)$$

and we find that the only non-vanishing Christoffel symbols of the type $\Gamma^r_{\mu\nu}$ are

$$\Gamma^r_{tt} = \frac{f'}{2g}, \quad \Gamma^r_{tr} = \Gamma^r_{rt} = \frac{\dot{g}}{2g},$$

$$\Gamma^r_{rr} = \frac{g'}{2g}, \quad \Gamma^r_{\theta\theta} = -\frac{r}{g}, \quad \Gamma^r_{\phi\phi} = -\frac{r\sin^2\theta}{g}. \qquad (8.15)$$

For $x^\mu = \theta$ we have

$$\frac{d}{d\lambda}\left(r^2\frac{d\theta}{d\lambda}\right) - r^2\sin\theta\cos\theta\left(\frac{d\phi}{d\lambda}\right)^2 = 0\,,$$

$$r^2\frac{d^2\theta}{d\lambda^2} + 2r\frac{dr}{d\lambda}\frac{d\theta}{d\lambda} - r^2\sin\theta\cos\theta\left(\frac{d\phi}{d\lambda}\right)^2 = 0\,,$$

$$\frac{d^2\theta}{d\lambda^2} + \frac{2}{r}\frac{dr}{d\lambda}\frac{d\theta}{d\lambda} - \sin\theta\cos\theta\left(\frac{d\phi}{d\lambda}\right)^2 = 0\,, \tag{8.16}$$

and the non-vanishing Christoffel symbols of the type $\Gamma^\theta_{\mu\nu}$ are

$$\Gamma^\theta_{r\theta} = \Gamma^\theta_{\theta r} = \frac{1}{r}\,, \quad \Gamma^\theta_{\phi\phi} = -\sin\theta\cos\theta\,. \tag{8.17}$$

Lastly, for $x^\mu = \phi$ the Euler–Lagrange equation reads

$$\frac{d}{d\lambda}\left(r^2\sin^2\theta\frac{d\phi}{d\lambda}\right) = 0\,,$$

$$r^2\sin^2\theta\frac{d^2\phi}{d\lambda^2} + 2r\sin^2\theta\frac{dr}{d\lambda}\frac{d\phi}{d\lambda} + 2r^2\sin\theta\cos\theta\frac{d\theta}{d\lambda}\frac{d\phi}{d\lambda} = 0\,,$$

$$\frac{d^2\phi}{d\lambda^2} + \frac{2}{r}\frac{dr}{d\lambda}\frac{d\phi}{d\lambda} + 2\cot\theta\frac{d\theta}{d\lambda}\frac{d\phi}{d\lambda} = 0\,, \tag{8.18}$$

and we find

$$\Gamma^\phi_{r\phi} = \Gamma^\phi_{\phi r} = \frac{1}{r}\,, \quad \Gamma^\phi_{\theta\phi} = \Gamma^\phi_{\phi\theta} = \cot\theta\,. \tag{8.19}$$

Now that we have all the Christoffel symbols, we can calculate the non-vanishing components of the Ricci tensor $R_{\mu\nu}$. The tt-component is

$$\begin{aligned}
R_{tt} &= \frac{\partial\Gamma^t_{tt}}{\partial t} + \frac{\partial\Gamma^r_{tt}}{\partial r} - \frac{\partial}{\partial t}\left(\Gamma^t_{tt} + \Gamma^r_{tr}\right) + \Gamma^t_{tt}\left(\Gamma^t_{tt} + \Gamma^r_{tr}\right)\\
&\quad + \Gamma^r_{tt}\left(\Gamma^t_{rt} + \Gamma^r_{rr} + \Gamma^\theta_{r\theta} + \Gamma^\phi_{r\phi}\right) - \Gamma^t_{tt}\Gamma^t_{tt} - \Gamma^t_{tr}\Gamma^t_{tt} - \Gamma^t_{tr}\Gamma^t_{tt} - \Gamma^r_{tr}\Gamma^r_{tr}\\
&= \frac{\partial\Gamma^r_{tt}}{\partial r} - \frac{\partial\Gamma^r_{tr}}{\partial t} + \Gamma^t_{tt}\Gamma^r_{tr} + \Gamma^r_{tt}\left(\Gamma^r_{rr} + \Gamma^\theta_{r\theta} + \Gamma^\phi_{r\phi}\right) - \Gamma^t_{tr}\Gamma^t_{tt} - \Gamma^r_{tr}\Gamma^r_{tr}\\
&= \frac{f''}{2g} - \frac{f'g'}{2g^2} - \frac{\ddot{g}}{2g} + \frac{\dot{g}^2}{2g^2} + \frac{\dot{f}}{2f}\frac{\dot{g}}{2g}\\
&\quad + \frac{f'}{2g}\left(\frac{g'}{2g} + \frac{1}{r} + \frac{1}{r}\right) - \frac{f'}{2f}\frac{f'}{2g} - \frac{\dot{g}}{2g}\frac{\dot{g}}{2g}\\
&= \frac{f''}{2g} - \frac{f'}{4g}\left(\frac{f'}{f} + \frac{g'}{g}\right) + \frac{f'}{rg} + \frac{\dot{f}\dot{g}}{4fg} + \frac{\dot{g}^2}{4g^2} - \frac{\ddot{g}}{2g}\,. \tag{8.20}
\end{aligned}$$

The tr-component of $R_{\mu\nu}$ is given by

$$R_{tr} = \frac{\partial \Gamma^t_{tr}}{\partial t} + \frac{\partial \Gamma^r_{tr}}{\partial r} - \frac{\partial}{\partial r}\left(\Gamma^t_{tt} + \Gamma^r_{tr}\right) + \Gamma^t_{tr}\left(\Gamma^t_{tt} + \Gamma^r_{tr}\right)$$

$$+ \Gamma^r_{tr}\left(\Gamma^t_{rt} + \Gamma^r_{rr} + \Gamma^\theta_{r\theta} + \Gamma^\phi_{r\phi}\right) - \Gamma^t_{tt}\Gamma^t_{rt} - \Gamma^t_{tr}\Gamma^r_{rt} - \Gamma^r_{tt}\Gamma^t_{rr} - \Gamma^r_{tr}\Gamma^r_{rr}$$

$$= \frac{\partial \Gamma^t_{tr}}{\partial t} - \frac{\partial \Gamma^t_{tt}}{\partial r} + \Gamma^r_{tr}\left(\Gamma^t_{rt} + \Gamma^\theta_{r\theta} + \Gamma^\phi_{r\phi}\right) - \Gamma^r_{tt}\Gamma^t_{rr}$$

$$= \frac{\dot{f}'}{2f} - \frac{f'\dot{f}}{2f^2} - \frac{\dot{f}'}{2f} + \frac{\dot{f}f'}{2f^2}$$

$$+ \frac{\dot{g}}{2g}\left(\frac{f'}{2f} + \frac{1}{r} + \frac{1}{r}\right) - \frac{f'}{2g}\frac{\dot{g}}{2f}$$

$$= \frac{\dot{g}^2}{rg}. \tag{8.21}$$

The *rr*-component is

$$R_{rr} = \frac{\partial \Gamma^t_{rr}}{\partial t} + \frac{\partial \Gamma^r_{rr}}{\partial r} - \frac{\partial}{\partial r}\left(\Gamma^t_{rt} + \Gamma^r_{rr} + \Gamma^\theta_{r\theta} + \Gamma^\phi_{r\phi}\right)$$

$$+ \Gamma^t_{rr}\left(\Gamma^t_{tt} + \Gamma^r_{tr}\right) + \Gamma^r_{rr}\left(\Gamma^t_{rt} + \Gamma^r_{rr} + \Gamma^\theta_{r\theta} + \Gamma^\phi_{r\phi}\right)$$

$$- \Gamma^t_{rt}\Gamma^t_{rt} - \Gamma^t_{rr}\Gamma^r_{rt} - \Gamma^r_{rt}\Gamma^t_{rr} - \Gamma^r_{rr}\Gamma^r_{rr} - \Gamma^\theta_{r\theta}\Gamma^\theta_{r\theta} - \Gamma^\phi_{r\phi}\Gamma^\phi_{r\phi}$$

$$= \frac{\partial \Gamma^t_{rr}}{\partial t} - \frac{\partial \Gamma^t_{rt}}{\partial r} - \frac{\partial \Gamma^\theta_{r\theta}}{\partial r} - \frac{\partial \Gamma^\phi_{r\phi}}{\partial r} + \Gamma^t_{rr}\Gamma^t_{tt} + \Gamma^r_{rr}\left(\Gamma^t_{rt} + \Gamma^\theta_{r\theta} + \Gamma^\phi_{r\phi}\right)$$

$$- \Gamma^t_{rt}\Gamma^t_{rt} - \Gamma^r_{rr}\Gamma^r_{rt} - \Gamma^\theta_{r\theta}\Gamma^\theta_{r\theta} - \Gamma^\phi_{r\phi}\Gamma^\phi_{r\phi}$$

$$= \frac{\ddot{g}}{2f} - \frac{\dot{f}\dot{g}}{2f^2} - \frac{f''}{2f} + \frac{f'^2}{2f^2} + \frac{1}{r^2} + \frac{1}{r^2} + \frac{\dot{g}}{2f}\frac{\dot{f}}{2f} + \frac{g'}{2g}\left(\frac{f'}{2f} + \frac{1}{r} + \frac{1}{r}\right)$$

$$- \frac{f'}{2f}\frac{f'}{2f} - \frac{\dot{g}}{2f}\frac{\dot{g}}{2g} - \frac{1}{r^2} - \frac{1}{r^2}$$

$$= -\frac{f''}{2f} + \frac{f'}{4f}\left(\frac{f'}{f} + \frac{g'}{g}\right) + \frac{g'}{rg} - \frac{\dot{f}\dot{g}}{4f^2} - \frac{\dot{g}^2}{4fg} + \frac{\ddot{g}}{2f}. \tag{8.22}$$

The $\theta\theta$-component reads

$$R_{\theta\theta} = \frac{\partial \Gamma^r_{\theta\theta}}{\partial r} - \frac{\partial \Gamma^\phi_{\theta\phi}}{\partial \theta} + \Gamma^r_{\theta\theta}\left(\Gamma^t_{rt} + \Gamma^r_{rr} + \Gamma^\theta_{r\theta} + \Gamma^\phi_{r\phi}\right) - \Gamma^r_{\theta\theta}\Gamma^\theta_{\theta r} - \Gamma^\theta_{\theta r}\Gamma^r_{\theta\theta} - \Gamma^\phi_{\theta\phi}\Gamma^\phi_{\theta\phi}$$

$$= \frac{\partial \Gamma^r_{\theta\theta}}{\partial r} - \frac{\partial \Gamma^\phi_{\theta\phi}}{\partial \theta} + \Gamma^r_{\theta\theta}\left(\Gamma^t_{rt} + \Gamma^r_{rr} + \Gamma^\phi_{r\phi}\right) - \Gamma^\theta_{\theta r}\Gamma^r_{\theta\theta} - \Gamma^\phi_{\theta\phi}\Gamma^\phi_{\theta\phi}$$

$$= -\frac{1}{g} + \frac{rg'}{g^2} + 1 + \cot^2\theta - \frac{r}{g}\left(\frac{f'}{2f} + \frac{g'}{2g} + \frac{1}{r}\right) + \frac{1}{r}\frac{r}{g} - \cot^2\theta$$

$$= 1 - \frac{1}{g} + \frac{r}{2g}\left(\frac{g'}{g} - \frac{f'}{f}\right). \tag{8.23}$$

The $\phi\phi$-component is

$$
\begin{aligned}
R_{\phi\phi} &= \frac{\partial \Gamma^r_{\phi\phi}}{\partial r} + \frac{\partial \Gamma^\theta_{\phi\phi}}{\partial \theta} + \Gamma^r_{\phi\phi}\left(\Gamma^t_{rt} + \Gamma^r_{rr} + \Gamma^\theta_{r\theta} + \Gamma^\phi_{r\phi}\right) + \Gamma^\theta_{\phi\phi}\Gamma^\phi_{\theta\phi} \\
&\quad - \Gamma^r_{\phi\phi}\Gamma^\phi_{\phi r} - \Gamma^\phi_{\phi r}\Gamma^r_{\phi\phi} - \Gamma^\theta_{\phi\phi}\Gamma^\phi_{\phi\theta} - \Gamma^\phi_{\phi\theta}\Gamma^\theta_{\phi\phi} \\
&= \frac{\partial \Gamma^r_{\phi\phi}}{\partial r} + \frac{\partial \Gamma^\theta_{\phi\phi}}{\partial \theta} + \Gamma^r_{\phi\phi}\left(\Gamma^t_{rt} + \Gamma^r_{rr} + \Gamma^\theta_{r\theta}\right) - \Gamma^\phi_{\phi r}\Gamma^r_{\phi\phi} - \Gamma^\phi_{\phi\theta}\Gamma^\theta_{\phi\phi} \\
&= -\frac{\sin^2\theta}{g} + \frac{r\sin^2\theta\, g'}{g^2} + \sin^2\theta - \cos^2\theta - \frac{r\sin^2\theta}{g}\left(\frac{f'}{2f} + \frac{g'}{2g} + \frac{1}{r}\right) \\
&\quad + \frac{1}{r}\frac{r\sin^2\theta}{g} + \sin\theta\cos\theta\cot\theta \\
&= \sin^2\theta\left[1 - \frac{1}{g} + \frac{r}{2g}\left(\frac{g'}{g} - \frac{f'}{f}\right)\right],
\end{aligned}
\tag{8.24}
$$

and it is equal to the $\theta\theta$-component multiplied by $\sin^2\theta$.

The other components of the Ricci tensor vanish. This can be seen noting that the transformations in (8.4) have the following effect:

$$
\begin{aligned}
R_{\theta\mu} \to R_{\tilde\theta\mu} &= \frac{\partial\theta}{\partial\tilde\theta}\frac{\partial x^\nu}{\partial x^\mu}R_{\theta\nu} = -R_{\theta\mu} \quad (\text{for } \mu \neq \theta), \\
R_{\phi\mu} &\to -R_{\phi\mu} \quad (\text{for } \mu \neq \phi).
\end{aligned}
\tag{8.25}
$$

Since the metric is invariant under such transformations, the components of the Ricci tensor should be invariant too, and therefore they must vanish.

Eventually, we have four independent equations

$$
R_{tt} = \frac{f''}{2g} - \frac{f'}{4g}\left(\frac{f'}{f} + \frac{g'}{g}\right) + \frac{f'}{rg} + \frac{\dot f\dot g}{4fg} + \frac{\dot g^2}{4g^2} - \frac{\ddot g}{2g} = 0,
\tag{8.26}
$$

$$
R_{tr} = \frac{\dot g^2}{rg} = 0,
\tag{8.27}
$$

$$
R_{rr} = -\frac{f''}{2f} + \frac{f'}{4f}\left(\frac{f'}{f} + \frac{g'}{g}\right) + \frac{g'}{rg} - \frac{\dot f\dot g}{4f^2} - \frac{\dot g^2}{4fg} + \frac{\ddot g}{2f} = 0,
\tag{8.28}
$$

$$
R_{\theta\theta} = 1 - \frac{1}{g} + \frac{r}{2g}\left(\frac{g'}{g} - \frac{f'}{f}\right) = 0,
\tag{8.29}
$$

From Eq. (8.27) we see that $g = g(r)$. Equation (8.29) can be written as

$$
\frac{f'}{f} = \frac{2g}{r} - \frac{2}{r} + \frac{g'}{g},
\tag{8.30}
$$

and we thus see that f'/f is a function of r only. This means that f can be written as

$$f(t,r) = f_1(t) f_2(r) .$$ (8.31)

With the following coordinate transformation

$$dt \rightarrow d\tilde{t} = \sqrt{f_1(t)} dt ,$$ (8.32)

we can always absorb $f_1(t)$ in the temporal coordinate. Eventually, the line element of the spacetime can be written in the following form

$$ds^2 = -f(r)c^2 dt^2 + g(r)dr^2 + r^2 \left(d\theta^2 + \sin^2\theta d\phi^2\right) ,$$ (8.33)

which shows that the metric is independent of t.

We combine Eqs. (8.26) and (8.28) in the following way

$$g R_{tt} + f R_{rr} = 0 ,$$ (8.34)

and we find

$$\frac{f''}{2} - \frac{f'}{4}\left(\frac{f'}{f} + \frac{g'}{g}\right) + \frac{f'}{r} - \frac{f''}{2} + \frac{f'}{4}\left(\frac{f'}{f} + \frac{g'}{g}\right) + \frac{fg'}{rg} = 0 ,$$

$$\frac{f'}{r} + \frac{fg'}{rg} = 0 ,$$

$$\frac{1}{rg}\frac{d}{dr}(fg) = 0$$

$$fg = \text{constant} .$$ (8.35)

Imposing the condition in Eq. (8.9), which is necessary because far from the source we want to recover the Minkowski spacetime, we find

$$g = \frac{1}{f} .$$ (8.36)

We can now rewrite Eq. (8.29) in terms of $f(r)$ only and solve the new differential equation

$$1 - f + \frac{rf}{2}\left(-f\frac{f'}{f^2} - \frac{f'}{f}\right) = 0 ,$$

$$1 - f - rf' = 0 ,$$

$$\frac{d}{dr}(rf) = 1 ,$$

$$f = 1 + \frac{C}{r} \, , \tag{8.37}$$

where C is a constant. The constant C can be inferred from the Newtonian limit. From Eq. (6.14), we know that

$$g_{tt} = -f = -\left(1 + \frac{2\Phi}{c^2}\right) . \tag{8.38}$$

For a spherically symmetric distribution of mass, the Newtonian potential reads

$$\Phi = -\frac{G_N M}{r} \, , \tag{8.39}$$

where M is the mass of the body generating the gravitational field. We thus find $C = -2G_N M/c^2$ and the line element of the spacetime reads

$$ds^2 = -\left(1 - \frac{2G_N M}{c^2 r}\right)c^2 dt^2 + \frac{dr^2}{1 - \frac{2G_N M}{c^2 r}} + r^2\left(d\theta^2 + \sin^2\theta d\phi^2\right) . \tag{8.40}$$

This concludes the proof of the theorem. The solution is called the Schwarzschild metric. It is remarkable that M is the only parameter that characterizes the spacetime metric in the exterior region; that is, the gravitational field in the exterior region is independent of the internal structure and composition of the massive body. As shown in Appendix F, M can be associated to the actual mass of the body only in the Newtonian limit. As discussed in Sect. 3.7 within the framework of special relativity, the total mass of a physical system is lower than the sum of the masses of its constituents because there is also a binding energy. The same is true here.

Note that the Schwarzschild metric has been derived only assuming that the space-time is spherically symmetric and solving the vacuum Einstein equations. The fact that the metric is independent of t is a consequence, it is not an assumption. This implies that the matter distribution does not have to be static, but it can move while maintaining the spherical symmetry, for example pulsating, and the vacuum solution is still described by the Schwarzschild metric. This implies that a spherically symmetric pulsating distribution of matter does not emit gravitational waves (this point will be discussed better in Chap. 12).

8.3 Schwarzschild Metric

In the Schwarzschild metric, the coordinates (ct, r, θ, ϕ) can assume the following values

$$t \in (-\infty, \infty) \, , \quad r \in [r_0, \infty) \, , \quad \theta \in (0, \pi) \, , \quad \phi \in [0, 2\pi) \, , \tag{8.41}$$

Table 8.1 Mass M, Schwarzschild radius r_S, and physical radius r_0 of the Sun, Earth, and a proton

Object	M (g)	r_S	r_0
Sun	$1.99 \cdot 10^{33}$	2.95 km	$6.97 \cdot 10^5$ km
Earth	$5.97 \cdot 10^{27}$	8.87 mm	$6.38 \cdot 10^3$ km
Proton	$1.67 \cdot 10^{-24}$	$2.48 \cdot 10^{-39}$ fm	0.8 fm

where r_0 is the radius of the body. The Schwarzschild metric is indeed valid in the vacuum only, namely in the "exterior" region. A different solution will describe the "interior" region $r < r_0$, where $T^{\mu\nu} \neq 0$. Note that the metric is ill-defined at the so-called *Schwarzschild radius* r_S

$$r_S = \frac{2G_N M}{c^2} . \tag{8.42}$$

This requires $r_0 > r_S$. In general, this is not a problem: as shown in Table 8.1, the Schwarzschild radius is typically much smaller than the radius of an object.

The relation between the temporal coordinate of the Schwarzschild metric, t, and the proper time of an observer at a point with fixed (r, θ, ϕ) is

$$d\tau = \sqrt{1 - \frac{r_S}{r}} dt < dt . \tag{8.43}$$

The relation between the space coordinates and the proper distance is more tricky. For simplicity, we consider the case of a distance in the radial direction. For a light signal, $ds^2 = 0$, and therefore in the case of pure radial motion we have

$$dt = \pm \frac{1}{c} \frac{dr}{1 - \frac{r_S}{r}} , \tag{8.44}$$

where the sign is $+$ $(-)$ if the light signal is moving to larger (smaller) radii. Instead of the temporal coordinate t, we rewrite Eq. (8.44) in terms of the proper time of an observer at a point with fixed (r, θ, ϕ)

$$d\tau = \pm \frac{1}{c} \frac{dr}{\sqrt{1 - \frac{r_S}{r}}} . \tag{8.45}$$

We can thus define the infinitesimal proper distance $d\rho$ as

$$d\rho = \frac{dr}{\sqrt{1 - \frac{r_S}{r}}} > dr . \tag{8.46}$$

This is indeed the infinitesimal distance along the radial direction measured by an observer at that point with the help of a light signal. The proper distance between the point (r_1, θ, ϕ) and the point (r_2, θ, ϕ) is obtained by integrating over the radial direction

$$\Delta\rho = \int_{r_1}^{r_2} \frac{dr}{\sqrt{1 - \frac{r_S}{r}}} \approx \int_{r_1}^{r_2} \left(1 + \frac{1}{2}\frac{r_S}{r}\right) dr$$

$$= (r_2 - r_1) + \frac{r_S}{2} \ln\frac{r_2}{r_1} . \tag{8.47}$$

In the case of the Minkowski spacetime, $r_S = 0$, and we recover the standard distance $r_2 - r_1$. For $r_S \neq 0$, there is a correction proportional to r_S.

Note that $d\tau \to dt$ and $d\rho \to dr$ for $r \to \infty$, which can be interpreted as the fact that the Schwarzschild coordinates correspond to the coordinates of an observer at infinity.

8.4 Motion in the Schwarzschild Metric

Let us now study the motion of test-particles in the Schwarzschild metric. The Lagrangian of the system is (for simplicity, we set $m = 0, 1$ the mass of, respectively, massless and massive particles)

$$L = \frac{1}{2}\left(-fc^2\dot{t}^2 + g\dot{r}^2 + r^2\dot{\theta} + r^2\sin^2\theta\dot{\phi}^2\right) , \tag{8.48}$$

where here the dot ˙ indicates the derivative with respect to the proper time/affine parameter λ and

$$f = \frac{1}{g} = 1 - \frac{r_S}{r} . \tag{8.49}$$

The Euler–Lagrange equation for the θ coordinate is equal to that in Newton's gravity met in Sect. 1.8. Without loss of generality, we can study the case of a particle moving in the equatorial plane $\theta = \pi/2$. The Lagrangian (8.48) thus simplifies to

$$L = \frac{1}{2}\left(-fc^2\dot{t}^2 + g\dot{r}^2 + r^2\dot{\phi}^2\right) . \tag{8.50}$$

There are three constants of motion: the energy (as measured at infinity) E, the angular momentum (as measured at infinity) L_z,[2] and the mass of the test-particle.

---

[2] As in Sect. 1.8, we use the notation L_z because this is also the axial component of the angular momentum (since $\theta = \pi/2$) and we do not want to call it L because it may generate confusion with the Lagrangian.

The conservation of the energy of the test-particle follows from the fact that the Lagrangian (8.50) is independent of the time coordinate t[3]

$$\frac{d}{d\lambda}\frac{\partial L}{\partial \dot{t}} - \frac{\partial L}{\partial t} = \frac{d}{d\lambda}\frac{\partial L}{\partial \dot{t}} = 0 \Rightarrow p_t = \frac{1}{c}\frac{\partial L}{\partial \dot{t}} = -fc\dot{t} = -\frac{E}{c} = \text{constant}. \quad (8.51)$$

The Lagrangian (8.50) is also independent of the coordinate ϕ, and we thus have the conservation of the angular momentum

$$\frac{d}{d\lambda}\frac{\partial L}{\partial \dot{\phi}} - \frac{\partial L}{\partial \phi} = \frac{d}{d\lambda}\frac{\partial L}{\partial \dot{\phi}} = 0 \Rightarrow p_\phi = \frac{\partial L}{\partial \dot{\phi}} = r^2\dot{\phi} = L_z = \text{constant}. \quad (8.52)$$

The conservation of the mass comes from the equation

$$g_{\mu\nu}\dot{x}^\mu \dot{x}^\nu = -fc^2\dot{t}^2 + g\dot{r}^2 + r^2\dot{\phi}^2 = -kc^2, \quad (8.53)$$

where $k = 0$ (for massless particles) and $k = 1$ (for massive particles).

From Eqs. (8.51) and (8.52) we find, respectively,

$$\dot{t} = \frac{E}{c^2 f}, \quad \dot{\phi} = \frac{L_z}{r^2}. \quad (8.54)$$

We plug these expressions of $\dot{t}$ and $\dot{\phi}$ into Eq. (8.53) and we find

$$g\dot{r}^2 + \frac{L_z^2}{r^2} - \frac{E^2}{c^2 f} = -kc^2. \quad (8.55)$$

If we multiply Eq. (8.55) by $f/2$ and we write the explicit form of f and g, we obtain

$$\frac{1}{2}\dot{r}^2 + \left(1 - \frac{r_S}{r}\right)\frac{L_z^2}{2r^2} - \frac{1}{2}\frac{E^2}{c^2} = -\frac{1}{2}\left(1 - \frac{r_S}{r}\right)kc^2. \quad (8.56)$$

This equation can be rewritten as

$$\frac{1}{2}\dot{r}^2 = \frac{E^2 - kc^4}{2c^2} - V_{\text{eff}}, \quad (8.57)$$

where

---

[3]Note that for a massive particle we choose $\lambda = \tau$ the particle proper time. In such a case, for a static particle at infinity we have $\dot{t} = 1$ and $E = c^2$ (because we are assuming $m = 1$, otherwise we would have $E = mc^2$); that is, the particle energy is just the rest mass. For a static particle at smaller radii, $E < c^2$ because the (Newtonian) gravitational potential energy is negative. Note also that $p_t = -E/c$ is conserved while the temporal component of the 4-momentum, p^t, is not a constant of motion. The same is true for p_ϕ and p^ϕ: only p_ϕ is conserved.

Fig. 8.1 Comparison between the effective potential in Eq. (8.58) for $k = 1$ valid in the Schwarzschild metric (red solid line) and the effective potential in Eq. (1.92) valid in Newton's gravity for the gravitational field generated by a point-like massive body (blue dashed line). We assume $G_N M = 1$, $L_z = 3.9$, and $c = 1$

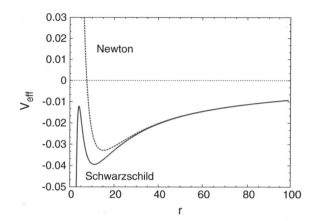

$$V_{\text{eff}} = -k\frac{G_N M}{r} + \frac{L_z^2}{2r^2} - \frac{G_N M L_z^2}{c^2 r^3}. \tag{8.58}$$

Equation (8.57) is the counterpart of Eq. (1.91) in Newton's gravity. The effective potential V_{eff} in Eq. (8.58) can be compared with the Newtonian effective potential in Eq. (1.92). For $k = 1$, we see that the first and the second terms on the right hand side in Eq. (8.58) are exactly those in Eq. (1.92). The third term is the correction to the Newtonian case and becomes important only at very small radii, because it scales as $1/r^3$. Figure 8.1 shows the difference between the effective potential in Eq. (8.58) and that in Eq. (1.92).

Let us now proceed as we did in Sect. 1.9 for the derivation of Kepler's Laws. We write

$$\frac{dr}{d\lambda} = \frac{dr}{d\phi}\frac{d\phi}{d\lambda} = \frac{L_z}{r^2}\frac{dr}{d\phi}, \tag{8.59}$$

and we remove the parameter λ in Eq. (8.57)

$$\frac{L_z^2}{2r^4}\left(\frac{dr}{d\phi}\right)^2 - k\frac{G_N M}{r} + \frac{L_z^2}{2r^2} - \frac{G_N M L_z^2}{c^2 r^3} = \frac{E^2 - kc^4}{2c^2}. \tag{8.60}$$

We introduce the variable $u = u(\phi)$

$$r = \frac{1}{u}, \quad u' = \frac{du}{d\phi}. \tag{8.61}$$

Equation (8.60) becomes

$$u'^2 - k\frac{2G_N M}{L_z^2}u + u^2 - \frac{2G_N M}{c^2}u^3 = \frac{E^2 - kc^4}{c^2 L_z^2}. \tag{8.62}$$

We derive this equation with respect to ϕ and we obtain

$$2u' \left(u'' - k\frac{G_N M}{L_z^2} + u - \frac{3G_N M}{c^2}u^2 \right) = 0 . \tag{8.63}$$

The equations of the orbits are thus

$$u' = 0 , \tag{8.64}$$

$$u'' - k\frac{G_N M}{L_z^2} + u - \frac{3G_N M}{c^2}u^2 = 0 . \tag{8.65}$$

From Eq. (8.64), we find circular orbits as in the Newtonian case from Eq. (1.100). Equation (8.65) is the relativistic generalization of Eq. (1.101).

8.5   Schwarzschild Black Holes

The Schwarzschild metric can describe the exterior region of a massive body, namely the region $r > r_0$, where $r_0 > r_S$ is the radius of the object. A different metric holds in the interior region $r < r_0$. As already discussed in Sect. 8.3, for typical bodies $r_0 \gg r_S$ and therefore it is irrelevant that the metric is not well-defined at $r = r_S$. If this is not the case and there is no interior solution, we have a black hole and the surface $r = r_S$ is the black hole *event horizon*. We will see in Sect. 10.5 how a similar object can be formed.

From Eq. (8.51), we can write

$$dt = \frac{E}{c^2 f}d\lambda = \frac{E}{c^2 f}\frac{d\lambda}{dr}dr = \frac{E}{c^2 f}\frac{dr}{\dot{r}} . \tag{8.66}$$

From Eq. (8.57), we have

$$\frac{1}{\dot{r}} = -\frac{c}{\sqrt{E^2 - kc^4 - 2V_{\text{eff}}c^2}} , \tag{8.67}$$

where the sign $-$ is chosen because we are interested in a particle moving to smaller radii. We replace $1/\dot{r}$ in Eq. (8.66) with the expression on the right hand side in Eq. (8.67). In the case of a massive particle ($k = 1$) with vanishing angular momentum ($L_z = 0$), we find

$$dt = -\frac{1}{c}\frac{E}{1 - \frac{r_S}{r}}\frac{dr}{\sqrt{E^2 - c^4 + \frac{r_S c^4}{r}}} , \tag{8.68}$$

which describes the motion of a massive particle falling onto the massive object with vanishing angular momentum. If we integrate both the left and the right hand sides in Eq. (8.68), we find that, according to our coordinate system corresponding to that of a distant observer, the particle takes the time Δt to move from the radius r_2 to the radius $r_1 < r_2$

$$\Delta t = \frac{1}{c} \int_{r_1}^{r_2} \frac{E}{1 - \frac{r_S}{r}} \frac{dr}{\sqrt{E^2 - c^4 + \frac{r_S c^4}{r}}} . \tag{8.69}$$

For $r_1 \rightarrow r_S$, $\Delta t \rightarrow \infty$ regardless of the value of E (for an explicit example, we can consider the case $E = c^2$ corresponding to a particle at rest at infinity); that is, for an observer at very large radii the particle takes an infinite time to reach the radial coordinate $r = r_S$.

Let us now calculate the time measured by the same particle. The relation between the proper time of the particle, τ, and its radial coordinate, r, can be inferred from Eq. (8.57), since $\dot{r} = dr/d\tau$. We have

$$d\tau = -\frac{cdr}{\sqrt{E^2 - c^4 - 2V_{\text{eff}}c^2}} .$$

For $L_z = 0$, we have

$$d\tau = -\frac{cdr}{\sqrt{E^2 - c^4 + \frac{r_S c^4}{r}}} .$$

Integrating we find

$$\Delta\tau = \int_{r_1}^{r_2} \frac{cdr}{\sqrt{E^2 - c^4 + \frac{r_S c^4}{r}}} .$$

For $r_1 \rightarrow r_S$, $\Delta\tau$ remains finite; that is, the particle can cross the surface at $r = r_S$, but the coordinate system of the distant observer can only describe the motion of the particle for $r > r_S$.

The radius $r = r_S$ is the black hole event horizon and causally disconnects the black hole ($r < r_S$) from the exterior region ($r > r_S$). A particle in the exterior region can cross the event horizon and enter the black hole (indeed it takes a finite time to reach and cross the surface at $r = r_S$) but then it cannot communicate with the exterior region any longer (this point will be more clear in the next section).

Note that the metric is ill-defined at $r = r_S$, but the spacetime is regular there. For instance, the *Kretschmann scalar* (as a scalar, it is an invariant) is

$$\mathcal{K} \equiv R^{\mu\nu\rho\sigma} R_{\mu\nu\rho\sigma} = \frac{48 G_N^2 M^2}{c^4 r^6} = \frac{12 r_S^2}{r^6} , \tag{8.70}$$

and does not diverge at $r = r_S$.

The singularity of the metric at $r = r_S$ depends on the choice of the coordinate system and can be removed by a coordinate transformation.[4] For instance, the Lemaitre coordinates (cT, R, θ, ϕ) are defined as

$$cdT = cdt + \left(\frac{r_S}{r}\right)^{1/2}\left(1 - \frac{r_S}{r}\right)^{-1} dr,$$

$$dR = cdt + \left(\frac{r}{r_S}\right)^{1/2}\left(1 - \frac{r_S}{r}\right)^{-1} dr. \tag{8.71}$$

In the Lemaitre coordinates, the line element of the Schwarzschild metric reads

$$ds^2 = -c^2 dT^2 + \frac{r_S}{r} dR^2 + r^2 d\theta^2 + r^2 \sin^2\theta d\phi^2, \tag{8.72}$$

where

$$r = (r_S)^{1/3}\left[\frac{3}{2}(R - cT)\right]^{2/3}. \tag{8.73}$$

The Schwarzschild radius $r = r_S$ in the new coordinates is

$$\frac{3}{2}(R - cT) = r_S, \tag{8.74}$$

and the metric is regular there. The Lemaitre coordinates can well describe both the black hole region $0 < r < r_S$ and the exterior region $r > r_S$.

The maximal analytic extension of the Schwarzschild spacetime is found when we employ the Kruskal–Szekeres coordinates. In these coordinates, the line element reads

$$ds^2 = \frac{4r_S^3}{r}e^{-r/r_S}(-d\tilde{t}^2 + d\tilde{r}^2) + r^2 d\theta^2 + r^2 \sin^2\theta d\phi^2, \tag{8.75}$$

where $\tilde{t}$ and $\tilde{r}$ are (dimensionless) coordinates defined as

$$\tilde{t} = \begin{cases} \left(\frac{r}{r_S} - 1\right)^{1/2} e^{r/(2r_S)} \sinh\left(\frac{ct}{2r_S}\right) & \text{if } r > r_S, \\ \left(1 - \frac{r}{r_S}\right)^{1/2} e^{r/(2r_S)} \cosh\left(\frac{ct}{2r_S}\right) & \text{if } 0 < r < r_S. \end{cases}$$

$$\tilde{r} = \begin{cases} \left(\frac{r}{r_S} - 1\right)^{1/2} e^{r/(2r_S)} \cosh\left(\frac{ct}{2r_S}\right) & \text{if } r > r_S, \\ \left(1 - \frac{r}{r_S}\right)^{1/2} e^{r/(2r_S)} \sinh\left(\frac{ct}{2r_S}\right) & \text{if } 0 < r < r_S. \end{cases} \tag{8.76}$$

---

[4]Note that the metric is ill-defined even at $r = 0$, which is a true spacetime singularity and cannot be removed by a coordinate transformation. The Kretschmann scalar diverges at $r = 0$.

The Schwarzschild solution in Kruskal–Szekeres coordinates includes also a white hole and a parallel universe, which are not present in the Schwarzschild spacetime in Schwarzschild coordinates. This will be briefly shown in the next section.

8.6 Penrose Diagrams

Penrose diagrams are 2-dimensional spacetime diagrams particularly suitable to studying the global properties and the causal structure of asymptotically flat spacetimes. Every point represents a 2-dimensional sphere of the original 4-dimensional spacetime. Penrose diagrams are obtained by a conformal transformation of the original coordinates such that the entire spacetime is transformed into a compact region. Since the transformation is conformal, angles are preserved. In this section, we employ units in which $c = 1$ and therefore null geodesics are lines at 45°. Time-like geodesics are inside the light-cone, space-like geodesics are outside. A more detailed discussion on the topic can be found, for instance, in [1, 2].

8.6.1 Minkowski Spacetime

The simplest example is the Penrose diagram of the Minkowski spacetime. In spherical coordinates (t, r, θ, ϕ), the line element of the Minkowski spacetime is $(c = 1)$

$$ds^2 = -dt^2 + dr^2 + r^2 d\theta^2 + r^2 \sin^2 \theta d\phi^2 . \tag{8.77}$$

With the following conformal transformation

$$t = \frac{1}{2} \tan \frac{T + R}{2} + \frac{1}{2} \tan \frac{T - R}{2} ,$$
$$r = \frac{1}{2} \tan \frac{T + R}{2} - \frac{1}{2} \tan \frac{T - R}{2} , \tag{8.78}$$

the line element becomes

$$ds^2 = \left(4 \cos^2 \frac{T + R}{2} \cos^2 \frac{T - R}{2} \right)^{-1} \left(-dT^2 + dR^2 \right)$$
$$+ r^2 d\theta^2 + r^2 \sin^2 \theta d\phi^2 . \tag{8.79}$$

Note that the transformation in (8.78) employs the tangent function, tan, in order to bring points at infinity to points at a finite value in the new coordinates.

The Penrose diagram for the Minkowski spacetime is shown in Fig. 8.2. The semi-infinite (t, r) plane is now a triangle. The dashed vertical line is the origin $r = 0$.

Every point corresponds to the 2-sphere (θ, ϕ). There are five different asymptotic regions. Without a rigorous treatment, they can be defined as follows[5]:

Future time-like infinity i^+: the region toward which time-like geodesics extend. It corresponds to the points at $t \to \infty$ with finite r.

Past time-like infinity i^-: the region from which time-like geodesics come. It corresponds to the points at $t \to -\infty$ with finite r.

Spatial infinity i^0: the region toward which space-like slices extend. It corresponds to the points at $r \to \infty$ with finite t.

Future null infinity $\mathscr{I}^+$: the region toward which outgoing null geodesics extend. It corresponds to the points at $t + r \to \infty$ with finite $t - r$.

Past null infinity $\mathscr{I}^-$: the region from which ingoing null geodesics come. It corresponds to the points at $t - r \to -\infty$ with finite $t + r$.

These five asymptotic regions are points or segments in the Penrose diagram. Their T and R coordinates are:

$$
\begin{aligned}
i^+ \quad & T = \pi, \quad R = 0. \\
i^- \quad & T = -\pi, \quad R = 0. \\
i^0 \quad & T = 0, \quad R = \pi,
\end{aligned}
\tag{8.80}
$$

and

$$
\begin{aligned}
\mathscr{I}^+ \quad & T + R = \pi, \quad T - R \in (-\pi; \pi). \\
\mathscr{I}^- \quad & T - R = -\pi, \quad T + R \in (-\pi; \pi).
\end{aligned}
\tag{8.81}
$$

8.6.2 Schwarzschild Spacetime

Penrose diagrams become a powerful tool to explore the global properties and the causal structure of more complicated spacetimes.

Let us now consider the Schwarzschild spacetime in Kruskal–Szekeres coordinates. The line element is given in Eq. (8.75). With the following coordinate transformation

---

[5]The symbol $\mathscr{I}$ is usually pronounced "scri".

Fig. 8.2 Penrose diagram
for the Minkowski spacetime

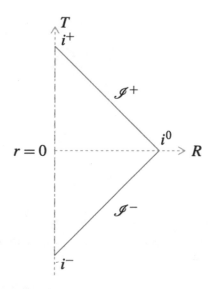

$$\tilde{t} = \frac{1}{2} \tan \frac{T + R}{2} + \frac{1}{2} \tan \frac{T - R}{2},$$
$$\tilde{r} = \frac{1}{2} \tan \frac{T + R}{2} - \frac{1}{2} \tan \frac{T - R}{2}, \qquad (8.82)$$

the line element becomes

$$ds^2 = \frac{32 G_N^3 M^3}{r} e^{-r/(2 G_N M)} \left(4 \cos^2 \frac{T + R}{2} \cos^2 \frac{T - R}{2} \right)^{-1} \left(-dT^2 + dR^2 \right)$$
$$+ r^2 d\theta^2 + r^2 \sin^2 \theta d\phi^2 . \qquad (8.83)$$

Figure 8.3 shows the Penrose diagram for the maximal extension of the Schwarzschild spacetime with its asymptotic regions i^+, i^-, i^0, $\mathscr{I}^+$, and $\mathscr{I}^-$. We can distinguish four regions, indicated, respectively, by I, II, III, and IV in the figure.

Region I corresponds to our universe, namely the exterior region of the Schwarzschild spacetime in Schwarzschild coordinates. Region II is the black hole, so the Schwarzschild spacetime in Schwarzschild coordinates has only regions I and II. The central singularity of the black hole at $r = 0$ is represented by the line with wiggles above region II. The event horizon of the black hole at $r = r_S$ is the red line at 45° separating regions I and II. Any ingoing light ray in region I is captured by the black hole, while any outgoing light ray in region I reaches future null infinity $\mathscr{I}^+$. Null and time-like geodesics in region II cannot exit the black hole and they necessarily fall to the singularity at $r = 0$.

Regions III and IV emerge from the extension of the Schwarzschild spacetime. Region III corresponds to another universe. The red line at 45° separating regions II and III is the event horizon of the black hole at $r = r_S$. Like in region I, any light ray in

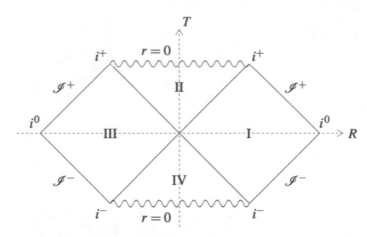

Fig. 8.3 Penrose diagram for the maximal extension of the Schwarzschild spacetime

region III can either cross the event horizon or escape to infinity. No future-oriented null or time-like geodesics can escape from region II. Our universe in region I and the other universe in region III cannot communicate: no null or time-like geodesic can go from one region to another.

Region IV is a *white hole*. If a black hole is a region of the spacetime where null and time-like geodesics can only enter and never exit, a white hole is a region where null and time-like geodesics can only exit and never enter. The red lines at 45° separating region IV from regions I and III are the horizons at $r = r_S$ of the white hole.

Problems

8.1 Let us consider a massive particle orbiting a geodesic circular orbit in the Schwarzschild spacetime. Calculate the relation between the particle proper time and the coordinate time t of the Schwarzschild metric.

8.2 Let us consider the Penrose diagram for the Minkowski spacetime, Fig. 8.2. We have a massive particle that emits an electromagnetic pulse at $t = 0$. Show the trajectories of the massive particle and of the electromagnetic pulse in the Penrose diagram.

8.3 Let us consider the Penrose diagram for the maximal extension of the Schwarzschild spacetime, Fig. 8.3. Show the future light-cone of an event in region I, of an event inside the black hole, and of an event inside the white hole.

References

1. C.W. Misner, K.S. Thorne, J.A. Wheeler, *Gravitation* (W. H. Freeman and Company, San Francisco, 1973)
2. P.K. Townsend, gr-qc/9707012

Chapter 9
Classical Tests of General Relativity

In 1916, Albert Einstein proposed three tests to verify his theory [2]:

1. The gravitational redshift of light.
2. The perihelion precession of Mercury.
3. The deflection of light by the Sun.

These three tests are today referred to as the *classical tests of general relativity*, even if, strictly speaking, the gravitational redshift of light is a test of the Einstein Equivalence Principle [6], while the other two are tests of the Schwarzschild solution in the weak field limit.[1] In 1964, Irwin Shapiro proposed another test [4], which is often called the *fourth classical test of general relativity*. Like the perihelion precession and the light deflection, even the test put forward by Shapiro is actually a test of the Schwarzschild solution in the weak field limit.

In the case of the perihelion precession of Mercury and of the deflection of light by the Sun, we can test, respectively, the trajectories of massive and massless test-particles in the Schwarzschild background. As discussed in Sect. 8.4, we have two differential equations, Eqs. (8.64) and (8.65), which we rewrite here for convenience

$$u' = 0,\tag{9.1}$$

$$u'' - k\frac{G_N M}{L_z^2} + u - \frac{3G_N M}{c^2}u^2 = 0,\tag{9.2}$$

where $k = 1$ for massive particles and $k = 0$ for massless particles. These two equations are the counterpart of Eqs. (1.100) and (1.101) of Newton's gravity. Equations (1.100) and (9.1) are identical and are simply the equation for a circle, so there

---

[1]Note, for example, that the Schwarzschild metric is a solution even in some alternative theories of gravity. Since the Mercury perihelion precession and the light deflection are only sensitive to the trajectories of particles, these tests can only verify the Schwarzschild metric (assuming geodesic motion), they cannot distinguish Einstein's gravity from those alternative theories of gravity in which the Schwarzschild metric is a solution of their field equations.

© Springer Nature Singapore Pte Ltd. 2018, corrected publication 2020
C. Bambi, *Introduction to General Relativity*, Undergraduate Lecture Notes in Physics, https://doi.org/10.1007/978-981-13-1090-4_9

are no interesting implications. Equation (1.101) has the solution (1.103), and the orbits are ellipses, parabolas, or hyperbolas according to the value of the constant A. Equation (9.2) has the last term proportional to u^2, which is absent in Newton's gravity and introduces relativistic corrections.

9.1   Gravitational Redshift of Light

Let us consider a static and spherically symmetric gravitational field. As already seen in Sect. 8.1, the line element can be written as

$$ds^2 = -f(r)c^2dt^2 + g(r)dr^2 + r^2 \left(d\theta^2 + \sin^2\theta d\phi^2\right) . \tag{9.3}$$

Let us also consider two observers at rest in this coordinate system. Observer A has spatial coordinates (r_A, θ, ϕ) and observer B has spatial coordinates (r_B, θ, ϕ); that is, the coordinates of the two observers differ only in the radial coordinate. At the location of observer A there is an emission of monochromatic electromagnetic radiation. Observer A measures the frequency ν_A for a time interval $\Delta\tau_A$. The number of wavefronts is

$$n = \nu_A \Delta\tau_A . \tag{9.4}$$

The radiation then reaches observer B, who measures the frequency ν_B for a time interval $\Delta\tau_B$, and therefore the number of wavefronts $n = \nu_B \Delta\tau_B$. Since the observers A and B must measure the same number of wavefronts n, we have

$$\frac{\nu_A}{\nu_B} = \frac{\Delta\tau_B}{\Delta\tau_A} . \tag{9.5}$$

For the electromagnetic signal propagating from the location of observer A to the location of observer B, $ds^2 = 0$. Moreover, since A and B have the same value for the coordinates θ and ϕ, along the trajectory of the signal we have

$$f(r)c^2dt^2 = g(r)dr^2 . \tag{9.6}$$

If we integrate over dt and dr, we find the time interval measured in the coordinate system (ct, r, θ, ϕ) that the first wavefront takes to go from the location of observer A to that of observer B

$$t_B^1 - t_A^1 = \int dt' = \frac{1}{c} \int_{r_A}^{r_B} dr' \sqrt{\frac{g(r')}{f(r')}} . \tag{9.7}$$

In the same way, we can compute the time interval that the last wavefront takes to go from observer A to observer B. Since the right hand side of Eq. (9.7) is independent of time, we have

$$t_B^n - t_A^n = t_B^1 - t_A^1 \,, \tag{9.8}$$

which we rewrite as

$$t_A^n - t_A^1 = t_B^n - t_B^1 \,. \tag{9.9}$$

Equation (9.9) shows that, as measured in the coordinate system (ct, r, θ, ϕ), the time intervals of the electromagnetic signal at the location of observer A and at the location of observer B are the same.

The time interval of the electromagnetic signal measured by observer A is [remember Eq. (8.43)]

$$\Delta \tau_A = \int_{t_A^1}^{t_A^n} dt \sqrt{f(r_A)} = \sqrt{f(r_A)} \left(t_A^n - t_A^1 \right) \,. \tag{9.10}$$

Observer B measures the time interval

$$\Delta \tau_B = \sqrt{f(r_B)} \left(t_B^n - t_B^1 \right) \,. \tag{9.11}$$

Employing Eq. (9.9), we find

$$\frac{\nu_A}{\nu_B} = \sqrt{\frac{f(r_B)}{f(r_A)}} \,. \tag{9.12}$$

In the Newtonian limit, $f = 1 + 2\Phi/c^2$, as we found in Eq. (6.14). We can thus rewrite Eq. (9.12) as

$$\frac{\nu_A}{\nu_B} = \sqrt{\frac{1 + 2\Phi_B/c^2}{1 + 2\Phi_A/c^2}} \approx 1 + \frac{\Phi_B}{c^2} - \frac{\Phi_A}{c^2} \,. \tag{9.13}$$

The relative variation of the frequency is

$$\frac{\Delta \nu}{\nu} = \frac{\nu_B - \nu_A}{\nu_B} = \frac{\Phi_A - \Phi_B}{c^2} = -\frac{\Delta \Phi}{c^2} \,. \tag{9.14}$$

This phenomenon is called the *gravitational redshift* of light. It was measured for the first time by Robert Pound and Glen Rebka in 1960 [3]. They used a moving atomic source such that the Doppler blueshift could exactly compensate the gravitational redshift that the photons experienced to reach the detector located at a height of 22.5 m.

For small distances, the Earth's gravitational field can be approximated as constant. If we send an electromagnetic signal from the ground to a detector at the height

h, the gravitational redshift is

$$\frac{\Delta \nu}{\nu} = -\frac{gh}{c^2}, \tag{9.15}$$

where g is the gravitational acceleration on Earth. For $g = 9.81 \, \text{m/s}^2$ and $h = 22.5 \, \text{m}$, we find

$$\frac{\Delta \nu}{\nu} = -2 \cdot 10^{-15}. \tag{9.16}$$

9.2   Perihelion Precession of Mercury

Let us study Eq. (9.2) for massive particles ($k = 1$). We introduce the dimensionless variable $y = Ru$, where R is the characteristic value of the radial coordinate of the orbit of the particle and therefore we expect a solution $y = O(1)$. Equation (9.2) becomes

$$y'' - \alpha + y - \varepsilon y^2 = 0, \tag{9.17}$$

where α and ε are

$$\alpha = \frac{G_N M R}{L_z^2}, \quad \varepsilon = \frac{3 G_N M}{c^2 R} = \frac{3}{2} \frac{r_S}{R}. \tag{9.18}$$

Note that $\varepsilon \ll 1$. For the Sun, the Schwarzschild radius is $r_S = 3 \, \text{km}$. The characteristic orbital radius of Mercury is $R \sim 5 \cdot 10^7 \, \text{km}$, and therefore $\varepsilon \sim 10^{-7}$. We can use ε as an expansion parameter and write y as

$$y = y_0 + \varepsilon y_1 + O(\varepsilon^2). \tag{9.19}$$

Equation (9.17) becomes

$$y_0'' + \varepsilon y_1'' + y_0 + \varepsilon y_1 = \alpha + \varepsilon y_0^2 + O\left(\varepsilon^2\right), \tag{9.20}$$

and we have to solve two differential equations

$$y_0'' + y_0 = \alpha, \tag{9.21}$$
$$y_1'' + y_1 = y_0^2. \tag{9.22}$$

Equation (9.21) is the equation of Newton's gravity (see Sect. 1.9) and the solution is

$$y_0 = \alpha + A \cos \phi \,, \tag{9.23}$$

We plug the solution (9.23) into Eq. (9.22) and we find

$$y_1'' + y_1 = \alpha^2 + A^2 \cos^2 \phi + 2\alpha A \cos \phi$$
$$= \alpha^2 + \frac{A^2}{2} + \frac{A^2}{2} \cos (2\phi) + 2\alpha A \cos \phi \,. \tag{9.24}$$

Let us now rewrite y_1 as the sum of three functions

$$y_1 = y_{11} + y_{12} + y_{13} \,, \tag{9.25}$$

and we split Eq. (9.24) into three parts as follows

$$y_{11}'' + y_{11} = \alpha^2 + \frac{A^2}{2} \,, \tag{9.26}$$

$$y_{12}'' + y_{12} = \frac{A^2}{2} \cos (2\phi) \,, \tag{9.27}$$

$$y_{13}'' + y_{13} = 2\alpha A \cos \phi \,. \tag{9.28}$$

The homogeneous solutions have the form $B \cos \phi$. The inhomogeneous solutions are

$$y_{11} = \alpha^2 + \frac{A^2}{2} \,, \tag{9.29}$$

$$y_{12} = -\frac{A^2}{6} \cos (2\phi) \,, \tag{9.30}$$

$$y_{13} = \alpha A \phi \sin \phi \,. \tag{9.31}$$

We can thus write the solution of Eq. (9.17) up to $O\left(\varepsilon^2\right)$:

$$y = \alpha + A \cos \phi + \varepsilon \left[\alpha^2 + \frac{A^2}{2} - \frac{A^2}{6} \cos (2\phi) + \alpha A \phi \sin \phi + B \cos \phi \right] + O\left(\varepsilon^2\right) \,. \tag{9.32}$$

Note that the corrections to the Newtonian orbit are proportional to ε, which is roughly the ratio between the Schwarzschild radius of the Sun and the characteristic radius of the orbit, and that the only term that grows with the number of orbital revolutions is $\alpha A \phi \sin \phi$; that is, this term becomes more and more important when we consider longer and longer time intervals. We rewrite the solution in (9.32) neglecting the terms that do not grow with the number of orbital revolutions and employing the relation

$$\cos(\phi - \varepsilon\alpha\phi) = \cos\phi \cos(\varepsilon\alpha\phi) + \sin\phi \sin(\varepsilon\alpha\phi)$$
$$= \cos\phi + \varepsilon\alpha\phi \sin\phi + O\left(\varepsilon^2\right) . \tag{9.33}$$

The result is

$$y = \alpha + A\cos(\phi - \varepsilon\alpha\phi) , \tag{9.34}$$

and, in terms of the coordinate r,

$$\frac{1}{r} = \frac{\alpha}{R} + \frac{A}{R}\cos(\phi - \varepsilon\alpha\phi) . \tag{9.35}$$

The function in (9.35) is periodic with period 2π with respect to the argument $\phi - \varepsilon\alpha\phi$. We can thus write

$$\phi(1 - \varepsilon\alpha) = 2\pi n , \tag{9.36}$$

and

$$\phi = 2\pi n(1 + \varepsilon\alpha) + O\left(\varepsilon^2\right) , \tag{9.37}$$

After n orbital revolutions, there is a shift $\delta\phi$ with respect to the Newtonian prediction

$$\delta\phi = \phi - 2\pi n = 2\pi\varepsilon\alpha n . \tag{9.38}$$

Comparing Eq. (9.35) with Eqs. (D.6) and (D.7) in Appendix D, we see that

$$\alpha = \frac{R}{a\left(1 - e^2\right)} , \tag{9.39}$$

and Eq. (9.38) can be rewritten as

$$\delta\phi = \frac{6\pi G_N M}{c^2}\frac{n}{a\left(1 - e^2\right)} . \tag{9.40}$$

Note that the perihelion precession of planets in the Solar System is already expected in Newtonian mechanics and the latter is larger than the relativistic contribution. Table 9.1 shows the case of the perihelion precession of Mercury. The dominant contribution comes from the equinox precession due to the fact that we observe Mercury from Earth. A minor but still large contribution comes from the perturbation of Mercury's orbit by other planets, in particular Venus, Earth, and Jupiter. In the end, the perihelion precession due to relativistic effects is a small contribution. Urbain Le Verrier was the first, in 1859, to point out an anomaly in the perihelion precession of Mercury within Newton's theory. The origin of this anomaly was under debate for a long time. It was the first test passed by Einstein's gravity, but the

Table 9.1 Contributions to the Mercury perihelion precession. $\delta\phi$ is in seconds of arc per century. Table readapted from [1]

Cause	$\delta\phi$ (as/100 years)
Mercury	0.03 ± 0.00
Venus	277.86 ± 0.68
Earth	90.04 ± 0.08
Mars	2.54 ± 0.00
Jupiter	153.58 ± 0.00
Saturn	7.30 ± 0.01
Uranus	0.14 ± 0.00
Neptune	0.04 ± 0.00
Solar oblateness	0.01 ± 0.02
Equinox precession	5025.65 ± 0.50
Sum	5557.18 ± 0.85
Observed	5599.74 ± 0.41
Difference	42.56 ± 0.94
Relativistic effect	43.03 ± 0.03

explanation was not immediately accepted, because of a certain skepticism towards this theory by a large fraction of the scientific community and the difficulties in the measurements of the Newtonian effects of the perihelion precession of Mercury.

The perihelion precession of other planets in the Solar System is smaller, essentially because their orbital radius is larger and their eccentricity is lower, but it can still be measured for Venus and Earth. In the case of binary pulsars, the relativistic contribution to the orbital precession can be much larger. For instance, in the binary PSR1913+16, the relativistic contribution to the perihelion precession is about $4°$ per year [5].

9.3 Deflection of Light

Let us now consider the case of massless particles ($k = 0$). As before, we introduce the variable $y = Ru$, where R is still the characteristic value of the radial coordinate of the particle. Equation (9.2) is now

$$y'' + y - \varepsilon y^2 = 0.\tag{9.41}$$

If R is the radius of the Sun, $R = 7 \cdot 10^5$ km, and $\varepsilon \sim 10^{-5}$. We proceed as in the previous section and we write y as an expansion in ε

$$y = y_0 + \varepsilon y_1 + O(\varepsilon^2).\tag{9.42}$$

Equation (9.41) becomes

$$y_0'' + \varepsilon y_1'' + y_0 + \varepsilon y_1 = \varepsilon y_0^2 + O\left(\varepsilon^2\right) , \tag{9.43}$$

and we have to solve the following equations

$$y_0'' + y_0 = 0 , \tag{9.44}$$
$$y_1'' + y_1 = y_0^2 . \tag{9.45}$$

Equation (9.44) provides the Newtonian solution

$$y_0 = A \cos \phi , \tag{9.46}$$

which can be rewritten in terms of the radial coordinate r

$$\frac{1}{r} = \frac{A}{R} \cos \phi . \tag{9.47}$$

This is the equation of a straight line in polar coordinates. For $\phi = 0$, we find the impact parameter $b = R/A$ (see Fig. 9.1).

We plug the solution (9.46) into Eq. (9.45) and we get

$$y_1'' + y_1 = \frac{R^2}{b^2} \cos^2 \phi = \frac{R^2}{2b^2} + \frac{R^2}{2b^2} \cos (2\phi) . \tag{9.48}$$

We write y_1 as the sum of two contributions

$$y_1 = y_{11} + y_{12} , \tag{9.49}$$

and Eq. (9.48) can be split into two parts

$$y_{11}'' + y_{11} = \frac{R^2}{2b^2} , \tag{9.50}$$

$$y_{12}'' + y_{12} = \frac{R^2}{2b^2} \cos (2\phi) , \tag{9.51}$$

The homogeneous solutions have the form $B \cos \phi$. The inhomogeneous solutions are

$$y_{11} = \frac{R^2}{2b^2} , \tag{9.52}$$

$$y_{12} = -\frac{R^2}{6b^2} \cos (2\phi) . \tag{9.53}$$

y_1 is thus given by

$$y_1 = \frac{R^2}{2b^2} - \frac{R^2}{6b^2} \cos(2\phi) + B \cos\phi$$

$$= \frac{2R^2}{3b^2} - \frac{R^2}{3b^2} \cos^2\phi + B \cos\phi, \qquad (9.54)$$

and the solution of Eq. (9.41) up to $O\left(\varepsilon^2\right)$ is

$$\frac{1}{r} = \frac{1}{b} \cos\phi + \varepsilon\left(\frac{2R}{3b^2} - \frac{R}{3b^2} \cos^2\phi + \frac{B}{R} \cos\phi\right) + O\left(\varepsilon^2\right). \qquad (9.55)$$

For $\phi = 0$, we find the minimum value of the radial coordinate of the orbit of the particle

$$\frac{1}{r_{\min}} = \frac{1}{b} + \varepsilon\left(\frac{R}{3b^2} + \frac{B}{R}\right). \qquad (9.56)$$

We define

$$\tilde{\varepsilon} = \frac{\varepsilon R}{b}, \quad \tilde{B} = \frac{B}{R}, \qquad (9.57)$$

and we consider the limit $r \to \infty$. Equation (9.55) becomes

$$-\frac{\tilde{\varepsilon}}{3} \cos^2\phi + \left(1 + \varepsilon\tilde{B}b\right) \cos\phi + \frac{2}{3}\tilde{\varepsilon} = 0,$$

$$\cos^2\phi - \frac{3\left(1 + \varepsilon\tilde{B}b\right)}{\tilde{\varepsilon}} \cos\phi - 2 = 0, \qquad (9.58)$$

which is a second order equation in $\cos\phi$. The solution is

$$\cos\phi = \frac{3\left(1 + \varepsilon\tilde{B}b\right)}{2\tilde{\varepsilon}}\left[1 \pm \sqrt{1 + \frac{8}{9}\frac{\tilde{\varepsilon}^2}{\left(1 + \varepsilon\tilde{B}b\right)^2}}\right]. \qquad (9.59)$$

The solution with the sign $+$ has no physical meaning, because $\cos\phi$ cannot exceed 1. The physical solution is that with the sign $-$ and we have

$$\cos\phi \approx \frac{3\left(1 + \varepsilon\tilde{B}b\right)}{2\tilde{\varepsilon}}\left[1 - 1 - \frac{4}{9}\frac{\tilde{\varepsilon}^2}{\left(1 + \varepsilon\tilde{B}b\right)^2}\right] = -\frac{2}{3}\frac{\tilde{\varepsilon}}{\left(1 + \varepsilon\tilde{B}b\right)}$$

$$= -\frac{2}{3}\tilde{\varepsilon} + O\left(\varepsilon^2\right) = -\frac{r_s}{b} + O\left(\varepsilon^2\right). \qquad (9.60)$$

For $\delta \equiv r_{\mathrm{S}}/b \ll 1$, we have

$$\cos\left(\pm\frac{\pi}{2}\pm\delta\right) = -\delta + O\left(\delta^2\right),\tag{9.61}$$

and we see that the solution of ϕ in Eq. (9.60) is

$$\phi = \pm\frac{\pi}{2}\pm\delta.\tag{9.62}$$

As we can see from Fig. 9.1, the total deflection of the light ray is Δ

$$\Delta = 2\delta = \frac{2r_{\mathrm{S}}}{b} = \frac{4G_{\mathrm{N}}M}{c^2b}.\tag{9.63}$$

If b is the radius of the Sun, we find $\Delta = 1.75$ as.

In reality, it is not possible to observe light rays with impact parameter b equal to the radius of the Sun $R_{\odot}$ because of the presence of the Solar corona. We have to consider photons with $b > R_{\odot}$ and in this case the deflection angle is

$$\Delta' = \Delta\frac{R_{\odot}}{b}.\tag{9.64}$$

The observation of the deflection of light rays by the Sun in 1919 by a group led by Arthur Eddington was the first test specifically done to verify Einstein's gravity. The observation was done during a Solar eclipse, when the light from the Sun was blocked by the Moon and it was thus possible to observe stars close to the Sun. From the comparison of photographs of the same region of the sky during the eclipse and when the Sun was not there, it was possible to measure the deflection angle Δ'

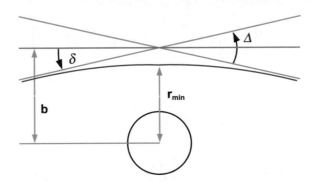

Fig. 9.1 Trajectory of a massless particle. The green line is the trajectory in Newton's gravity and is described by $r = b/\cos\phi$, where r is the radial coordinate of the polar coordinate system centered at the center of the massive body. The blue line is the trajectory in the Schwarzschild metric, Eq. (9.55). At $r \to \infty$, the polar angle is $\phi = \pm\pi/2 \pm \delta$. The total deflection angle is thus $\Delta = 2\delta$

and then infer Δ. This kind of optical observations is quite challenging and the final measurement is affected by systematic effects. Today, radio observations can provide more reliable and precise measurements.

9.4 Shapiro's Effect

In the perihelion precession of Mercury and in the deflection of light, we see how relativistic effects change the trajectories predicted in Newton's gravity. In 1964, Irwin Shapiro proposed a new test, often called the fourth classical test of general relativity, which is based on the measurement of the time delay of an electromagnetic signal to move from one point to another in the Solar System with respect to the time that the same signal would take in a flat spacetime [4].

Figure 9.2 is a sketch to illustrate Shapiro's effect. The Sun is at the point O. We want to calculate the time that an electromagnetic signal takes to go from point A with spatial coordinates $(r_A, \pi/2, \phi_A)$ to point B with spatial coordinates $(r_B, \pi/2, \phi_B)$. As before, without loss of generality, we consider the equatorial plane $\theta = \pi/2$. C is the point of the trajectory of the electromagnetic signal with the smallest value of the radial coordinate, which we call r_C.

For massless particles, Eq. (8.55) is

$$g \left(\frac{dr}{d\lambda} \right)^2 + \frac{L_z^2}{r^2} - \frac{E^2}{c^2 f} = 0 . \tag{9.65}$$

Employing Eq. (8.51), we can write

$$\frac{dr}{d\lambda} = \frac{dr}{dt} \frac{dt}{d\lambda} = \frac{dr}{dt} \frac{E}{c^2 f} , \tag{9.66}$$

and we remove the affine parameter λ in Eq. (9.65)

$$\frac{E^2}{c^4 f^3} \left(\frac{dr}{dt} \right)^2 + \frac{L_z^2}{r^2} - \frac{E^2}{c^2 f} = 0 , \tag{9.67}$$

where we have also exploited the fact that $g = 1/f$ in the Schwarzschild metric. At $r = r_C$, we have $dr/dt = 0$, and therefore

$$L_z^2 = \frac{E^2 r_C^2}{c^2 f(r_C)} . \tag{9.68}$$

With this expression we can rewrite Eq. (9.67) without the constants that depend on the parametrization of the orbit

Fig. 9.2 Shapiro's effect.
The Sun is at point O and we
want to measure the time that
an electromagnetic signal
takes to go from point A to B
and return to A

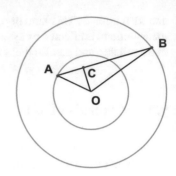

$$\frac{1}{f^3}\left(\frac{dr}{dt}\right)^2 + \frac{c^2}{f(r_C)}\frac{r_C^2}{r^2} - \frac{c^2}{f} = 0. \tag{9.69}$$

We find

$$dt = \pm\frac{1}{c}\frac{dr}{\sqrt{\left[1 - \frac{f(r)}{f(r_C)}\frac{r_C^2}{r^2}\right]f^2(r)}}. \tag{9.70}$$

The time, as measured by the coordinate system (ct, r, θ, ϕ), that an electromagnetic signal takes to go from point A to point C is

$$t_{AC} = \frac{1}{c}\int_{r_C}^{r_A}\frac{dr}{\sqrt{\left[1 - \frac{f(r)}{f(r_C)}\frac{r_C^2}{r^2}\right]f^2(r)}}. \tag{9.71}$$

We only consider the first order terms in the expansion in r_S/r and r_S/r_C

$$\left[1 - \frac{f(r)}{f(r_C)}\frac{r_C^2}{r^2}\right]f^2(r) = \left[1 - \frac{\left(1 - \frac{r_S}{r}\right)}{\left(1 - \frac{r_S}{r_C}\right)}\frac{r_C^2}{r^2}\right]\left(1 - \frac{r_S}{r}\right)^2$$

$$= \left[1 - \left(1 - \frac{r_S}{r} + \frac{r_S}{r_C}\right)\frac{r_C^2}{r^2}\right]\left(1 - \frac{2r_S}{r}\right)$$

$$= 1 - \frac{r_C^2}{r^2} - \frac{2r_S}{r} + \frac{3r_S}{r}\frac{r_C^2}{r^2} - \frac{r_S}{r_C}\frac{r_C^2}{r^2}, \tag{9.72}$$

and therefore

$$\frac{1}{\sqrt{\left[1 - \frac{f(r)}{f(r_C)} \frac{r_C^2}{r^2}\right] f^2(r)}} = \frac{1}{\sqrt{1 - \frac{r_C^2}{r^2}}} \left[1 - \frac{1}{2} \frac{1}{1 - \frac{r_C^2}{r^2}} \left(\frac{3r_S}{r} \frac{r_C^2}{r^2} - \frac{2r_S}{r} - \frac{r_S}{r_C} \frac{r_C^2}{r^2}\right)\right]$$

$$= \sqrt{\frac{r^2}{r^2 - r_C^2} - \frac{1}{2}\left(\frac{r^2}{r^2 - r_C^2}\right)^{3/2} \left(\frac{3r_S}{r} \frac{r_C^2}{r^2} - \frac{2r_S}{r} - \frac{r_S}{r_C} \frac{r_C^2}{r^2}\right)}$$

$$= \sqrt{\frac{r^2}{r^2 - r_C^2}} + \frac{1}{2} \frac{r_S}{(r^2 - r_C^2)^{3/2}} \left(2r^2 + r_C r - 3r_C^2\right). \quad (9.73)$$

The leading order term of the integral (9.71) is

$$\frac{1}{c} \int_{r_C}^{r_A} \frac{r\,dr}{\sqrt{r^2 - r_C^2}} = \frac{1}{c}\sqrt{r_A^2 - r_C^2}. \quad (9.74)$$

The next-to-leading order terms are

$$\frac{1}{c}\int_{r_C}^{r_A} \frac{r_S r^2\,dr}{(r^2 - r_C^2)^{3/2}} = \frac{1}{c}\left[-\frac{r_S r}{\sqrt{r^2 - r_C^2}} + r_S \ln\left(r + \sqrt{r^2 - r_C^2}\right)\right]_{r_C}^{r_A}, \quad (9.75)$$

$$\frac{1}{2c}\int_{r_C}^{r_A} \frac{r_S r_C r\,dr}{(r^2 - r_C^2)^{3/2}} = \frac{1}{2c}\left[-\frac{r_S r_C}{\sqrt{r^2 - r_C^2}}\right]_{r_C}^{r_A}, \quad (9.76)$$

$$\frac{1}{2c}\int_{r_C}^{r_A} \frac{-3r_S r_C^2\,dr}{(r^2 - r_C^2)^{3/2}} = \frac{1}{2c}\left[\frac{3r_S r}{\sqrt{r^2 - r_C^2}}\right]_{r_C}^{r_A}. \quad (9.77)$$

These integrals diverge at $r = r_C$, but this is because we have expanded in r_S/r and r_S/r_C and we see that the sum of those integrals with $r \to r_C$ is finite

$$(9.75) + (9.76) + (9.77) = \frac{r_S}{2c} \frac{r_A - r_C}{\sqrt{r_A^2 - r_C^2}} + \frac{r_S}{c} \ln\left(r_A + \sqrt{r_A^2 - r_C^2}\right) - \frac{r_S}{c} \ln(r_C)$$

$$= \frac{r_S}{2c}\sqrt{\frac{r_A - r_C}{r_A + r_C}} + \frac{r_S}{c} \ln\left(\frac{r_A + \sqrt{r_A^2 - r_C^2}}{r_C}\right). \quad (9.78)$$

Eventually, the time that the electromagnetic signal takes to go from point A to point C turns out to be

$$t_{AC} = \frac{1}{c}\sqrt{r_A^2 - r_C^2} + \frac{r_S}{c}\ln\left(\frac{r_A + \sqrt{r_A^2 - r_C^2}}{r_C}\right) + \frac{r_S}{2c}\sqrt{\frac{r_A - r_C}{r_A + r_C}}. \quad (9.79)$$

In the case of a flat spacetime, the time that the electromagnetic signal would take to go from point A to point C is

$$\tilde{t}_{AC} = \frac{1}{c}\sqrt{r_A^2 - r_C^2}, \quad (9.80)$$

and corresponds to the leading order term in (9.79). The total time that the electromagnetic signal takes to go from A to B and come back to A is

$$t_{\text{tot}} = 2t_{AC} + 2t_{BC}, \quad (9.81)$$

while in flat spacetime it would be

$$\tilde{t}_{\text{tot}} = 2\tilde{t}_{AC} + 2\tilde{t}_{BC} = \frac{2}{c}\sqrt{r_A^2 - r_C^2} + \frac{2}{c}\sqrt{r_B^2 - r_C^2}. \quad (9.82)$$

The maximum time delay with respect to the flat spacetime is when $r_C = R_\odot$, where $R_\odot$ is the radius of the surface of the Sun. The result is $(R_\odot \ll r_A, r_B)$

$$\delta t_{\text{max}} = \frac{4G_N M}{c^3}\left[1 + \ln\left(\frac{4r_A r_B}{R_\odot^2}\right)\right]. \quad (9.83)$$

Note that we are using the coordinate time t, not the proper time of the observer in A. However, the correction is of the order of r_S/r_A.

The phenomenon of time delay of an electromagnetic signal passing near a massive body is commonly called *Shapiro's effect*. Current measurements are in perfect agreement with the theoretical predictions. To have an idea of the magnitude of the effect, let us estimate the time delay of an electromagnetic signal to go from Earth to Mercury and return to Earth when the two planets are at opposite sides of the Sun, so r_C is the radius of the Sun. We plug the following values into Eq. (9.83)

$$r_A = 150 \cdot 10^6 \text{ km}, \quad r_B = 58 \cdot 10^6 \text{ km}, \quad R_\odot = 0.7 \cdot 10^6 \text{ km},$$
$$r_S = 2.95 \text{ km}. \quad (9.84)$$

The result is

$$\delta t_{\text{max}} = 0.24 \text{ ms}. \quad (9.85)$$

9.5 Parametrized Post-Newtonian Formalism

The Parametrized Post-Newtonian (PPN) formalism is a convenient approach to test the solutions of Einstein's gravity in the weak field regime. The basic idea is to write the metric as an expansion about the Minkowski metric in terms of certain gravitational potentials. If we want to test the Schwarzschild metric in the Solar System, we write the most general static and spherically symmetric line element as an expansion in r_S/r, where r_S is the Schwarzschild radius of the Sun. The approach is traditionally formulated in isotropic coordinates and, as seen in Sect. 8.1, the most general static and spherically symmetric line element reads[2]

$$
ds^2 = -\left(1 - \frac{2G_N M}{c^2 R} + \beta \frac{2G_N^2 M^2}{c^4 R^2} + \cdots \right) c^2 dt^2
$$
$$
+ \left(1 + \gamma \frac{2G_N M}{c^2 R} + \cdots \right) \left(dx^2 + dy^2 + dz^2\right), \qquad (9.87)
$$

where β and γ are free parameters to be determined by observations.

In spherical-like coordinates, the line element (9.87) is

$$
ds^2 = -\left(1 - \frac{2G_N M}{c^2 R} + \beta \frac{2G_N^2 M^2}{c^4 R^2} + \cdots \right) c^2 dt^2
$$
$$
+ \left(1 + \gamma \frac{2G_N M}{c^2 R} + \cdots \right) \left(dR^2 + R^2 d\theta^2 + R^2 \sin^2\theta d\phi^2\right). \quad (9.88)
$$

With the transformation

$$
R \to r = R\left(1 + \gamma \frac{G_N M}{c^2 R} + \cdots \right), \qquad (9.89)
$$

with inverse transformation

$$
R = r\left(1 - \gamma \frac{G_N M}{c^2 r} + \cdots \right), \qquad (9.90)
$$

we write the line element in the more conventional Schwarzschild coordinates

---

[2]In order to have the correct Newtonian limit, the line element must have the form

$$
ds^2 = -\left(1 - \frac{2G_N M}{c^2 R} + ... \right) c^2 dt^2 + (1 + ...) \left(dx^2 + dy^2 + dz^2\right). \qquad (9.86)
$$

For higher order terms, there are no theoretical requirements, and therefore we introduce β and γ.

$$ds^2 = -\left[1 - \frac{2G_N M}{c^2 r} + (\beta - \gamma)\frac{2G_N^2 M^2}{c^4 r^2} + \cdots\right]c^2 dt^2$$

$$+ \left(1 + \gamma\frac{2G_N M}{c^2 r} + \cdots\right)dr^2 + r^2\left(d\theta^2 + \sin^2\theta\, d\phi^2\right). \quad (9.91)$$

Now we can easily see that the Schwarzschild metric is recovered when $\beta = \gamma = 1$.

If we employ the metric in (9.91) in place of the Schwarzschild one in the discussion of the Mercury precession, Eq. (9.40) becomes

$$\delta\phi = \frac{6\pi G_N M}{c^2}\frac{n}{a\left(1 - e^2\right)}\left(\frac{2 - \beta + 2\gamma}{3}\right). \quad (9.92)$$

If we do the same for the problem of light bending, Eq. (9.63) becomes

$$\Delta = \frac{4G_N M}{c^2 b}\left(\frac{1 + \gamma}{2}\right). \quad (9.93)$$

For the Shapiro time delay, Eq. (9.83) becomes

$$\delta t_{\max} = \frac{4G_N M}{c^3}\left[1 + \left(\frac{1 + \gamma}{2}\right)\ln\left(\frac{4r_A r_B}{R_\odot^2}\right)\right]. \quad (9.94)$$

From the measurements of observational effects, like the perihelion precession of Mercury $\delta\phi$, the light bending Δ, and the Shapiro time delay δt, we can constrain the PPN parameters β and γ to check whether they are consistent with 1, as required by the Schwarzschild metric. The current most stringent constraints are [6]

$$|\beta - 1| < 8 \cdot 10^{-5}, \quad (9.95)$$

$$|\gamma - 1| < 2.3 \cdot 10^{-5}. \quad (9.96)$$

Current observations thus confirm the Schwarzschild solution within their precision.

References

1. G.M. Clemence, Rev. Mod. Phys. **19**, 361 (1947)
2. A. Einstein, Ann. Phys. **49**, 769 (1916). [Annalen Phys. **14**, 517 (2005)]
3. R.V. Pound, G.A. Rebka Jr., Phys. Rev. Lett. **4**, 337 (1960)
4. I.I. Shapiro, Phys. Rev. Lett. **13**, 789 (1964)
5. J.H. Taylor, R.A. Hulse, L.A. Fowler, G.E. Gullahorn, J.M. Rankin, Astrophys. J. **206**, L53 (1976)
6. C.M. Will, Living Rev. Relativ. **17**, 4 (2014), arXiv:1403.7377[gr-qc]

Chapter 10
Black Holes

The aim of this chapter is to provide a general overview on black holes, and on black holes in 4-dimensional Einstein's gravity in particular. Contrary to the previous chapters, often we will only present the final result, without providing all the calculations. Throughout the chapter, unless stated otherwise, we will employ natural units in which $G_N = c = 1$. Such a choice significantly simplifies all the formulas and represents the standard convention in this line of research. The black hole mass M sets the size of the system and the associated length and time scales are, respectively,

$$\frac{G_N M}{c^2} = 1.48 \left(\frac{M}{M_\odot}\right) \text{ km}, \quad \frac{G_N M}{c^3} = 4.92 \left(\frac{M}{M_\odot}\right) \text{ μs}, \quad (10.1)$$

10.1 Definition

Roughly speaking, a black hole is a region of spacetime in which gravity is so strong that it is impossible to escape or send information to the exterior region. With the concepts introduced in Sect. 8.6, we can provide the following definition of black hole in an asymptotically flat spacetime[1]:

Black hole. A *black hole* in an asymptotically flat spacetime $\mathcal{M}$ is the set of events that do not belong to the causal past of the future null infinity $J^-(\mathscr{I}^+)$, namely

$$\mathscr{B} = \mathcal{M} - J^-(\mathscr{I}^+) \neq \emptyset. \quad (10.2)$$

The *event horizon* is the boundary of the region $\mathscr{B}$.

---

[1]Note that such a definition of black hole is not limited to Einstein's gravity and can be applied whenever it is possible to define $J^-(\mathscr{I}^+)$.

© Springer Nature Singapore Pte Ltd. 2018
C. Bambi, *Introduction to General Relativity*, Undergraduate Lecture Notes in Physics, https://doi.org/10.1007/978-981-13-1090-4_10

$\mathscr{I}^+$ is the future null infinity and was already discussed in Sect. 8.6. $J^-(\mathscr{P})$ is called the causal past of the region $\mathscr{P}$ and is the set of all events that causally precede $\mathscr{P}$; that is, for every element in $J^-(\mathscr{P})$ there exists at least one smooth future-directed time-like or light-like curve extending to $\mathscr{P}$. All future-directed curves (either time-like or light-like) starting from the region $\mathscr{B}$ fail to reach null infinity $\mathscr{I}^+$. A black hole is thus a one-way membrane: if something crosses the event horizon it can no longer send any signal to the asymptotically flat region. For more details, see for instance Ref. [9].

The Penrose diagram of the Schwarzschild spacetime was presented in Fig. 8.3. We can see that there the event horizon is a line at 45°, and therefore no time-like and light-like trajectories can escape from the interior region (the black hole) and reach $\mathscr{I}^+$. The black hole does not belong to $J^-(\mathscr{I}^+)$.

10.2 Reissner–Nordström Black Holes

In Einstein's gravity, the simplest black hole solution is represented by the Schwarzschild spacetime met in Chap. 8. It describes an uncharged and spherically symmetric black hole and is completely characterized by only one parameter, the black hole mass M. The next-to-simplest black hole solution is the *Reissner–Nordström spacetime*. It describes a non-rotating black hole with a non-vanishing electric charge. Now the solution is specified by two parameters, namely the black hole mass M and the black hole electric charge Q.

As a useful recipe to remember, the Reissner–Nordström line element can be obtained from the Schwarzschild one with the substitution $M \to M - Q^2/(2r)$. The result is[2]

$$ds^2 = -\left(1 - \frac{2M}{r} + \frac{Q^2}{r^2}\right)dt^2 + \left(1 - \frac{2M}{r} + \frac{Q^2}{r^2}\right)^{-1} dr^2 + r^2 d\theta^2$$
$$+r^2 \sin^2\theta d\phi^2 . \tag{10.4}$$

The solution of $g^{rr} = 0$ is[3]

---

[2]In the international system of units, the line element reads

$$ds^2 = -\left(1 - \frac{2G_N M}{c^2 r} + \frac{G_N Q^2}{4\pi\varepsilon_0 c^4 r^2}\right)dt^2 + \left(1 - \frac{2G_N M}{c^2 r} + \frac{G_N Q^2}{4\pi\varepsilon_0 c^4 r^2}\right)^{-1} dr^2 + r^2 d\theta^2 \tag{10.3}$$
$$+r^2 \sin^2\theta d\phi^2 ,$$

where $1/(4\pi\varepsilon_0)$ is the Coulomb force constant.

[3]The definition of event horizon is provided in Sect. 10.1, but it cannot be directly applied for determining the coordinates of the event horizon from a known metric. For the black hole solutions discussed in this textbook (Schwarzschild, Reissner–Nordström, Kerr), the radial coordinate of the event horizon corresponds to the larger root of $g^{rr} = 0$. For more general cases, the interested reader can refer to [1] and references therein. Note that the issue of the existence of an event horizon in these solutions is non-trivial. This is also proved by the fact that the Schwarzschild metric was found

$$r_\pm = M \pm \sqrt{M^2 - Q^2}\,, \tag{10.5}$$

where the larger root, r_+, is the event horizon, while the smaller root, r_-, is the inner horizon. The horizons only exist for $|Q| \le M$. For $|Q| > M$, there is no horizon, the singularity at $r = 0$ is "naked",[4] and the Reissner–Nordström solution describes the spacetime of a naked singularity rather than that of a black hole.

10.3 Kerr Black Holes

Albert Einstein proposed his theory at the end of 1915 and the Schwarzschild solution was found immediately after, in 1916, by Karl Schwarzschild. Still in 1916, Hans Reissner solved the Einstein equations for a point-like charged mass, and, in 1918, Gunnar Nordström found the metric for a spherically symmetric charged mass. It was only in 1963 that Roy Kerr found the solution for a rotating black hole [8]. The *Kerr solution* describes a rotating uncharged black hole in 4-dimensional Einstein's gravity and is specified by two parameters; that is, the black hole mass M and the black hole spin angular momentum J. In Boyer–Lindquist coordinates, the line element is

$$ds^2 = -\left(1 - \frac{2Mr}{\Sigma}\right)dt^2 - \frac{4aMr\sin^2\theta}{\Sigma}dt d\phi + \frac{\Sigma}{\Delta}dr^2 + \Sigma d\theta^2$$
$$+ \left(r^2 + a^2 + \frac{2a^2 Mr\sin^2\theta}{\Sigma}\right)\sin^2\theta d\phi^2\,, \tag{10.6}$$

where

$$\Sigma = r^2 + a^2\cos^2\theta\,, \quad \Delta = r^2 - 2Mr + a^2\,, \tag{10.7}$$

and $a = J/M$ is the *specific spin*. It is often convenient to introduce the dimensionless *spin parameter*[5] $a_* = a/M = J/M^2$.

If we expand the line element in Eq. (10.6) in a/r and M/r, we find

---

in 1916 and it was only in 1958 that David Finkelstein realized that there was an event horizon with specific properties.

[4]A *naked singularity* is a singularity of the spacetime that is not inside a black hole and thus belongs to the causal past of future null infinity.

[5]Reintroducing the speed of light c and Newton's gravitational constants G_N, we have

$$a = \frac{J}{cM}\,, \quad a_* = \frac{a}{r_g} = \frac{cJ}{G_N M^2}\,, \tag{10.8}$$

where $r_g = G_N M/c^2$ is the *gravitational radius*.

$$ds^2 = -\left(1 - \frac{2M}{r} + \frac{2a^2 M \cos^2 \theta}{r^3} + \cdots \right) dt^2 - \left(\frac{4aM \sin^2 \theta}{r} + \cdots \right) dt d\phi$$

$$+ \left[1 + \frac{a^2 \left(\cos^2 \theta - 1\right)}{r^2} + \cdots \right] \frac{dr^2}{1 - 2M/r} + \left(r^2 + a^2 \cos^2 \theta\right) d\theta^2$$

$$+ \left(r^2 + a^2 + \frac{2a^2 M \sin^2 \theta}{r} + \cdots \right) \sin^2 \theta d\phi^2 . \tag{10.9}$$

From this expression, we can see that the Kerr metric reduces to the Schwarzschild solution for $a = 0$. Moreover, by comparing this line element with that around a slow-rotating massive body (see Appendix G), we can realize that $a = J/M$, as we have already asserted without any proof.

As in the Reissner–Nordström metric, there are two solutions for the equation $g^{rr} = 0$; that is

$$r_{\pm} = M \pm \sqrt{M^2 - a^2} . \tag{10.10}$$

r_+ is the radius of the event horizon, which requires $|a| \leq M$. For $|a| > M$ there is no horizon and the spacetime has a naked singularity at $r = 0$. Note that the topology of the spacetime singularity in the Kerr solution is different from those in the Schwarzschild and Reissner–Nordström spacetimes. The singularity is only in the equatorial plane. For example, the Kretschmann scalar $\mathscr{K}$ is

$$\mathscr{K} = \frac{48M^2}{\Sigma^6} \left(r^6 - 15a^2 r^4 \cos^2 \theta + 15a^4 r^2 \cos^4 \theta - a^6 \cos^6 \theta\right) , \tag{10.11}$$

and diverges at $r = 0$ only for $\theta = \pi/2$.

It is possible to show that geodesics outside the equatorial plane can reach the singularity and extend to another universe. Let us consider the Kerr–Schild coordinates (t', x, y, z), which are related to the Boyer–Lindquist ones by

$$x + iy = (r + ia) \sin \theta \exp \left[i \int d\phi + i \int \frac{a}{\Delta} dr\right] ,$$

$$z = r \cos \theta ,$$

$$t' = \int dt - \int \frac{r^2 + a^2}{\Delta} dr - r , \tag{10.12}$$

where i is the imaginary unit, namely $i^2 = -1$. r is implicitly given by

$$r^4 - \left(x^2 + y^2 + z^2 - a^2\right) r^2 - a^2 z^2 = 0 . \tag{10.13}$$

The singularity at $r = 0$ and $\theta = \pi/2$ corresponds to $z = 0$ and $x^2 + y^2 = a^2$, namely it is a ring. It is possible to extend the spacetime to negative r, and the ring connects two universes. However, the region $r < 0$ posses closed time-like curves.[6] More details can be found, for instance, in [5].

In what follows, we will discuss a number of properties of the Kerr metric. However, the results can be easily extended to more general stationary, axisymmetric, and asymptotically flat spacetimes. We will present both general equations (that can be easily applied to other spacetimes) as well as the expressions for the Kerr metric in Boyer–Lindquist coordinates. More details can be found in Ref. [1].

10.3.1 Equatorial Circular Orbits

Time-like circular orbits in the equatorial plane of the Kerr metric are particularly important [4]. For example, they are relevant for the structure of accretion disks around astrophysical black holes [1].

Let us write the line element in the so-called *canonical form*

$$ds^2 = g_{tt}dt^2 + 2g_{t\phi}dtd\phi + g_{rr}dr^2 + g_{\theta\theta}d\theta^2 + g_{\phi\phi}d\phi^2 , \qquad (10.14)$$

where the metric coefficients are independent of t and ϕ. For example, this is the case of the Boyer–Lindquist coordinates in (10.6). As we know, the Lagrangian for a free point-like particle is that in Eq. (3.1) and it is convenient to parametrize the trajectory with the particle proper time τ. For simplicity, we set the rest-mass of the particle $m = 1$.

Since the metric is independent of the coordinates t and ϕ, we have two constants of motion, namely the specific energy at infinity E and the axial component of the specific angular momentum at infinity L_z

$$\frac{d}{d\tau}\frac{\partial L}{\partial \dot{t}} - \frac{\partial L}{\partial t} = 0 \Rightarrow p_t \equiv \frac{\partial L}{\partial \dot{t}} = g_{tt}\dot{t} + g_{t\phi}\dot{\phi} = -E , \qquad (10.15)$$

$$\frac{d}{d\tau}\frac{\partial L}{\partial \dot{\phi}} - \frac{\partial L}{\partial \phi} = 0 \Rightarrow p_\phi \equiv \frac{\partial L}{\partial \dot{\phi}} = g_{t\phi}\dot{t} + g_{\phi\phi}\dot{\phi} = L_z . \qquad (10.16)$$

The term "specific" is used to indicate that E and L_z are, respectively, the energy and the angular momentum per unit rest-mass. We remind the reader that here we use the convention of a metric with signature $(-+++)$. For a metric with signature $(+---)$, we would have $p_t = E$ and $p_\phi = -L_z$. Equations (10.15) and (10.16) can be solved to find the t and the ϕ components of the 4-velocity of the test-particle

---

[6]A closed time-like curve is a closed time-like trajectory; if there are closed time-like curves in a spacetime, massive particles can travel backwards in time. While the Einstein equations have solutions with similar properties, they are thought to be non-physical and therefore they are usually not taken into account.

$$\dot{t} = \frac{Eg_{\phi\phi} + L_z g_{t\phi}}{g_{t\phi}^2 - g_{tt}g_{\phi\phi}}, \tag{10.17}$$

$$\dot{\phi} = -\frac{Eg_{t\phi} + L_z g_{tt}}{g_{t\phi}^2 - g_{tt}g_{\phi\phi}}. \tag{10.18}$$

From the conservation of the rest-mass, $g_{\mu\nu}\dot{x}^\mu\dot{x}^\nu = -1$, we can write the equation

$$g_{rr}\dot{r}^2 + g_{\theta\theta}^2\dot{\theta}^2 = V_{\text{eff}}(r, \theta), \tag{10.19}$$

where V_{eff} is the effective potential

$$V_{\text{eff}} = \frac{E^2 g_{\phi\phi} + 2EL_z g_{t\phi} + L_z^2 g_{tt}}{g_{t\phi}^2 - g_{tt}g_{\phi\phi}} - 1. \tag{10.20}$$

Circular orbits in the equatorial plane are located at the zeros and the turning points of the effective potential: $\dot{r} = \dot{\theta} = 0$, which implies $V_{\text{eff}} = 0$, and $\ddot{r} = \ddot{\theta} = 0$, which requires, respectively, $\partial_r V_{\text{eff}} = 0$ and $\partial_\theta V_{\text{eff}} = 0$. From these conditions, we could obtain the specific energy E and the axial component of the specific angular momentum L_z of a test-particle in equatorial circular orbits. However, it is faster to proceed in the following way. We write the geodesic equations as

$$\frac{d}{d\tau}\left(g_{\mu\nu}\dot{x}^\nu\right) = \frac{1}{2}\left(\partial_\mu g_{\nu\rho}\right)\dot{x}^\nu\dot{x}^\rho. \tag{10.21}$$

In the case of equatorial circular orbits, $\dot{r} = \dot{\theta} = \ddot{r} = 0$, and the radial component of Eq. (10.21) reduces to

$$\left(\partial_r g_{tt}\right)\dot{t}^2 + 2\left(\partial_r g_{t\phi}\right)\dot{t}\dot{\phi} + \left(\partial_r g_{\phi\phi}\right)\dot{\phi}^2 = 0. \tag{10.22}$$

The angular velocity $\Omega = \dot{\phi}/\dot{t}$ is

$$\Omega_\pm = \frac{-\partial_r g_{t\phi} \pm \sqrt{\left(\partial_r g_{t\phi}\right)^2 - \left(\partial_r g_{tt}\right)\left(\partial_r g_{\phi\phi}\right)}}{\partial_r g_{\phi\phi}}, \tag{10.23}$$

where the upper (lower) sign refers to corotating (counterrotating) orbits, namely orbits with angular momentum parallel (antiparallel) to the spin of the central object. From $g_{\mu\nu}\dot{x}^\mu\dot{x}^\nu = -1$ with $\dot{r} = \dot{\theta} = 0$, we can write

$$\dot{t} = \frac{1}{\sqrt{-g_{tt} - 2\Omega g_{t\phi} - \Omega^2 g_{\phi\phi}}}. \tag{10.24}$$

Equation (10.15) becomes

Fig. 10.1 Specific energy E of a test-particle in equatorial circular orbits in the Kerr spacetime as a function of the Boyer–Lindquist radial coordinate r. Every curve corresponds to the spacetime with a particular value of the spin parameter a_*. The values are $a_* = 0, 0.5, 0.8, 0.9, 0.95, 0.99,$ and 0.999

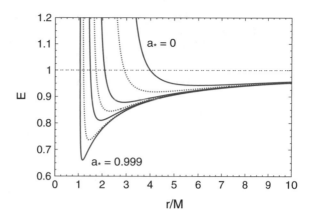

$$E = -\left(g_{tt} + \Omega g_{t\phi}\right) \dot{t}$$

$$= -\frac{g_{tt} + \Omega g_{t\phi}}{\sqrt{-g_{tt} - 2\Omega g_{t\phi} - \Omega^2 g_{\phi\phi}}} . \tag{10.25}$$

Equation (10.16) becomes

$$L_z = \left(g_{t\phi} + \Omega g_{\phi\phi}\right) \dot{t}$$

$$= \frac{g_{t\phi} + \Omega g_{\phi\phi}}{\sqrt{-g_{tt} - 2\Omega g_{t\phi} - \Omega^2 g_{\phi\phi}}} . \tag{10.26}$$

If we employ the metric coefficients of the Kerr solution in Boyer–Lindquist coordinates, Eqs. (10.25) and (10.26) become, respectively, [4]

$$E = \frac{r^{3/2} - 2Mr^{1/2} \pm aM^{1/2}}{r^{3/4}\sqrt{r^{3/2} - 3Mr^{1/2} \pm 2aM^{1/2}}} , \tag{10.27}$$

$$L_z = \pm\frac{M^{1/2}\left(r^2 \mp 2aM^{1/2}r^{1/2} + a^2\right)}{r^{3/4}\sqrt{r^{3/2} - 3Mr^{1/2} \pm 2aM^{1/2}}} , \tag{10.28}$$

and the angular velocity of the test-particle is

$$\Omega_\pm = \pm\frac{M^{1/2}}{r^{3/2} \pm aM^{1/2}} \tag{10.29}$$

where, again, the upper sign refers to corotating orbits, the lower sign to counterrotating ones. Figure 10.1 shows E as a function of the radial coordinate r for different values of the spin parameter of the black hole. Figure 10.2 shows the profile of L_z.

As we can see from Eqs. (10.25) and (10.26), as well as from Eqs. (10.27) and (10.28), E and L_z diverge when their denominator vanishes. This happens at the radius of the *photon orbit* r_γ

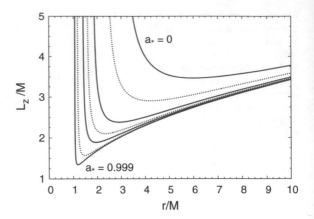

$$g_{tt} + 2\Omega g_{t\phi} + \Omega^2 g_{\phi\phi} = 0 \quad \Rightarrow \quad r = r_\gamma \,. \tag{10.30}$$

In Boyer–Lindquist coordinates, the equation for the radius of the photon orbit is

$$r^{3/2} - 3Mr^{1/2} \pm 2aM^{1/2} = 0 \,. \tag{10.31}$$

The solution is [4]

$$r_\gamma = 2M \left\{ 1 + \cos \left[\frac{2}{3} \arccos \left(\mp \frac{a}{M} \right) \right] \right\} \,. \tag{10.32}$$

For $a = 0, r_\gamma = 3M$. For $a = M, r_\gamma = M$ for corotating orbits and $4M$ for counter-rotating orbits.

The radius of the *marginally bound orbit* r_{mb} is defined as

$$E = -\frac{g_{tt} + \Omega g_{t\phi}}{\sqrt{-g_{tt} - 2\Omega g_{t\phi} - \Omega^2 g_{\phi\phi}}} = 1 \quad \Rightarrow \quad r = r_{\mathrm{mb}} \,. \tag{10.33}$$

The orbit is marginally bound, which means that the test-particle has sufficient energy to escape to infinity (the test-particle cannot reach infinity if $E < 1$, and can reach infinity with a finite velocity if $E > 1$). In the Kerr metric in Boyer–Lindquist coordinates, we find [4]

$$r_{\mathrm{mb}} = 2M \mp a + 2\sqrt{M(M \mp a)} \,. \tag{10.34}$$

For $a = 0, r_{\mathrm{mb}} = 4M$. For $a = M, r_{\mathrm{mb}} = M$ for corotating orbits and

$$r_{\mathrm{mb}} = \left(3 + 2\sqrt{2} \right) M \approx 5.83M \tag{10.35}$$

for counterrotating orbits.

The radius of the *marginally stable orbit*, r_{ms}, more often called the radius of the *innermost stable circular orbit* (ISCO), r_{ISCO}, is defined by

$$\partial_r^2 V_{eff} = 0 \quad \text{or} \quad \partial_\theta^2 V_{eff} = 0 \quad \Rightarrow \quad r = r_{ISCO}. \tag{10.36}$$

In the Kerr metric, equatorial circular orbits turn out to always be vertically stable, so the ISCO radius is determined by $\partial_r^2 V_{eff}$ only. In Boyer–Lindquist coordinates, we have [4]

$$r_{ISCO} = 3M + Z_2 \mp \sqrt{(3M - Z_1)(3M + Z_1 + 2Z_2)},$$
$$Z_1 = M + \left(M^2 - a^2\right)^{1/3} \left[(M + a)^{1/3} + (M - a)^{1/3}\right],$$
$$Z_2 = \sqrt{3a^2 + Z_1^2}. \tag{10.37}$$

For $a = 0$, $r_{ISCO} = 6M$. For $a = M$, $r_{ISCO} = M$ for corotating orbits and $r_{ISCO} = 9M$ for counterrotating orbits. The derivation can be found, for instance, in [5].

Note that the innermost stable circular orbit is located at the minimum of the energy E. r_{ISCO} can thus be obtained even from the equation $dE/dr = 0$. After some manipulation, one finds

$$\frac{dE}{dr} = \frac{r^2 - 6Mr \pm 8aM^{1/2}r^{1/2} - 3a^2}{2r^{7/4}\left(r^{3/2} - 3Mr^{1/2} \pm aM^{1/2}\right)^{3/2}} M, \tag{10.38}$$

and $r^2 - 6Mr \pm 8aM^{1/2}r^{1/2} - 3a^2 = 0$ is the same equation as that we can obtain from Eq. (10.36), see [5]. Moreover, the minimum of the energy E and of the axial component of the angular momentum L_z are located at the same radius. After some manipulation, dL_z/dr is

$$\frac{dL_z}{dr} = \frac{r^2 - 6Mr \pm aM^{1/2}r^{1/2} - 3a^2}{2r^{7/4}\left(r^{3/2} - 3Mr^{1/2} \pm aM^{1/2}\right)^{3/2}} \left(M^{1/2}r^{3/2} \pm aM\right), \tag{10.39}$$

and we see that $dE/dr = 0$ and $dL_z/dr = 0$ have the same solution.

The concept of innermost stable circular orbit has no counterpart in Newtonian mechanics. As we have already seen in Sect. 8.4 for the Schwarzschild metric, and we can do the same for the Kerr one, the equations of motion of a test-particle can be written as the equations of motion in Newtonian mechanics with a certain effective potential. At small radii, the effective potential is dominated by an attractive term that is absent in Newtonian mechanics and is responsible for the existence of the innermost stable circular orbit.

Figure 10.3 shows the radius of the event horizon r_+, of the photon orbit r_γ, of the marginally bound circular orbit r_{mb}, and of the innermost stable circular orbit r_{ISCO} in the Kerr metric in Boyer–Lindquist coordinates as functions of a_*.

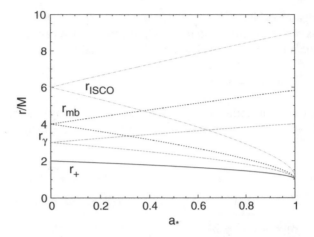

Fig. 10.3 Radial coordinates of the event horizon r_+ (red solid curve), of the photon orbit r_γ (green dashed curve), of the marginally bound circular orbit r_{mb} (blue dotted curve), and of the innermost stable circular orbit r_{ISCO} (cyan dashed-dotted curve) in the Kerr metric in Boyer–Lindquist coordinates as functions of the spin parameter a_*. For every radius, the upper curve refers to the counterrotating orbits, the lower curve to the corotating ones

In the end, in the Kerr metric we have the following picture for E and L_z. At large radii, we recover the Newtonian limit for E and L_z for a particle in the gravitational field of a point-like massive body. As the radial coordinate decreases, E and L_z monotonically decrease as well, up to when we reach a minimum, which is at the same radius for E and L_z. This is the radius of the innermost stable circular orbit. If we move to smaller radii, E and L_z increase. First, we find the radius of the marginally bound circular orbit, defined by the condition $E = 1$. As the radial coordinate decreases, E and L_z continue increasing, and they diverge at the photon orbit. There are no circular orbits with radial coordinate smaller than the photon orbit and massive particles can reach the photon orbit in the limit of infinite energy.

Note that, in the case of an extremal Kerr black hole with $a = M$, one finds $r_+ = r_\gamma = r_{mb} = r_{ISCO} = M$ for corotating orbits. However, the Boyer–Lindquist coordinates are not well-defined at the event horizon and these special orbits do not coincide [4]. If we write $a = M(1 - \varepsilon)$ with $\varepsilon \to 0$, we find

$$r_+ = M\left(1 + \sqrt{2\varepsilon} + \cdots\right), \qquad r_\gamma = M\left(1 + 2\sqrt{\frac{2\varepsilon}{3}} + \cdots\right),$$

$$r_{mb} = M\left(1 + 2\sqrt{\varepsilon} + \cdots\right), \qquad r_{ISCO} = M\left[1 + (4\varepsilon)^{1/3} + \cdots\right]. \quad (10.40)$$

We can then evaluate the proper radial distance between r_+ and the other radii [4]. Sending ε to zero, the result is

$$\int_{r_+}^{r_\gamma} \frac{r'\,dr'}{\sqrt{\Delta}} \rightarrow \frac{1}{2} M \ln 3\,,$$

$$\int_{r_+}^{r_{mb}} \frac{r'\,dr'}{\sqrt{\Delta}} \rightarrow M \ln\left(1 + \sqrt{2}\right)\,,$$

$$\int_{r_+}^{r_{ISCO}} \frac{r'\,dr'}{\sqrt{\Delta}} \rightarrow \frac{1}{6} M \ln\left(\frac{2^7}{\varepsilon}\right)\,, \tag{10.41}$$

which clearly shows that these orbits do not coincide even if in Boyer–Lindquist coordinates they have the same value. The energy E and the axial component of the angular momentum L_z are

$$\begin{aligned}
r &= r_\gamma & E &\rightarrow \infty, & L_z &\rightarrow 2EM\,, \\
r &= r_{mb} & E &\rightarrow 1, & L_z &\rightarrow 2M\,, \\
r &= r_{ISCO} & E &\rightarrow \tfrac{1}{\sqrt{3}}, & L_z &\rightarrow \tfrac{2}{\sqrt{3}}M\,.
\end{aligned} \tag{10.42}$$

10.3.2 Fundamental Frequencies

Equatorial circular orbits are characterized by three fundamental frequencies:

1. *Orbital frequency* (or Keplerian frequency) ν_ϕ: it is the inverse of the orbital period.
2. *Radial epicyclic frequency* ν_r: it is the frequency of radial oscillations around the mean orbit.
3. *Vertical epicyclic frequency* ν_θ: it is the frequency of vertical oscillations around the mean orbit.

These three frequencies only depend on the metric of the spacetime and on the radius of the orbit.

Let us start by considering a point-like massive body in Newtonian gravity. The gravitational potential is $V = -M/r$. The three fundamental frequencies are given by

$$\nu_\phi = \nu_r = \nu_\theta = \frac{1}{2\pi} \frac{M^{1/2}}{r^{3/2}}\,, \tag{10.43}$$

and have the same value. The profile of these frequencies is shown in the top left panel in Fig. 10.4.

In the Schwarzschild metric, we have the innermost stable circular orbit and the photon radius, both absent in Newtonian gravity. Circular orbits with radii smaller than r_{ISCO} are radially unstable. The radial epicyclic frequency ν_r must thus reach a maximum at some radius $r_{max} > r_{ISCO}$ and then vanish at r_{ISCO}. The orbital and the vertical epicyclic frequencies are instead defined up to the innermost circular orbit, the so-called photon orbit (see Sect. 10.3.1). There are no circular orbits with

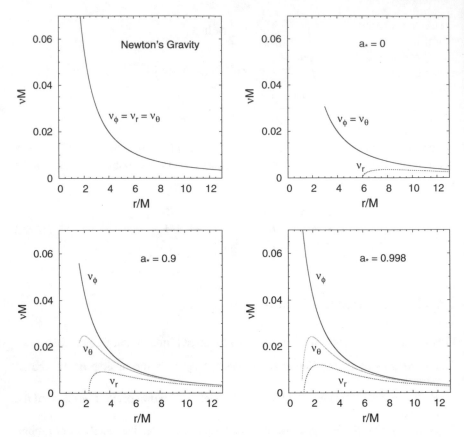

Fig. 10.4 Fundamental frequencies of a test-particle. Top left panel: Newtonian gravity with the potential $V = -M/r$; the three fundamental frequencies have the same value. Top right panel: Schwarzschild metric; the orbital and the vertical epicyclic frequencies have the same value, the radial epicyclic frequency vanishes at the radius of the innermost stable circular orbit. Bottom left panel: Kerr metric with the spin parameter $a_* = 0.9$. Bottom right panel: Kerr metric with the spin parameter $a_* = 0.998$

radius smaller than that of the photon orbit. The three fundamental frequencies as a function of the radial coordinate r in the Schwarzschild metric are shown in the top right panel in Fig. 10.4. We always have $\nu_\phi = \nu_\theta > \nu_r$, see below Eq. (10.54).

In the Kerr metric, $\nu_\theta > \nu_r$ is still true and, for corotating orbits, $\nu_\phi \geq \nu_\theta$. The three fundamental frequencies as a function of the radial coordinate r in the Kerr spacetime with the spin parameter $a_* = 0.9$ and 0.998 and for corotating orbits are shown, respectively, in the bottom left and bottom right panels in Fig. 10.4.

Let us now see how we can calculate the fundamental frequencies ν_ϕ, ν_r, and ν_θ. The orbital angular velocity was already found in Eqs. (10.23) and (10.29). The orbital frequency is $\nu_\phi = \Omega_\phi/2\pi$.

For the calculation of the radial and the vertical epicyclic frequencies, we can start from Eq. (10.19). In the linear regime, we can consider separately small perturbations around circular equatorial orbits along, respectively, the radial and the vertical directions. For the radial direction, we assume $\dot{\theta} = 0$, we write $\dot{r} = i(dr/dt)$, and we find

$$\left(\frac{dr}{dt}\right)^2 = \frac{1}{g_{rr}\dot{t}^2}V_{\text{eff}}\,. \tag{10.44}$$

We derive Eq. (10.44) with respect to the coordinate t and we obtain

$$\begin{aligned}
\frac{d^2r}{dt^2} &= \frac{1}{2}\frac{\partial}{\partial r}\left(\frac{1}{g_{rr}\dot{t}^2}V_{\text{eff}}\right) \\
&= \frac{V_{\text{eff}}}{2}\frac{\partial}{\partial r}\left(\frac{1}{g_{rr}\dot{t}^2}\right) + \frac{1}{2g_{rr}\dot{t}^2}\frac{\partial V_{\text{eff}}}{\partial r}\,.
\end{aligned} \tag{10.45}$$

If δ_r is a small displacement around the mean orbit, i.e. $r = r_0 + \delta_r$, we have

$$\frac{d^2r}{dt^2} = \frac{d^2\delta_r}{dt^2}\,,$$

$$V_{\text{eff}}(r_0 + \delta_r) = V_{\text{eff}}(r_0) + \left(\frac{\partial V_{\text{eff}}}{\partial r}\right)_{r=r_0}\delta_r + O(\delta_r^2) = O(\delta_r^2)\,,$$

$$\begin{aligned}
\left(\frac{\partial V_{\text{eff}}}{\partial r}\right)_{r=r_0+\delta_r} &= \left(\frac{\partial V_{\text{eff}}}{\partial r}\right)_{r=r_0} + \left(\frac{\partial^2 V_{\text{eff}}}{\partial r^2}\right)_{r=r_0}\delta_r + O(\delta_r^2) \\
&= \left(\frac{\partial^2 V_{\text{eff}}}{\partial r^2}\right)_{r=r_0}\delta_r + O(\delta_r^2)\,.
\end{aligned} \tag{10.46}$$

A similar expression can be derived for the coordinate θ to find the vertical epicyclic frequency introducing a small displacement around the mean orbit δ_θ, i.e. $\theta = \pi/2 + \delta_\theta$. Neglecting, respectively, terms $O(\delta_r^2)$ and $O(\delta_\theta^2)$, we find the following differential equations

$$\frac{d^2\delta_r}{dt^2} + \Omega_r^2\delta_r = 0\,, \tag{10.47}$$

$$\frac{d^2\delta_\theta}{dt^2} + \Omega_\theta^2\delta_\theta = 0\,, \tag{10.48}$$

where

$$\Omega_r^2 = -\frac{1}{2g_{rr}\dot{t}^2}\frac{\partial^2 V_{\text{eff}}}{\partial r^2}\,, \tag{10.49}$$

$$\Omega_\theta^2 = -\frac{1}{2g_{\theta\theta}\dot{t}^2}\frac{\partial^2 V_{\text{eff}}}{\partial\theta^2}\,. \tag{10.50}$$

The radial epicyclic frequency is $\nu_r = \Omega_r/2\pi$ and the vertical one is $\nu_\theta = \Omega_\theta/2\pi$.

In the Kerr metric in Boyer–Lindquist coordinates, the three fundamental frequencies can be written in an analytic and compact form as follows

$$\nu_\phi = \frac{1}{2\pi} \frac{M^{1/2}}{r^{3/2} \pm a M^{1/2}}, \tag{10.51}$$

$$\nu_r = \nu_\phi \sqrt{1 - \frac{6M}{r} \pm \frac{8a M^{1/2}}{r^{3/2}} - \frac{3a^2}{r^2}}, \tag{10.52}$$

$$\nu_\theta = \nu_\phi \sqrt{1 \mp \frac{4a M^{1/2}}{r^{3/2}} + \frac{3a^2}{r^2}}. \tag{10.53}$$

The Schwarzschild limit is recovered by imposing $a = 0$ and we obtain that the orbital and the vertical epicyclic frequencies coincide

$$\nu_\phi = \nu_\theta = \frac{1}{2\pi} \frac{M^{1/2}}{r^{3/2}}, \quad \nu_r = \nu_\phi \sqrt{1 - \frac{6M}{r}}. \tag{10.54}$$

The Newtonian limit can be quickly recovered from the Schwarzschild case by considering only the leading order term in M/r and the result is given in Eq. (10.43). In the Kerr spacetime $\nu_\theta \geq \nu_r$.

It may be useful to have an estimate of the order of magnitude of these frequencies. For a Schwarzschild black hole, the orbital frequency is

$$\nu_\phi(a_* = 0) = 220 \left(\frac{10 \, M_\odot}{M} \right) \left(\frac{6 \, M}{r} \right)^{3/2} \text{Hz}. \tag{10.55}$$

10.3.3   Frame Dragging

In Newton's gravity, only the mass of a body is responsible for the gravitational force. On the contrary, its angular momentum has no gravitational effects. In Einstein's gravity, even the angular momentum alters the geometry of the spacetime. *Frame dragging* refers to the capability of a spinning massive body of "dragging" the spacetime.

In Eq. (10.16), L_z is the specific angular momentum of a test-particle as measured at infinity. However, even if $L_z = 0$, the angular velocity of a test-particle $\Omega = \dot{\phi}/\dot{t}$ may be non-vanishing if $g_{t\phi} \neq 0$. In other words, the spinning body generating the gravitational field forces the test-particle to orbit with a non-vanishing angular velocity. If $L_z = 0$, from Eq. (10.16) we find

$$\Omega = -\frac{g_{t\phi}}{g_{\phi\phi}}. \tag{10.56}$$

In the Kerr metric in Boyer–Lindquist coordinates, we have

$$\Omega = \frac{2Mar}{\left(r^2 + a^2\right)\Sigma + 2Ma^2r\sin^2\theta}.$$
(10.57)

The *angular velocity of the event horizon* is the angular velocity at the event horizon of a test-particle with vanishing angular momentum at infinity:

$$\Omega_H = -\left(\frac{g_{t\phi}}{g_{\phi\phi}}\right)_{r=r_+} = \frac{a_*}{2r_+},$$
(10.58)

where r_+ is the radius of the event horizon.

The phenomenon of frame dragging is particularly strong in the *ergoregion*, which is the exterior region of the spacetime in which $g_{tt} > 0$. In the Schwarzschild spacetime, there is no ergoregion. In the Kerr spacetime, the ergoregion is between the event horizon and the *static limit*; that is

$$r_+ < r < r_{sl},$$
(10.59)

where r_{sl} the radius of the static limit and in Boyer–Lindquist coordinates is

$$r_{sl} = M + \sqrt{M^2 - a^2\cos^2\theta}.$$
(10.60)

The static limit is simply the surface $g_{tt} = 0$, while $g_{tt} < 0\,(> 0)$ for $r > r_{sl}\,(< r_{sl})$. Ergoregions may exist even inside spinning compact bodies like neutron stars, so it is not a concept related to black holes only.

In the ergoregion, frame dragging is so strong that static test-particles are not possible, namely test-particles with constant spatial coordinates. Everything must rotate. The line element of a static test-particle is ($dr = d\theta = d\phi = 0$)

$$ds^2 = g_{tt}dt^2.$$
(10.61)

Outside the ergoregion, $g_{tt} < 0$, and therefore a static test-particle follows a time-like geodesic. Inside the ergoregion, $g_{tt} > 0$, and a static test-particle would correspond to a space-like trajectory, which is not allowed.

10.4 No-Hair Theorem

Rotating black holes with a non-vanishing electric charge are described by the *Kerr–Newman solution*, which is the natural generalization of the Reissner–Nordström and Kerr spacetimes. The black hole is now completely specified by three parameters: the black hole mass M, the black hole electric charge Q, and the black hole spin angular momentum J. The line element of the Kerr–Newman solution can be obtained from that of the Kerr metric with the substitution $M \rightarrow M - Q^2/(2r)$. In Boyer–Lindquist coordinates, the radius of the event horizon is at

$$r_+ = M + \sqrt{M^2 - Q^2 - a^2} \,, \tag{10.62}$$

where $a = J/M$ as in the Kerr metric. Equation (10.62) reduces to Eq. (10.5) for $a = 0$ and to Eq. (10.10) for $Q = 0$. The condition for the existence of the event horizon is

$$\sqrt{Q^2 + a^2} \leq M \,. \tag{10.63}$$

If Eq. (10.63) is not satisfied, there is no black hole and the singularity at $r = 0$ is naked.

It turns out that the Kerr–Newman metric is the only stationary, axisymmetric, asymptotically flat, and regular (i.e. without singularities or closed time-like curves on or outside the event horizon) solution of the 4-dimensional electro-vacuum[7] Einstein equations. This is essentially the conclusion of the so-called *no-hair theorem*, which is actually a family of theorems because there are many different versions and it can also be extended beyond Einstein's gravity. The name "no-hair" is to indicate that black holes have no features, even if, strictly speaking, they would have three hairs, namely M, Q, and J. Violations of the no-hair theorem are possible if we relax some hypotheses or in some extensions of Einstein's gravity. For more details, see [1] and references therein.

10.5   Gravitational Collapse

When a star exhausts all its nuclear fuel, the gas pressure cannot balance the star's own weight, and the body shrinks to find a new equilibrium configuration. For most stars, the pressure of degenerate electrons stops the collapse and the star becomes a white dwarf. However, if the collapsing part of the star is too heavy, the mechanism does not work, matter reaches higher densities, and protons and electrons transform into neutrons. If the pressure of degenerate neutrons stops the collapse, the star becomes a neutron star. If the collapsing core is still too massive and even the neutron pressure cannot stop the process, there is no known mechanism capable of finding a new equilibrium configuration, and the body should undergo a complete collapse. In this case, the final product is a black hole.

The aim of this section is to present the simplest gravitational collapse model. This solution is analytic and nicely shows how the gravitational collapse of a spherically symmetric cloud of dust creates a spacetime singularity and an event horizon. For

---

[7]For electro-vacuum, we mean that the energy-momentum tensor on the right hand side of the Einstein equations either vanishes or is that of the electromagnetic field.

a review, see e.g. [7]. Numerical simulations can treat more realistic models, where the final product is still a black hole [2, 3].

We want to consider a spherically symmetric collapse, so the spacetime must be spherically symmetric. As we saw in Sect. 8.1, the most general line element of a spherically symmetric spacetime can always be written as (here we employ a different notation with respect to Sect. 8.1)

$$ds^2 = -e^{2\lambda}dt^2 + e^{2\psi}dr^2 + R^2\left(d\theta^2 + \sin^2\theta d\phi^2\right), \qquad (10.64)$$

where λ, ψ, and R are functions of t and r only.

Let us assume that the collapsing body can be described by a perfect fluid with energy density ρ and pressure P. The coordinate system of the line element in Eq. (10.64) is called comoving because the coordinates t and r are "attached" to every collapsing particle. This is the rest-frame of the collapsing fluid and therefore the fluid 4-velocity is $u^\mu = (e^{-\lambda}, \mathbf{0})$. The energy-momentum tensor is $T^\mu_\nu = \mathrm{diag}(-\rho, P, P, P)$.

With the line element in Eq. (10.64), the Einstein tensor reads

$$G^t_t = -\frac{F'}{R^2 R'} + \frac{2\dot{R}e^{-2\lambda}}{RR'}\left(\dot{R}' - \dot{R}\lambda' - \dot{\psi}R'\right), \qquad (10.65)$$

$$G^r_r = -\frac{\dot{F}}{R^2\dot{R}} - \frac{2R'e^{-2\psi}}{R\dot{R}}\left(\dot{R}' - \dot{R}\lambda' - \dot{\psi}R'\right), \qquad (10.66)$$

$$G^t_r = -e^{2\psi-2\lambda}G^r_t = \frac{2e^{-2\lambda}}{R}\left(\dot{R}' - \dot{R}\lambda' - \dot{\psi}R'\right), \qquad (10.67)$$

$$G^\theta_\theta = G^\phi_\phi = \frac{e^{-2\psi}}{R}\left[\left(\lambda'' + \lambda'^2 - \lambda'\psi'\right)R + R'' + R'\lambda' - R'\psi'\right] + \qquad (10.68)$$

$$-\frac{e^{-2\lambda}}{R}\left[\left(\ddot{\psi} + \dot{\psi}^2 - \dot{\lambda}\dot{\psi}\right)R + \ddot{R} + \dot{R}\dot{\psi} - \dot{R}\dot{\lambda}\right]. \qquad (10.69)$$

The Einstein equations are

$$G^t_r = 0 \Rightarrow \dot{R}' - \dot{R}\lambda' - \dot{\psi}R' = 0, \qquad (10.70)$$

$$G^t_t = 8\pi T^t_t \Rightarrow \frac{F'}{R^2 R'} = 8\pi\rho, \qquad (10.71)$$

$$G^r_r = 8\pi T^r_r \Rightarrow \frac{\dot{F}}{R^2\dot{R}} = -8\pi P, \qquad (10.72)$$

where the prime $'$ and the dot $\dot{}$ denote, respectively, the derivative with respect to r and t. F is called the Misner–Sharp mass

$$F = R\left(1 - e^{-2\psi}R'^2 + e^{-2\lambda}\dot{R}^2\right), \qquad (10.73)$$

which is defined by the relation

$$1 - \frac{F}{R} = g_{\mu\nu} (\partial^\mu R)(\partial^\nu R) \,. \tag{10.74}$$

From Eq. (10.71), we can see that the Misner–Sharp mass is proportional to the gravitational mass within the radius r at the time t

$$F(r) = \int_0^r F' d\tilde{r} = 8\pi \int_0^r \rho R^2 R' d\tilde{r} = 2M(r) \,. \tag{10.75}$$

A fourth relation can be obtained from the covariant conservation of the matter energy-momentum tensor

$$\nabla_\mu T^\mu_\nu = 0 \Rightarrow \lambda' = -\frac{P'}{\rho + P} \,. \tag{10.76}$$

10.5.1   Dust Collapse

Let us consider the case in which the equation of state is $P = 0$ (dust). Equations (10.70)–(10.72), and (10.76) become

$$\dot{R}' - \dot{R}\lambda' - \dot{\psi} R' = 0 \,, \tag{10.77}$$

$$\frac{F'}{R^2 R'} = 8\pi\rho \,, \tag{10.78}$$

$$\frac{\dot{F}}{R^2 \dot{R}} = 0 \,, \tag{10.79}$$

$$\lambda' = 0 \,. \tag{10.80}$$

Equation (10.79) shows that, in the case of dust, F is independent of t, namely there is no inflow or outflow through any spherically symmetric shell with radial coordinate r. This means that the exterior spacetime is described by the Schwarzschild solution. Note that for $P \neq 0$ this may not be true, and the interior region should be matched with a non-vacuum spacetime. If r_b is the comoving radial coordinate of the boundary of the cloud of dust, $F(r_b) = 2M$, where M is the mass parameter appearing in the Schwarzschild exterior solution.

Equation (10.80) implies that $\lambda = \lambda(t)$, and we can choose the time gauge in such a way that $\lambda = 0$. It is indeed always possible to define a new time coordinate $\tilde{t}$ such that $d\tilde{t} = e^\lambda dt$ and therefore $g_{\tilde{t}\tilde{t}} = -1$.

Equation (10.77) becomes $\dot{R}' - \dot{\psi} R' = 0$, and we can write

$$R' = e^{g(r)+\psi} \,. \tag{10.81}$$

We introduce the function $f(r) = e^{2g(r)} - 1$ and Eq. (10.73) becomes

$$\dot{R}^2 = \frac{F}{R} + f \, . \tag{10.82}$$

The line element can now be written as

$$ds^2 = -dt^2 + \frac{R'^2}{1+f} dr^2 + R^2 \left(d\theta^2 + \sin^2 \theta d\phi^2 \right) . \tag{10.83}$$

This is called the *Lemaitre–Tolman–Bondi metric*.

The Kretschmann scalar of the line element in (10.83) is

$$\mathscr{K} = 12 \frac{F'^2}{R^4 R'^2} - 32 \frac{F F'}{R^5 R'} + 48 \frac{F^2}{R^6} , \tag{10.84}$$

and diverges if $R = 0$. The system has a gauge degree of freedom that can be fixed by setting the scale at a certain time. It is common to set the area radius $R(t, r)$ to the comoving radius r at the initial time $t_i = 0$, namely $R(0, r) = r$, and introduce the scale factor a

$$R(t, r) = r a(t, r) \, . \tag{10.85}$$

We have thus $a = 1$ at $t = t_i$ and $a = 0$ at the time of the formation of the singularity. The condition for collapse is $\dot{a} < 0$. From Eq. (10.78), the regularity of the energy density at the initial time t_i requires us to write the Misner–Sharp mass as $F(r) = r^3 m(r)$, where $m(r)$ is a sufficiently regular function of r in the interval $[0, r_b]$. Equation (10.78) becomes

$$\rho = \frac{3m + r m'}{a^2 (a + r a')} \, . \tag{10.86}$$

The function $m(r)$ is usually written as a polynomial expansion around $r = 0$

$$m(r) = \sum_{k=0}^{\infty} m_k r^k , \tag{10.87}$$

where $\{m_k\}$ are constants. Requiring that the energy density ρ has no cusps at $r = 0$, $m_1 = 0$.

From Eq. (10.84), we see that the Kretschmann scalar diverges even when $R' = 0$ if $m' \neq 0$. However, the nature of these singularities is different: they arise from the overlapping of radial shells and are called shell crossing singularities [6]. Here the radial geodesic distance between shells with radial coordinates r and $r + dr$ vanishes, but the spacetime may be extended through the singularity by a suitable redefinition of the coordinates. To avoid any problem, it is common to impose that the collapse model has no shell crossing singularities, for instance requiring that $R' \neq 0$ or that m'/R' does not diverge.

At the initial time t_i, Eq. (10.82) becomes

$$\dot{a}(t_i, r) = -\sqrt{m + \frac{f}{r^2}}, \qquad (10.88)$$

and we can see that the choice of f corresponds to the choice of the initial velocity profile of the particles in the cloud. In order to have a finite velocity at all radii, it is necessary to impose some conditions on f. It is common to write $f(r) = r^2 b(r)$ and $b(r)$ as a polynomial expansion around $r = 0$:

$$b(r) = \sum_{k=0}^{\infty} b_k r^k. \qquad (10.89)$$

10.5.2   Homogeneous Dust Collapse

The simplest model of gravitational collapse is the Oppenheimer–Snyder one [10]. It describes the collapse of a homogeneous and spherically symmetric cloud of dust. In this case, $\rho = \rho(t)$ is independent of r, so $m = m_0$ and $b = b_0$. The line element reads[8]

$$ds^2 = -dt^2 + a^2 \left(\frac{dr^2}{1 + b_0 r^2} + r^2 d\theta^2 + r^2 \sin^2 \theta d\phi^2 \right). \qquad (10.90)$$

$b_0 = 0$ corresponds to a marginally bound collapse, namely the scenario in which the falling particles have vanishing velocity at infinity. Equation (10.82) becomes

$$\dot{a} = -\sqrt{\frac{m_0}{a} + b_0}. \qquad (10.91)$$

For $b_0 = 0$, the solution is

$$a(t) = \left(1 - \frac{3\sqrt{m_0}}{2} t \right)^{2/3}. \qquad (10.92)$$

The formation of the singularity occurs at the time

$$t_s = \frac{2}{3\sqrt{m_0}}. \qquad (10.93)$$

---

[8]Note that the interior metric is the time reversal of the Friedmann–Robertson–Walker solution, which will be discussed in the next chapter.

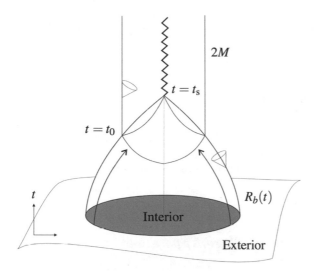

Fig. 10.5 Finkelstein diagram for the gravitational collapse of a homogeneous and spherically symmetric cloud of dust. $R_b(t)$ is the radius of the cloud (in the Schwarzschild coordinates of the exterior region) and separates the interior from the vacuum exterior. The cloud collapses as t increases and at the time $t = t_0$ the horizon forms at the boundary when $R_b(t_0) = 2M$. In the interior, the apparent horizon propagates inwards and reaches the center of symmetry at the time of the formation of the singularity $t = t_s$. For $t > t_s$, the spacetime has settled down to the usual Schwarzschild solution. Figure courtesy of Daniele Malafarina

Without a rigorous definition, the *apparent horizon* can be introduced as the surface defining the boundary between outward-pointing light rays moving outwards (outside the apparent horizon) and inwards (inside the apparent horizon). More details can be found in [1] and references therein. The curve $t_{ah}(r)$ describing the time at which the shell r crosses the apparent horizon can be obtained from

$$1 - \frac{F}{R} = 1 - \frac{r^2 m_0}{a} = 0. \tag{10.94}$$

For $b_0 = 0$, the solution is

$$t_{ah}(r) = t_s - \frac{2}{3}F = \frac{2}{3\sqrt{m_0}} - \frac{2}{3}r^3 m_0. \tag{10.95}$$

The *Finkelstein diagram* of the gravitational collapse of a homogeneous and spherically symmetric cloud of dust is sketched in Fig. 10.5. At the time $t = t_0$, the radius of the surface of the cloud crosses the Schwarzschild radius. We have the formation of both the event horizon in the exterior region and the apparent horizon at the boundary $r = r_b$, i.e. $t_0 = t_{ah}(r_b)$. As shown in Fig. 10.5, the exterior region is now settled down to the static Schwarzschild spacetime, while the radius of the apparent horizon propagates to smaller radii and reaches $r = 0$ at the time of the formation of the singularity t_s.

10.6   Penrose Diagrams

In this section, we will present (without any derivation) the Penrose diagrams for the Reissner–Nordström spacetime, the Kerr spacetime, and for the spacetime of the Oppenheimer–Snyder model. We will discuss their basic properties at a qualitative level.

10.6.1   Reissner–Nordström Spacetime

The Penrose diagram for the maximal extension of the Reissner–Nordström spacetime is illustrated in Fig. 10.6. As we can immediately realize, there are some similarities and some differences with respect to the Schwarzschild case. Region I is our universe outside the black hole and region III is the parallel universe, as in the maximal extension of the Schwarzschild spacetime. The red solid lines separating region II from regions I and III are still the event horizon. However, region II is now different. The Reissner–Nordström solution indeed has an event horizon at the radial coordinate r_+ and an inner horizon at r_-. Region II is now the region between r_+ and r_-.

The orange curve describes the trajectory of a hypothetical massive particle. The particle is initially in region I, crosses the event horizon r_+ and thus enters the black hole. Once in region II, the particle must inevitably cross the inner horizon as well. Inside the inner horizon, the particle has two options: either falling to the singularity at $r = 0$ or crossing again the inner horizon r_- to go to region IV'. Note that in the Schwarzschild spacetime it is impossible to avoid the singularity when we are inside the black hole. This is not true here. Note that this is due to the difference between the singularity in the Schwarzschild spacetime, which is represented by a horizontal line, and the singularity in the Reissner–Nordström spacetime, described by a vertical line. In the case of the orange trajectory in Fig. 10.6, the particle crosses the inner horizon and goes to region IV'. The latter is like the white hole of the Schwarzschild spacetime. The particle can exit the while hole and go either to universe I' or universe III'.

As we can see from Fig. 10.6, the maximal extension of the Reissner–Nordström spacetime possesses infinite "copies" of our universe, of the parallel universe, and of the interior region.

Fig. 10.6 Penrose diagram
for the maximal extension of
the Reissner–Nordström
spacetime

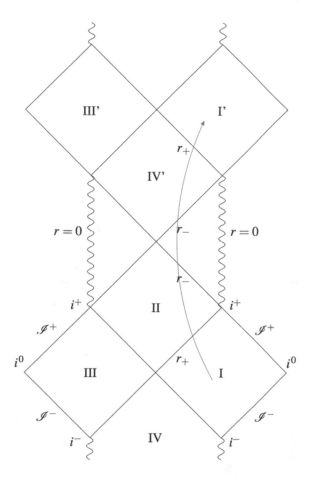

10.6.2 Kerr Spacetime

As shown in Fig. 10.7, the Penrose diagram for the maximal extension of the Kerr spacetime is quite similar to that of the Reissner–Nordström one: region II is now the region between the event horizon r_+ and the inner horizon r_-, the central black hole singularity is a vertical line and can be avoided, there is an infinite number of copies of our universe, of the parallel universe, and of the black hole region.

The main difference between the Kerr and the Reissner–Nordström spacetime is at the singularity. As we have already mentioned in Sect. 10.3, in the Kerr metric the spacetime is singular only for trajectories in the equatorial plane. The point $r = 0$ has the topology of a ring and it is possible to extend the spacetime to negative values of the radial coordinate r. This is exactly what is illustrated in Fig. 10.7. The green curve describes the trajectory of a massive particle that starts from region I, enters the black hole, crosses the inner horizon of the black hole, but then enters the gate

Fig. 10.7 Penrose diagram
for the maximal extension of
the Kerr spacetime

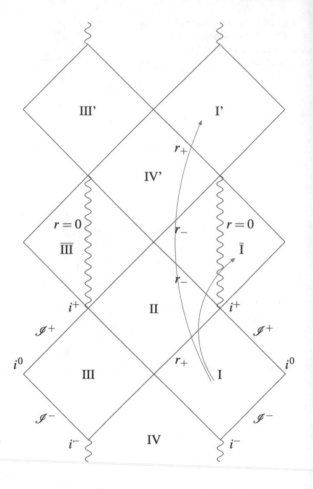

at $r = 0$ and reaches region $\bar{\mathrm{I}}$. Such a region exists in the Kerr solution because it is possible to extend the spacetime to $r < 0$, but it does not in the Reissner–Nordström solution. Note also that it is possible to return to the region with $r > 0$, because the point $r = 0$ is a vertical line.

10.6.3   Oppenheimer–Snyder Spacetime

The Penrose diagram for the maximal extension of the Schwarzschild spacetime was discussed in Sect. 8.6.2. Note that the Schwarzschild solution is static, while astrophysical black holes should be created from the gravitational collapse of massive bodies. The Penrose diagram of the Oppenheimer–Snyder model can thus better illustrate the properties of the spacetime with a real black hole.

Fig. 10.8 Penrose diagram for the gravitational collapse of a homogeneous and spherically symmetric cloud of dust (Oppenheimer–Snyder model). The letter S indicates the interior region of the collapsing body and the black arc extending from i^- to the singularity $r = 0$ (the horizontal line with wiggles) is its boundary

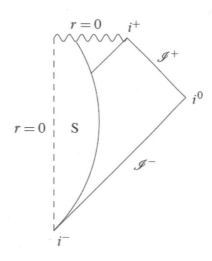

Figure 10.8 shows the Penrose diagram for the gravitational collapse of a homogeneous and spherically symmetric cloud of dust, and it is evident that it is substantially different from that discussed in Sect. 8.6.2. The Penrose diagram for a static spacetime with a massive body (e.g. a star) would be equivalent to that of the Minkowski spacetime illustrated in Fig. 8.2. The diagram changes when we have the creation of an event horizon, because the latter causally disconnects the interior region from the exterior. When the radius of the collapsing cloud crosses the corresponding Schwarzschild radius $r_S = 2M$, we have the formation of the event horizon represented by the red solid line at $45°$. At this point, the exterior region looks like region I in the Penrose diagram of the Schwarzschild spacetime. In the interior region, the cloud collapses to the center $r = 0$ and the whole region looks like region II in the Penrose diagram of the Schwarzschild spacetime. There is neither a white hole nor a parallel universe.

Problems

10.1 Write the inverse metric $g^{\mu\nu}$ of the Reissner–Nordström solution.

10.2 Write the inverse metric $g^{\mu\nu}$ of the Kerr solution in Boyer–Lindquist coordinates.

10.3 Check that Eq. (10.21) is a different form of the geodesic equations.

10.4 For a Kerr black hole, $|a_*| \leq 1$. Show that the spin parameter of Earth is $|a_*| \gg 1$.

References

1. C. Bambi, *Black Holes: A Laboratory for Testing Strong Gravity* (Springer, Singapore, 2017)
2. L. Baiotti, I. Hawke, P.J. Montero, F. Loffler, L. Rezzolla, N. Stergioulas, J.A. Font, E. Seidel, Phys. Rev. D **71**, 024035 (2005). [gr-qc/0403029]
3. L. Baiotti, L. Rezzolla, Phys. Rev. Lett. **97**, 141101 (2006). [gr-qc/0608113]
4. J.M. Bardeen, W.H. Press, S.A. Teukolsky, Astrophys. J. **178**, 347 (1972)
5. S. Chandrasekhar, *The Mathematical Theory of Black Holes* (Clarendon Press, Oxford, 1998)
6. C. Hellaby, K. Lake, Astrophys. J. **290**, 381 (1985)
7. P.S. Joshi, D. Malafarina, Int. J. Mod. Phys. D **20**, 2641 (2011), arXiv:1201.3660 [gr-qc]
8. R.P. Kerr, Phys. Rev. Lett. **11**, 237 (1963)
9. C.W. Misner, K.S. Thorne, J.A. Wheeler, *Gravitation* (W. H. Freeman and Company, San Francisco, 1973)
10. J.R. Oppenheimer, H. Snyder, Phys. Rev. **56**, 455 (1939)

Chapter 11
Cosmological Models

In this chapter, we will show how we can construct some simple models for the description of the Universe.[1] More details on cosmological models and on the evolution of the Universe can be found in standard textbooks on cosmology, like [1].

11.1 Friedmann–Robertson–Walker Metric

Our starting point to construct some simple cosmological models is the so-called *Cosmological Principle*:

> **Cosmological Principle**. The Universe is homogeneous and isotropic.

As a matter of fact, the Universe is far from being homogeneous and isotropic. We observe a lot of structures around us (stars, galaxies, clusters of galaxies). However, we can expect that the Universe can be well approximated as homogeneous and isotropic if we average over large volumes. The Cosmological Principle can be seen as a sort of Copernican Principle: there are no preferred points or preferred directions in the Universe. Our current model for the description of the Universe is called the Standard Model of Cosmology and is based on the Cosmological Principle. Nevertheless, today there is debate on the applicability of this principle. Since there are structures, the assumptions of homogeneity and isotropy inevitably introduce systematic effects in the measurements of the properties of the Universe, and the impact of these systematic effects on current measurements of cosmological parameters, which are more and more precise, is not clear.

---

[1]Note that it is common to use the initial capital letter in Universe only when we refer to our Universe. If we mean a generic universe/cosmological model, we write universe.

© Springer Nature Singapore Pte Ltd. 2018
C. Bambi, *Introduction to General Relativity*, Undergraduate Lecture Notes in Physics, https://doi.org/10.1007/978-981-13-1090-4_11

The Cosmological Principle requires that there are no preferred points (homogeneity, i.e. invariance under spatial translations) and no preferred directions (isotropy, i.e. invariance under spatial rotations) in the 3-dimensional spacetime. The spacetime geometry is still allowed to depend on time.[2] These assumptions strongly constrain the metric of the spacetime. The only background compatible with the Cosmological Principle is the *Friedmann–Robertson–Walker metric*. Its derivation requires some calculations, which are outlined in Appendix H. The line element reads

$$ds^2 = -c^2dt^2 + a^2 \left(\frac{dr^2}{1 - kr^2} + r^2d\theta^2 + r^2 \sin^2 \theta d\phi^2 \right) , \tag{11.1}$$

where $a = a(t)$ is the *scale factor*, which depends on the temporal coordinate t and is independent of the spatial coordinates (r, θ, ϕ), and k is a constant. In general, k can be positive, zero, or negative. Nevertheless, it is always possible to rescale the radial coordinate r and have $k = 1, 0$, or -1. If $k = 1$, we have a *closed universe*. If $k = 0$, we have a *flat universe*. If $k = -1$, we have an *open universe*. Note that, in general, a flat universe is not a flat spacetime, namely the Minkowski spacetime of special relativity. The spacetime is flat only when a is independent of t and $k = 0$. In such a case, we can redefine the radial coordinate to absorb the scale factor and the line element (11.1) becomes that of the Minkowski spacetime in spherical coordinates

$$ds^2 = -c^2dt^2 + dr^2 + r^2d\theta^2 + r^2 \sin^2 \theta d\phi^2 . \tag{11.2}$$

It is straightforward to compute some invariants of the Friedmann–Robertson–Walker spacetime with specific Mathematica packages (see Appendix E). For instance, the scalar curvature is

$$R = 6 \, \frac{kc^2 + \dot{a}^2 + \ddot{a}a}{a^2c^2} . \tag{11.3}$$

When $k = 0$ and a is constant, R vanishes, which is indeed the case of the Minkowski spacetime (but R may vanish even if the spacetime is curved; in Einstein's gravity $R = 0$ if $T^\mu_\mu = 0$). Note also that R diverges for $a \to 0$. Typical cosmological models start from a singularity, namely $a = 0$ at the initial time (*Big Bang*). The Kretschmann scalar is

$$\mathcal{K} = 12 \, \frac{k^2c^4 + 2kc^2\dot{a}^2 + \dot{a}^4 + \ddot{a}^2a^2}{a^4c^4} , \tag{11.4}$$

which also diverges for $a \to 0$.

Note that the Cosmological Principle is quite a strong assumption only based on the requirement that the Universe is homogeneous and isotropic. It is completely independent of the Einstein equations. If we impose the Cosmological Principle in a

---

[2]If we impose that the spacetime geometry is also independent of time ("Perfect" Cosmological Principle), we find cosmological models in disagreement with observations.

4-dimensional spacetime, the geometry is described by the Friedmann–Robertson–Walker metric in (11.1). If we specify that we are in Einstein's gravity and we know the matter content of the Universe, we can find $a(t)$ and k. In another gravity theory and/or for a different matter content we would find, in general, different solutions for $a(t)$ and k.

Note that the Cosmological Principle cannot determine some global properties of the Universe. For example, the Friedmann–Robertson–Walker metric can describe topologically different universes. If we assume that the Universe has a trivial topology, the 3-volume V is finite if $k = 1$ and infinite for $k = 0$ and -1:

$$V = \int_V \sqrt{^3g}\, d^3x = a^3 \int_0^{2\pi} d\phi \int_0^\pi \sin\theta d\theta \int_0^{r_k} \frac{r^2 dr}{\sqrt{1 - kr^2}}\,, \qquad (11.5)$$

where ^3g is the determinant of the spatial 3-metric,[3] $r_k = 1$ for $k = 1$, and $r_k = \infty$ for $k = 0$ and -1. If we integrate Eq. (11.5), we find

$$V = \begin{cases} \pi^2 a^3 & \text{for } k = 1 \\ \infty & \text{for } k = 0,\ -1 \end{cases}. \qquad (11.7)$$

In the case of a universe with non-trivial topology, the picture is more complicated, and depends on the specific configuration. Closed universes always have a finite volume, but in the case of flat and open universes the volume can be either finite or infinite. Current astrophysical data suggest that the Universe is almost flat, and we cannot say if $k = 1, 0$, or -1. In such a situation, even assuming that the Universe has a trivial topology we cannot say whether its volume is finite or infinite. In the case of a universe with non-trivial topology, the measurement of k is not enough to assert if it is finite or infinite. In principle, in the case of a non-trivial topology, the volume of the Universe may be evaluated by looking for "ghost images" of astronomical sources,[4] namely images of the same source coming from different directions. Since we currently have no evidence of the presence of ghost images due to the possible non-trivial topology of the Universe, we only have lower bounds of the possible finite size of the Universe in the case of non-trivial topologies.

---

[3]The line element of the spatial 3-metric of the Friedmann–Robertson–Walker spacetime reads

$$dl^2 = a^2 \left(\frac{dr^2}{1 - kr^2} + r^2 d\theta^2 + r^2 \sin^2\theta d\phi^2 \right). \qquad (11.6)$$

[4]In general, ghost images can be potentially seen in any universe (with either trivial or non-trivial topology) with at least one space dimension of finite size. However, it is when the topology is non-trivial that the detection of ghost images seems to be the simplest way to infer if the Universe has at least one space dimension of finite size.

11.2   Friedmann Equations

If we assume the Cosmological Principle, the metric of the spacetime must be described by the Friedmann–Robertson–Walker solution. The only undetermined quantities are the scale factor $a(t)$ and the constant k, which can be obtained once we specify the gravity theory (e.g. Einstein's gravity) and the matter content, and we solve the corresponding field equations.

The simplest cosmological models are constructed assuming that the matter in the Universe can be described by the energy-momentum tensor of a perfect fluid

$$T^{\mu\nu} = (\rho + P)\,\frac{u^\mu u^\nu}{c^2} + Pg^{\mu\nu}\,, \tag{11.8}$$

where ρ and P are, respectively, the energy density and the pressure of the fluid, and u^μ is the fluid 4-velocity. In the coordinate system of the Friedmann–Robertson–Walker metric, the Universe is manifestly homogeneous and isotropic. It corresponds to the rest-frame of the fluid, where the fluid 4-velocity becomes $u^\mu = (c, \mathbf{0})$. Let us note that the expression in Eq. (11.8) can be employed even when there are several matter components. In such a case, ρ and P are the total energy density and the total pressure, namely

$$\rho = \sum_i \rho_i\,, \quad P = \sum_i P_i\,, \tag{11.9}$$

where ρ_i and P_i are, respectively, the energy density and the pressure of the component i.

If we plug the Friedmann–Robertson–Walker metric (11.1) and the energy-momentum tensor of a perfect fluid (11.8) with $u^\mu = (c, \mathbf{0})$ into the Einstein equations, we find the field equations to solve. The tt component of the Einstein equations gives the *first Friedmann equation*:

$$H^2 = \frac{8\pi G_N}{3c^2}\rho - \frac{kc^2}{a^2}\,, \tag{11.10}$$

where $H = \dot{a}/a$ is the *Hubble parameter*. The rr, $\theta\theta$, and $\phi\phi$ components of the Einstein equations provide the same equation, which is called the *second Friedmann equation* and reads

$$\frac{\ddot{a}}{a} = -\frac{4\pi G_N}{3c^2}(\rho + 3P)\,. \tag{11.11}$$

Instead of the Einstein equations, we may use the covariant conservation of the energy-momentum tensor, $\nabla_\mu T^{\mu\nu} = 0$, which is a consequence of the Einstein equations. In such a case, we find

$$\dot{\rho} = -3H\left(\rho + P\right).$$ (11.12)

Note that Eqs. (11.10), (11.11), and (11.12) are not three independent equations. Only two equations are independent, and it is possible to obtain the third equation from the other two. For instance, if we derive the first Friedmann equation with respect to t, we have

$$\frac{2\dot{a}\ddot{a}a^2 - 2a\dot{a}^3}{a^4} = \frac{8\pi G_N}{3c^2}\dot{\rho} + \frac{2\dot{a}kc^2}{a^3},$$

$$H\frac{\ddot{a}}{a} - H\frac{\dot{a}^2}{a^2} = \frac{4\pi G_N}{3c^2}\dot{\rho} + \frac{kc^2}{a^2}H.$$ (11.13)

We replace the term $\dot{a}^2/a^2$ with the expression on the right hand side of the first Friedmann equation and $\dot{\rho}$ with the expression in Eq. (11.12)

$$H\frac{\ddot{a}}{a} - H\frac{8\pi G_N}{3c^2}\rho + H\frac{kc^2}{a^2} = \frac{4\pi G_N}{3c^2}[-3H\left(\rho + P\right)] + \frac{kc^2}{a^2}H,$$ (11.14)

and eventually we recover the second Friedmann equation.

At this point, we have two independent equations and three unknown functions of time (a, ρ, and P). In order to close the system and find a, ρ, and P, we need another equation. We can introduce the *equation of state* of the matter in the Universe. The simplest form is

$$P = w\rho,$$ (11.15)

where w is a constant. While this is a very simple equation of state, it includes the main physically relevant cases: dust ($w = 0$), radiation ($w = 1/3$), and vacuum energy ($w = -1$).

The first Friedmann equation for $k = 0$ reads

$$H^2 = \frac{8\pi G_N}{3c^2}\rho_c,$$ (11.16)

and defines the *critical density* ρ_c as the energy density of a flat universe. The value of the critical energy density today is

$$\rho_c^0 = \frac{3H_0^2 c^2}{8\pi G_N} = 1.88 \cdot 10^{-29}\ h_0^2\ c^2\ \text{g} \cdot \text{cm}^{-3}$$

$$= 11\ h_0^2\ \text{protons} \cdot \text{m}^{-3},$$ (11.17)

where H_0 is the *Hubble constant*, namely the value of the Hubble parameter today, and can be written as[5]

$$H_0 = 100 \, h_0 \, \frac{\text{km}}{\text{s} \cdot \text{Mpc}} \, . \tag{11.18}$$

h_0 is a dimensionless parameter of order 1. Since the value of the Hubble constant was not known with good precision in the past, it was common to use the expression in (11.18) and keep the parameter h_0 in all the equations. Today we know that $h_0 \approx 0.7$.

11.3 Cosmological Models

If we plug the equation of state (11.15) into Eq. (11.12), we have

$$\frac{\dot{\rho}}{\rho} = -3 \, (1 + w) \, \frac{\dot{a}}{a} \, . \tag{11.19}$$

The solution of this equation is

$$\rho \propto a^{-3(1+w)} \, . \tag{11.20}$$

In particular, we have the following relevant cases

$$\begin{array}{llll}
w = 0 & \rightarrow & \rho \propto 1/a^3 & \text{(dust)}, \\
w = 1/3 & \rightarrow & \rho \propto 1/a^4 & \text{(radiation)}, \\
w = -1 & \rightarrow & \rho = \text{constant} & \text{(vacuum energy)},
\end{array} \tag{11.21}$$

If we plug (11.20) into the first Friedmann equation, we neglect the term kc^2/a^2 (whose contribution can always be ignored at sufficiently early times in a universe made of dust or radiation because the term with ρ is dominant with respect to kc^2/a^2 for $a \rightarrow 0$), and we write $a \propto t^\alpha$, the first Friedmann equation reads

$$t^{-2} \propto t^{-3\alpha(1+w)} \, , \tag{11.22}$$

and we find

$$\alpha = \frac{2}{3 \, (1 + w)} \, . \tag{11.23}$$

---

[5]The parsec (pc) is a common unit of length in astronomy and cosmology. 1 pc $= 3.086 \cdot 10^{16}$ m. 1 Mpc $= 10^6$ pc.

11.3.1 Einstein Universe

For ordinary matter $\rho + 3P > 0$, and therefore the second Friedmann equation implies that $\ddot{a} < 0$; that is, the Universe cannot be static. However, this was against the common belief at the beginning of the 20th century. This apparent problem led Einstein to introduce the cosmological constant Λ into the theory, replacing the field equations (7.6) with (7.8). In the presence of Λ, the first and the second Friedmann equations read

$$H^2 = \frac{8\pi G_N}{3c^2}\rho + \frac{\Lambda c^2}{3} - \frac{kc^2}{a^2}, \tag{11.24}$$

$$\frac{\ddot{a}}{a} = -\frac{4\pi G_N}{3c^2}(\rho + 3P) + \frac{\Lambda c^2}{3}. \tag{11.25}$$

The so-called Einstein universe is a cosmological model in which matter is described by dust ($P = 0$) and there is a non-vanishing cosmological constant to make the universe static. If we require that $\dot{a} = \ddot{a} = 0$, from Eqs. (11.24) and (11.25) we find

$$\rho = \frac{\Lambda c^4}{4\pi G_N}, \quad a = \frac{1}{\sqrt{\Lambda}}, \quad k = 1. \tag{11.26}$$

The Einstein universe is unstable, namely small perturbations make it either collapse or expand. After the discovery of the expansion of the Universe by Hubble in 1929, the cosmological constant was (temporarily) removed from the Einstein equations.

11.3.2 Matter Dominated Universe

Let us now consider a universe filled with dust. The equation of state is $P = 0$, namely $w = 0$. The energy density scales as $1/a^3$, so we can write

$$\rho a^3 = \text{constant} \equiv C_1. \tag{11.27}$$

The first Friedmann equation becomes

$$\dot{a}^2 = \frac{8\pi G_N}{3c^2}\frac{C_1}{a} - kc^2. \tag{11.28}$$

In order to solve Eq. (11.28), we introduce the variable η

$$\frac{d\eta}{dt} = \frac{1}{a}. \tag{11.29}$$

Fig. 11.1 Scale factor a as a function of the cosmological time t for matter dominated universes. The scale factor is expressed in units in which $8\pi G_N C_1/c^2 = A = 1$

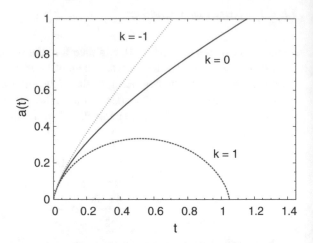

We replace t with η in Eq. (11.28) and the first Friedmann equation reads

$$a'^2 = \frac{8\pi G_N}{3c^2} C_1 a - kc^2 a^2 \,, \tag{11.30}$$

where the prime $'$ indicates the derivative with respect to η, i.e. $' = d/d\eta$. With the initial condition $a = 0$ at $t = 0$, we get the following parametric solutions for a and t. In the case of closed universes ($k = 1$), we have

$$a = \frac{4\pi G_N}{3c^2} C_1 A \left(1 - \cos\frac{\eta}{\sqrt{A}}\right) \,, \quad t = \frac{4\pi G_N}{3c^2} C_1 A \left(\eta - \sqrt{A}\sin\frac{\eta}{\sqrt{A}}\right) . \tag{11.31}$$

For flat universes ($k = 0$), we have

$$a = \frac{2\pi G_N}{3c^2} C_1 \eta^2 \,, \quad t = \frac{2\pi G_N}{9c^2} C_1 \eta^3 \,. \tag{11.32}$$

Lastly, for open universes ($k = -1$), we have

$$a = \frac{4\pi G_N}{3c^2} C_1 A \left(\cosh\frac{\eta}{\sqrt{A}} - 1\right) \,, \quad t = \frac{4\pi G_N}{3c^2} C_1 A \left(\sqrt{A}\sinh\frac{\eta}{\sqrt{A}} - \eta\right) . \tag{11.33}$$

$A = 1/(|k|c^2)$ has dimensions of time squared.

The scale factor a as a function of the cosmological time t for matter dominated universes is shown in Fig. 11.1. At $t = 0$, $a = 0$ and the matter density diverges. This is the Big Bang and the universe starts expanding. A closed universe expands up to a critical point and then recollapses. An open universe expands forever. The flat universe expands forever and represents the critical case separating closed and open universes.

Fig. 11.2 Scale factor a as a function of the cosmological time t for radiation dominated universes. The scale factor is expressed in units in which $8\pi G_N C_2/c^2 = c = 1$

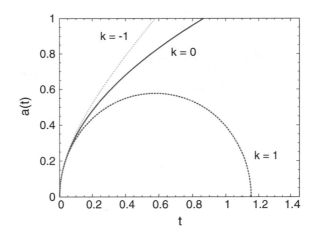

11.3.3 *Radiation Dominated Universe*

If the universe is filled with radiation, the energy density scales as $1/a^4$, and we can write

$$\rho a^4 = \text{constant} \equiv C_2 .\tag{11.34}$$

The first Friedmann equation becomes

$$\dot{a}^2 = \frac{8\pi G_N}{3c^2}\frac{C_2}{a^2} - kc^2 .\tag{11.35}$$

With the initial condition $a = 0$ at $t = 0$, we find the following solution

$$a = \left[\sqrt{\frac{32\pi G_N C_2}{3c^2}}\, t - kc^2 t^2\right]^{1/2} .\tag{11.36}$$

where $k = 1$, 0, or -1 for closed, flat, or open universes, respectively.

Figure 11.2 shows the scale factor a as a function of the cosmological time t for radiation dominated universes. As in the matter dominated universes, closed universes expand up to a critical point and then recollapse, while flat and open universes expand forever.

11.3.4 Vacuum Dominated Universe

In the case of vacuum energy, the equation of state is $P = -\rho$, and we find that ρ is constant. Equivalently, we can write the Friedmann equations without matter ($\rho = P = 0$) and with a non-vanishing cosmological constant

$$H^2 = \frac{\Lambda c^2}{3} - \frac{kc^2}{a^2} \, , \quad \frac{\ddot{a}}{a} = \frac{\Lambda c^2}{3} \, . \tag{11.37}$$

For $\Lambda > 0$, the solutions for closed, flat, and open universes are

$$k = 1 \rightarrow a = \sqrt{\frac{3k}{\Lambda}} \cosh\left(\sqrt{\frac{\Lambda}{3}} \, ct\right) . \tag{11.38}$$

$$k = 0 \rightarrow a = a(t = 0) \, \exp\left(\sqrt{\frac{\Lambda}{3}} \, ct\right) . \tag{11.39}$$

$$k = -1 \rightarrow a = \sqrt{\frac{-3k}{\Lambda}} \sinh\left(\sqrt{\frac{\Lambda}{3}} \, ct\right) . \tag{11.40}$$

All universes expand forever. For $k = 1$ and 0, the scale factor never vanishes, so there is no Big Bang.

If $\Lambda < 0$, we can only find a solution for $k = -1$

$$k = -1 \rightarrow a = \sqrt{\frac{3k}{\Lambda}} \cos\left(\sqrt{-\frac{\Lambda}{3}} \, ct\right) . \tag{11.41}$$

If $\Lambda = 0$, we have only the trivial case $k = 0$ and a constant corresponding to the Minkowski spacetime.

11.4 Properties of the Friedmann–Robertson–Walker Metric

11.4.1 Cosmological Redshift

Now we want to study the geodesic motion of test-particles in the Friedmann–Robertson–Walker metric. The Lagrangian is

$$L = \frac{1}{2} g_{\mu\nu} x'^{\mu} x'^{\nu} , \tag{11.42}$$

where $g_{\mu\nu}$s are the metric coefficients of the Friedmann–Robertson–Walker solution and the prime $'$ is used to indicate the derivative with respect to the proper time/affine parameter of the trajectory (in this chapter the dot $\cdot$ is used for the derivative of the coordinate t; e.g. $da/dt = \dot{a}$). Since in the Friedmann–Robertson–Walker metric the scale factor a depends on the time t, it is clear that the energy of the particle is not a constant of motion.

In the case of a photon, $g_{\mu\nu}x'^{\mu}x'^{\nu} = 0$. If we employ a coordinate system in which the motion is only along the radial direction, we have

$$c^2 t'^2 - a^2 \frac{r'^2}{1 - kr^2} = 0. \tag{11.43}$$

The Euler-Lagrange equation for the t coordinate reads

$$t'' = -\frac{a\dot{a}}{c^2}\frac{r'^2}{1 - kr^2}. \tag{11.44}$$

If we use Eq. (11.43) into Eq. (11.44), we find

$$t'' = -\frac{\dot{a}}{a}t'^2 = -\frac{a'}{a}t'. \tag{11.45}$$

Since t' is proportional to the energy of the photon, say E, $t''/t' = E'/E$. Equation (11.45) tells us that the the energy of a photon propagating in a Friedmann–Robertson–Walker spacetime scales as the inverse of the scale factor

$$E \propto 1/a. \tag{11.46}$$

This is the *cosmological redshift* and it is due to the expansion of the Universe. It is different from the Doppler redshift, which is due to the relative motion between the source and the observer and is present already in special relativity. It is also different from the gravitational redshift, which is due to climbing in a gravitational potential. It is instead connected to the behavior of the energy density of radiation, which scales as $1/a^4$: the photon number density scales as the inverse of the volume, $1/a^3$, while the photon energy scales as $1/a$, so the result is that the energy density scales as $1/a^4$.

11.4.2 Particle Horizon

Within a cosmological model with a Big Bang, the *particle horizon* at the time t is defined as the distance covered by a photon from the time of the Big Bang to the time t. It is an important concept because it defines the causally connected regions at any time: two points at a distance larger than the particle horizon have never exchanged any information. Let us consider a flat universe ($k = 0$), which can be seen as a valid approximation for any matter or radiation dominated universe for sufficiently early times. For simplicity, we choose a coordinate system in which the photon is at the origin $r = 0$ at the time $t = 0$ (Big Bang). From $ds^2 = 0$, we have

$$r = \int_0^r d\tilde{r} = c \int_0^t \frac{d\tilde{t}}{a} . \tag{11.47}$$

If we write the scale factor as $a \propto t^\alpha$ (see Sect. 11.3) and we integrate in $d\tilde{t}$, we find

$$r = \frac{ct}{a\,(1 - \alpha)} . \tag{11.48}$$

The proper distance at the time t between the origin and a point with the radial coordinate r is $d = a\,r$. The particle horizon is thus

$$d = \frac{ct}{1 - \alpha} . \tag{11.49}$$

If we have a universe filled with dust, $\alpha = 2/3$ and $d = 3ct$. If the universe is filled with radiation, $\alpha = 1/2$ and $d = 2ct$. For ordinary matter $w \geq 0$, the particle horizon grows linearly with time, and the scale factor grows slower, because $a \propto t^\alpha$ and $\alpha < 1$. In such a case, more and more regions in the universe get causally connected at later times.

11.5 Primordial Plasma

Let us consider a gas of particles in thermal equilibrium. We assume that the space is homogeneous and isotropic. The particle number density n and the particle energy density ρ at the temperature T are, respectively,

$$n = g \int \frac{d^3\mathbf{p}}{(2\pi\hbar)^3} f(\mathbf{p}), \quad \rho = g \int \frac{d^3\mathbf{p}}{(2\pi\hbar)^3} E(\mathbf{p}) f(\mathbf{p}), \tag{11.50}$$

where g is the number of internal degrees of freedom, $\mathbf{p}$ is the particle 3-momentum, $E = \sqrt{m^2c^4 + \mathbf{p}^2c^2}$ is the particle energy, and $f(\mathbf{p})$ is the Bose-Einstein distribution (in the case of bosons, namely for particles with integer spin) or the Fermi-Dirac distribution (for fermions, namely for particles with half-integer spin)

$$f(\mathbf{p}) = \begin{cases} \frac{1}{e^{(E-\mu)/k_{\mathrm{B}}T} - 1} & \text{Bose-Einstein distribution} \\ \frac{1}{e^{(E-\mu)/k_{\mathrm{B}}T} + 1} & \text{Fermi-Dirac distribution} \end{cases} . \tag{11.51}$$

Here μ is the chemical potential and k_B is the Boltzmann constant. The reader not familiar with these concepts may refer to any textbook on statistical mechanics.

For a non-degenerate ($\mu \ll T$) relativistic ($m \ll T$) gas, the particle number density turns out to be

$$
n = \frac{g}{2\pi^2 c^3 \hbar^3} \int \frac{E^2\, dE}{e^{E/k_B T} \pm 1} = \begin{cases} \frac{\zeta(3)}{\pi^2} \frac{g}{c^3 \hbar^3} (k_B T)^3 & \text{for bosons} \\[2ex] \frac{3}{4} \frac{\zeta(3)}{\pi^2} \frac{g}{c^3 \hbar^3} (k_B T)^3 & \text{for fermions} \end{cases}
\tag{11.52}
$$

where $\zeta(3) = 1.20206...$ is the Riemann Zeta function. The particle energy density is

$$
\rho = \frac{g}{2\pi^2 c^3 \hbar^3} \int \frac{E^3\, dE}{e^{E/k_B T} \pm 1} = \begin{cases} \frac{\pi^2}{30} \frac{g}{c^3 \hbar^3} (k_B T)^4 & \text{for bosons} \\[2ex] \frac{7}{8} \frac{\pi^2}{30} \frac{g}{c^3 \hbar^3} (k_B T)^4 & \text{for fermions} \end{cases}
\tag{11.53}
$$

For non-relativistic particles ($m \gg T$) with arbitrary chemical potential μ, we find (in the non-relativistic limit, there is no difference between bosons and fermions)

$$
n = g \left(\frac{m k_B T}{2\pi \hbar^2} \right)^{3/2} e^{-(mc^2 - \mu)/k_B T}, \quad \rho = mc^2 n.
\tag{11.54}
$$

If the gas is made of different species of particles, the total number density and the total energy density will be given by, respectively, the sum of all number densities and the sum of all energy densities, as in Eq. (11.9). Since the energy density of non-relativistic particles in thermal equilibrium is exponentially suppressed with respect to that of relativistic particles, their contribution may be ignored and we can write the total energy density of the universe as

$$
\rho = \frac{\pi^2}{30} \frac{g_{\text{eff}}}{c^3 \hbar^3} (k_B T)^4,
\tag{11.55}
$$

where g_{eff} is the effective number of light degrees of freedom

$$
g_{\text{eff}} = \sum_{\text{bosons}} g_b + \frac{7}{8} \sum_{\text{fermions}} g_f.
\tag{11.56}
$$

In general, g_{eff} will depend on the plasma temperature, because some particles may be relativistic above some temperature and become non-relativistic at lower temperatures.

If we plug the expression in Eq. (11.55) into the first Friedmann equation, we get

$$
H^2 = \frac{4\pi^3 G_N}{45 c^5 \hbar^3} g_{\text{eff}} (k_B T)^4 - \frac{kc^2}{a^2}.
\tag{11.57}
$$

In a radiation dominated universe, we have $a \propto t^{1/2}$ at sufficiently early times, and therefore $H = 1/2t$. If we plug such an expression for the Hubble parameter into Eq. (11.57), we can get a relation between the time t of the universe and the temperature T of the plasma. Within the Standard Model of particle physics, at $T > 1$ MeV we have $g_{\text{eff}} \sim 10 - 100$ [1]. The relation between time t and plasma temperature T is

$$t \sim 1 \left(\frac{1 \, \text{MeV}}{T} \right)^2 \text{s} .$$

(11.58)

11.6 Age of the Universe

The cosmological models discussed in Sect. 11.3 are very simple and it is possible to obtain a compact analytic expression for their scale factor a. In more realistic cosmological models, the Universe is filled with different components. However, if we know the contribution of each component and the value of the Hubble parameter at a certain time (for example today), we can calculate the evolution in time of the scale factor. In the cosmological models starting from a vanishing scale factor, like those in Sects. 11.3.2 and 11.3.3, we can define the age of the Universe as the time interval measured with respect to the temporal coordinate of the Friedmann–Robertson–Walker metric between the Big Bang ($a = 0$) and today.

To evaluate the age of the Universe today we can proceed as follows. First, we define the effective energy density associated to a possible non-vanishing k as

$$\rho_k = -\frac{3c^4}{8\pi G_{\text{N}}} \frac{k}{a^2} ,$$

(11.59)

and we rewrite the first Friedmann equation as

$$H^2 = \frac{8\pi G_{\text{N}}}{3c^2} \rho_{\text{c}}^0 \sum_i \frac{\rho_i}{\rho_{\text{c}}^0} ,$$

(11.60)

where ρ_{c}^0 is the value of the critical energy density today and the sum is over all the different components filling the Universe, including also ρ_k.

We define the *redshift factor* z as

$$1 + z \equiv \frac{a_0}{a} ,$$

(11.61)

where a_0 is the scale factor today and a is the scale factor at the redshift z. Today $z = 0$. We know from observations that the Universe is expanding, namely z increases as we go backwards in time. From Sect. 11.3 and Eq. (11.59), we know how the energy densities of different components evolve with the scale factor, and therefore with the redshift factor. If we restrict attention to non-relativistic matter (dust), vacuum energy, and the effective energy associated to k (curvature), we have

$$\rho_m = \rho_m^0 \left(1 + z\right)^3 , \quad \rho_\Lambda = \rho_\Lambda^0 , \quad \rho_k = \rho_k^0 \left(1 + z\right)^2 . \tag{11.62}$$

where ρ_m, ρ_Λ, and ρ_k are, respectively, the energy densities of non-relativistic matter, vacuum energy, and curvature at redshift z, and ρ_m^0, ρ_Λ^0, and ρ_k^0 are the same energy densities today. We plug the expressions in Eq. (11.62) into Eq. (11.60) and we find

$$H^2 = H_0^2 \left[\Omega_m^0 \left(1 + z\right)^3 + \Omega_\Lambda^0 + \Omega_k^0 \left(1 + z\right)^2\right] , \tag{11.63}$$

where $\Omega_i = \rho_i / \rho_c$ and the index 0 is to indicate their value today.

From the definition of the Hubble parameter, we find

$$H = \frac{\dot{a}}{a} = \frac{d}{dt} \ln \frac{a}{a_0} = \frac{d}{dt} \ln \frac{1}{1 + z} = -\frac{1}{1 + z} \frac{dz}{dt} , \tag{11.64}$$

and we can thus rewrite Eq. (11.63) as

$$\frac{dt}{dz} = -\frac{1}{1 + z} \frac{1}{H_0 \sqrt{\Omega_m^0 \left(1 + z\right)^3 + \Omega_\Lambda^0 + \Omega_k^0 \left(1 + z\right)^2}} . \tag{11.65}$$

Since $\Omega_m^0 + \Omega_\Lambda^0 + \Omega_k^0 = 1$ by definition, we can write $\Omega_k^0 = 1 - \Omega_m^0 - \Omega_\Lambda^0$ and remove Ω_k^0 in Eq. (11.65). Integrating by parts, we can find the time difference between today ($z = 0$) and the time at which the redshift of the Universe was z in terms of H_0, Ω_m^0, and Ω_Λ^0

$$\Delta t = \frac{1}{H_0} \int_0^z \frac{d\tilde{z}}{1 + \tilde{z}} \frac{1}{\sqrt{\left(1 + \Omega_m^0 \tilde{z}\right) \left(1 + \tilde{z}\right)^2 - \tilde{z} \left(2 + \tilde{z}\right) \Omega_\Lambda^0}} . \tag{11.66}$$

The age of the Universe is obtained when $z \to \infty$ (corresponding to $a = 0$)

$$\tau = \frac{1}{H_0} \int_0^\infty \frac{d\tilde{z}}{1 + \tilde{z}} \frac{1}{\sqrt{\left(1 + \Omega_m^0 \tilde{z}\right) \left(1 + \tilde{z}\right)^2 - \tilde{z} \left(2 + \tilde{z}\right) \Omega_\Lambda^0}} . \tag{11.67}$$

The integral in Eq. (11.67) is typically of order unity, so the age of the Universe is roughly given by $1/H_0 \approx 14$ Gyr and is not too sensitive to the exact matter content. For instance, in the simple case of a flat Universe without vacuum energy (i.e. $\Omega_m^0 = 1$ and $\Omega_\Lambda^0 = 0$), we find

$$\tau = \frac{1}{H_0} \int_0^\infty \frac{d\tilde{z}}{(1+\tilde{z})^{5/2}} = \frac{2}{3}\frac{1}{H_0} \approx 10 \text{ Gyr}. \tag{11.68}$$

In more general cases, it is necessary to integrate Eq. (11.67) numerically. We should also take into account the contribution of relativistic matter in order to get a more accurate result, but in the case of our Universe it only introduces a small correction.

11.7 Destiny of the Universe

For a universe filled with ordinary matter, there is a simple relation between its geometry (given by the value of the constant k) and its destiny (recollapse or eternal expansion): a closed universe must recollapse, while flat or open universes expand forever. In the presence of vacuum energy, this is not true any longer. A positive cosmological constant makes a universe without matter expand for any value of k. If the universe is filled with dust and vacuum energy, its destiny is determined by their relative contribution. If there is enough dust to stop the expansion, then the universe starts recollapsing to a new singular configuration with $a = 0$. If vacuum energy starts driving the expansion before the recollapse, we are in the opposite scenario and dust becomes less and less important in the evolution of the universe.

If we consider only dust and vacuum energy, the universe is flat if

$$\Omega_\mathrm{m} + \Omega_\Lambda = 1, \tag{11.69}$$

and it is closed (open) if $\Omega_\mathrm{m} + \Omega_\Lambda > 1 \, (< 1)$.

The curve separating eternally expanding universes from universes initially having an expanding phase followed by a contraction is given by

$$\Omega_\Lambda = \begin{cases} 0 & \text{for } \Omega_\mathrm{m} \le 1, \\ 4\Omega_\mathrm{m} \sin^3 \left[\frac{1}{3} \arcsin \left(\frac{\Omega_\mathrm{m}-1}{\Omega_\mathrm{m}} \right) \right] & \text{for } \Omega_\mathrm{m} > 1. \end{cases} \tag{11.70}$$

From the second Friedmann equation, we see that the universe's expansion is accelerating (decelerating) if $\rho + 3P < 0 \, (> 0)$. If we write $\rho = \rho_\mathrm{m} + \rho_\Lambda$ and $P = P_\Lambda = -\rho_\Lambda$, we find

$$\ddot{a} > 0 \Rightarrow \Omega_\mathrm{m} < 2\Omega_\Lambda \quad (\ddot{a} < 0 \Rightarrow \Omega_\mathrm{m} > 2\Omega_\Lambda). \tag{11.71}$$

Figure 11.3 shows the curves in Eqs. (11.69), (11.70), and (11.71) on the plane $(\Omega_\mathrm{m}, \Omega_\Lambda)$

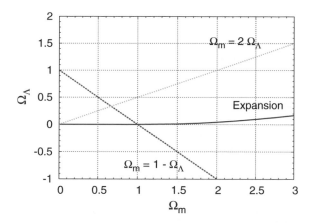

Fig. 11.3 Cosmological models with dust and vacuum energy. The line $\Omega_m = 1 - \Omega_\Lambda$ separates closed universes ($\Omega_m > 1 - \Omega_\Lambda$) from open universes ($\Omega_m < 1 - \Omega_\Lambda$). The line "Expansion" (red solid line) separates universes that expand forever (above) from universes that first expand and then recollapse (below). The line $\Omega_m = 2\Omega_\Lambda$ separates accelerating ($\ddot{a} > 0$) from decelerating ($\ddot{a} < 0$) universes

Problems

11.1 Write the tt component of the Einstein equations for the Friedmann–Robertson–Walker metric and a perfect fluid and derive the first Friedmann equation.

11.2 With the help of some Mathematica package, verify Eqs. (11.3) and (11.4).

11.3 Check that the Einstein universe is unstable.

11.4 Repeat the discussion in Sect. 11.6 about the age of the Universe in the case of a non-negligible contribution from a radiation component.

Reference

1. C. Bambi, A.D. Dolgov, *Introduction to Particle Cosmology: The Standard Model of Cosmology and its Open Problems* (Springer-Verlag, Berlin Heidelberg, 2016)

Chapter 12
Gravitational Waves

Gravitational waves are a hot topic today. It is now possible to directly detect gravitational waves from astrophysical sources, and we expect an impressive amount of completely new data in the next 10–20 years. The aim of this chapter is to provide an introductory overview on the topic. Contrary to the other chapters of the book, the discussion will not be at a purely theoretical level, and some sections will be devoted to observations and experimental facilities.

12.1 Historical Overview

Generally speaking, gravitational waves should be a prediction of any relativistic theory of gravity. Matter makes the spacetime curved, and therefore the motion of matter can alter the spacetime metric. Gravitational waves are like "ripples" in the curvature of the spacetime propagating at a finite velocity. Consistently with the Einstein Principle of Relativity, no signal can propagate (locally) at a velocity higher than the speed of light in vacuum.

Gravitational waves were predicted by Albert Einstein immediately after the formulation of his theory. However, the gravitational wave signal produced by typical astrophysical sources is extremely weak, and therefore the detection of gravitational waves is very challenging.

The first observational evidence for the existence of gravitational waves followed the discovery of the binary pulsar PSR 1913+16 by Russell Alan Hulse and Joseph Hooton Taylor in 1974. PSR 1913+16 is a binary system of two neutron stars, and one of them is seen as a pulsar, which makes this system a perfect laboratory for testing the predictions of Einstein's gravity. Since its discovery, the orbital period of PSR 1913+16 has decayed in agreement with the predictions of Einstein's equations for the emission of gravitational waves. From the radio data covering about 40 years of observations, we have [13]

© Springer Nature Singapore Pte Ltd. 2018, corrected publication 2020
C. Bambi, *Introduction to General Relativity*, Undergraduate Lecture Notes
in Physics, https://doi.org/10.1007/978-981-13-1090-4_12

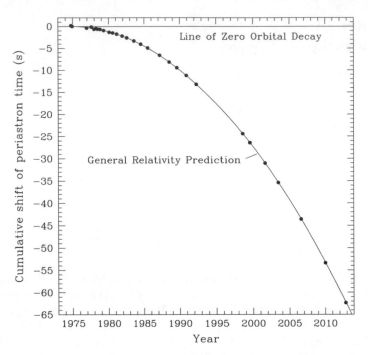

Fig. 12.1 Cumulative shift of the periastron time of PSR 1913+16 over about 40 years, from 1974 to 2012. The dots (with error bars too small to show) are the data and the solid curve is the prediction of Einstein's gravity for the emission of gravitational waves. From [13]. ©AAS. Reproduced with permission

$$\frac{\dot{P}_{\text{corrected}}}{\dot{P}_{\text{GR}}} = 0.9983 \pm 0.0016, \tag{12.1}$$

where $\dot{P}_{\text{corrected}}$ is the (corrected) observed orbital decay[1] and $\dot{P}_{\text{GR}}$ is the orbital decay due to gravitational waves expected in Einstein's gravity. Figure 12.1 shows the perfect agreement between the data (the black dots) and the theoretical prediction (the solid line).

Attempts of direct detection of gravitational waves started in the 1960s, with the resonant bar constructed by Joseph Weber. It was a 2 m aluminum cylinder held at room temperature and isolated from vibrations in a vacuum chamber. In the 1990s, a new generation of resonant detectors became operative, as well as the first generation of laser interferometers.

The first direct detection of gravitational waves was announced by the LIGO-Virgo collaboration in February 2016 [1]. The event, called GW150914 because it was detected on 14 September 2015, was the coalescence of two stellar-mass black

---

[1]One has to remove the effect due to the relative acceleration between us and the pulsar caused by the differential rotation of the Galaxy.

holes, both of about 30 $M_\odot$. They formed a black hole of about 60 $M_\odot$, releasing an energy of about 3 $M_\odot$ in the form of gravitational waves.

12.2 Gravitational Waves in Linearized Gravity

Let us consider a spacetime where we can write, not necessarily over the whole spacetime but at least on a sufficiently large region, the metric $g_{\mu\nu}$ as the Minkowski metric $\eta_{\mu\nu}$ plus a small perturbation $h_{\mu\nu}$

$$g_{\mu\nu} = \eta_{\mu\nu} + h_{\mu\nu}, \quad |h_{\mu\nu}| \ll 1. \tag{12.2}$$

$h_{\mu\nu}$ is called the *metric perturbation*. In linearized gravity, we neglect terms of second or higher order in $h_{\mu\nu}$. Indices of "tensors" of order $h_{\mu\nu}$ are raised and lowered using the Minkowski metric $\eta_{\mu\nu}$. The inverse metric $g^{\mu\nu}$ is

$$g^{\mu\nu} = \eta^{\mu\nu} - h^{\mu\nu} \tag{12.3}$$

Indeed, if we write $g^{\mu\nu} = \eta^{\mu\nu} + H^{\mu\nu}$, we see that

$$\begin{aligned}
g_{\mu\nu}g^{\nu\rho} &= \left(\eta_{\mu\nu} + h_{\mu\nu}\right)\left(\eta^{\nu\rho} + H^{\nu\rho}\right) \\
&= \eta_\mu^\rho + h_\mu^\rho + H_\mu^\rho + O\left(h^2\right),
\end{aligned} \tag{12.4}$$

and therefore $H^{\mu\nu} = -h^{\mu\nu}$. Note that $h_{\mu\nu}$ transforms as a tensor only under Lorentz transformations, but not under general coordinate transformations [see later Eqs. (12.18) and (12.19)].

Let us now write the Einstein equations in linearized gravity. From Eq. (5.76), the Riemann tensor $R_{\mu\nu\rho\sigma}$ is given by

$$R_{\mu\nu\rho\sigma} = \frac{1}{2}\left(\partial_\nu\partial_\rho h_{\mu\sigma} + \partial_\mu\partial_\sigma h_{\nu\rho} - \partial_\nu\partial_\sigma h_{\mu\rho} - \partial_\mu\partial_\rho h_{\nu\sigma}\right). \tag{12.5}$$

The Ricci tensor is

$$\begin{aligned}
R_{\nu\sigma} &= g^{\mu\rho}R_{\mu\nu\rho\sigma} = \eta^{\mu\rho}R_{\mu\nu\rho\sigma} \\
&= \frac{1}{2}\left(\partial_\nu\partial^\mu h_{\mu\sigma} + \partial^\rho\partial_\sigma h_{\nu\rho} - \partial_\nu\partial_\sigma h - \Box_\eta h_{\nu\sigma}\right)
\end{aligned} \tag{12.6}$$

where $h = \eta^{\mu\nu}h_{\mu\nu}$ is the trace of the metric perturbation and $\Box_\eta = \eta^{\mu\nu}\partial_\mu\partial_\nu$ is the d'Alembertian of flat spacetime. Lastly, the scalar curvature is

$$\begin{aligned}
R &= g^{\nu\sigma}R_{\nu\sigma} = \frac{1}{2}\left(\partial^\sigma\partial^\mu h_{\mu\sigma} + \partial^\rho\partial^\nu h_{\nu\rho} - \Box_\eta h - \Box_\eta h\right) \\
&= \partial^\mu\partial^\nu h_{\mu\nu} - \Box_\eta h.
\end{aligned} \tag{12.7}$$

With Eqs. (12.6) and (12.7) we can write the Einstein equations

$$\frac{1}{2}\left(\partial_\mu\partial^\sigma h_{\sigma\nu} + \partial^\sigma\partial_\nu h_{\mu\sigma} - \partial_\mu\partial_\nu h - \Box_\eta h_{\mu\nu}\right)$$

$$-\frac{1}{2}\eta_{\mu\nu}\left(\partial^\sigma\partial^\rho h_{\sigma\rho} - \Box_\eta h\right) = \frac{8\pi G_N}{c^4}T_{\mu\nu}. \tag{12.8}$$

Let us now simplify Eq. (12.8) by changing variables and coordinate system. First, we change variables. We define the *trace-reversed perturbation* as

$$\tilde{h}_{\mu\nu} = h_{\mu\nu} - \frac{1}{2}\eta_{\mu\nu}h. \tag{12.9}$$

The trace of $\tilde{h}_{\mu\nu}$ is

$$\tilde{h} = \eta^{\mu\nu}\tilde{h}_{\mu\nu} = \eta^{\mu\nu}\left(h_{\mu\nu} - \frac{1}{2}\eta_{\mu\nu}h\right) = h - 2h = -h, \tag{12.10}$$

and hence the name "trace-reversed". Note that

$$h_{\mu\nu} = \tilde{h}_{\mu\nu} - \frac{1}{2}\eta_{\mu\nu}\tilde{h}, \tag{12.11}$$

If we plug the expression (12.11) into Eq. (12.8), we find

$$\frac{1}{2}\left(\partial_\mu\partial^\sigma\tilde{h}_{\sigma\nu} - \frac{1}{2}\partial_\mu\partial_\nu\tilde{h} + \partial^\sigma\partial_\nu\tilde{h}_{\mu\sigma} - \frac{1}{2}\partial_\mu\partial_\nu\tilde{h} + \partial_\mu\partial_\nu\tilde{h} - \Box_\eta\tilde{h}_{\mu\nu} + \frac{1}{2}\eta_{\mu\nu}\Box_\eta\tilde{h}\right)$$

$$-\frac{1}{2}\eta_{\mu\nu}\left(\partial^\sigma\partial^\rho\tilde{h}_{\sigma\rho} - \frac{1}{2}\Box_\eta\tilde{h} + \Box_\eta\tilde{h}\right) = \frac{8\pi G_N}{c^4}T_{\mu\nu}, \tag{12.12}$$

and the Einstein equations in linearized gravity in terms of the trace-reversed perturbation are

$$\partial_\mu\partial^\sigma\tilde{h}_{\sigma\nu} + \partial^\sigma\partial_\nu\tilde{h}_{\mu\sigma} - \Box_\eta\tilde{h}_{\mu\nu} - \eta_{\mu\nu}\partial^\sigma\partial^\rho\tilde{h}_{\sigma\rho} = \frac{16\pi G_N}{c^4}T_{\mu\nu}. \tag{12.13}$$

12.2.1   Harmonic Gauge

Now we change coordinate system. We choose the coordinate system $\{x^\mu\}$ such that

$$\partial^\mu\tilde{h}_{\mu\nu} = 0. \tag{12.14}$$

The condition (12.14) is called the *harmonic gauge* (or the Hilbert gauge or the de Donder gauge). The choice of the harmonic gauge is similar to the choice of the

Lorentz gauge in Maxwell's theory, $\partial_\mu A^\mu = 0$ (see Chap. 4). In the harmonic gauge, Eq. (12.13) becomes

$$\Box_\eta \tilde{h}_{\mu\nu} = -\frac{16\pi G_N}{c^4} T_{\mu\nu} \,. \tag{12.15}$$

It is easy to see that we can always choose the harmonic gauge. Let us consider the coordinate transformation

$$x^\mu \to x'^\mu = x^\mu + \xi^\mu \,, \tag{12.16}$$

where ξ^μs are four functions of x^μ of the same order as $h_{\mu\nu}$. The inverse is

$$x^\mu = x'^\mu - \xi^\mu \,, \tag{12.17}$$

and the spacetime metrics in the two coordinates systems are related by

$$\begin{aligned}
g_{\mu\nu} \to g'_{\mu\nu} &= \frac{\partial x^\alpha}{\partial x'^\mu} \frac{\partial x^\beta}{\partial x'^\nu} g_{\alpha\beta} \\
&= \left(\delta^\alpha_\mu - \partial_\mu \xi^\alpha\right) \left(\delta^\beta_\nu - \partial_\nu \xi^\beta\right) \left(\eta_{\alpha\beta} + h_{\alpha\beta}\right) \\
&= \eta_{\mu\nu} + h_{\mu\nu} - \partial_\mu \xi_\nu - \partial_\nu \xi_\mu \,.
\end{aligned} \tag{12.18}$$

Since $g'_{\mu\nu} = \eta_{\mu\nu} + h'_{\mu\nu}$, the relation between the two metric perturbations is

$$h'_{\mu\nu} = h_{\mu\nu} - \partial_\mu \xi_\nu - \partial_\nu \xi_\mu \,, \tag{12.19}$$

and we see that $h_{\mu\nu}$ does not transform as a tensor under general coordinate transformations.[2] The relation between the two trace-reversed perturbations is

$$\tilde{h}'_{\mu\nu} = h'_{\mu\nu} - \frac{1}{2}\eta_{\mu\nu}h' = \tilde{h}_{\mu\nu} - \partial_\mu \xi_\nu - \partial_\nu \xi_\mu + \eta_{\mu\nu}\partial^\sigma \xi_\sigma \,. \tag{12.21}$$

If we are not in the harmonic gauge, we can perform the coordinate transformation in (12.16) to have $\partial^\mu \tilde{h}'_{\mu\nu} = 0$

$$\partial^\mu \tilde{h}'_{\mu\nu} = \partial^\mu \tilde{h}_{\mu\nu} - \Box_\eta \xi_\nu = 0 \,, \tag{12.22}$$

and therefore we need ξ_ν such that

---

[2]$h_{\mu\nu}$ transforms as a tensor under Lorentz transformations. If the transformation is $x^\mu \to x'^\mu = \Lambda^\mu_\nu x^\nu$, we have

$$g_{\mu\nu} = \eta_{\mu\nu} + h_{\mu\nu} \to g'_{\mu\nu} = \Lambda^\alpha_\mu \Lambda^\beta_\nu \left(\eta_{\alpha\beta} + h_{\alpha\beta}\right) = \eta_{\mu\nu} + \Lambda^\alpha_\mu \Lambda^\beta_\nu h_{\alpha\beta} \,, \tag{12.20}$$

and we see that $h'_{\mu\nu} = \Lambda^\alpha_\mu \Lambda^\beta_\nu h_{\alpha\beta}$.

$$\Box_\eta \xi_v = \partial^\mu \tilde{h}_{\mu v} \,. \tag{12.23}$$

Note that if we are in the harmonic gauge and we consider a new transformation such that $\Box_\eta \xi_\mu = 0$, we remain in the harmonic gauge. This is the counterpart of the transformation $A_\mu \to A_\mu + \partial_\mu \Lambda$ in Maxwell's theory, where the Lorentz gauge is preserved if $\Box_\eta \Lambda = 0$.

The formal solution of Eq. (12.15) is

$$\tilde{h}_{\mu v} (t, \mathbf{x}) = \frac{4G_N}{c^4} \int d^3x' \frac{T_{\mu v} \left(t - |\mathbf{x} - \mathbf{x}'|/c, \mathbf{x}'\right)}{|\mathbf{x} - \mathbf{x}'|} \,, \tag{12.24}$$

where the integral is performed in the flat 3-dimensional space and $|\mathbf{x} - \mathbf{x}'|$ is the Euclidean distance between the point $\mathbf{x}$ and the point $\mathbf{x}'$. In Cartesian coordinates (x, y, z), we have

$$|\mathbf{x} - \mathbf{x}'| = \sqrt{(x - x')^2 + (y - y')^2 + (z - z')^2} \,. \tag{12.25}$$

12.2.2   Transverse-Traceless Gauge

$\tilde{h}_{\mu v}$ is symmetric and therefore has ten independent components. The harmonic gauge (12.14) provides four conditions and reduces the number of independent components to six. However, we still have the freedom to choose four arbitrary functions ξ_μ satisfying the equation $\Box_\eta \xi^\mu = 0$.

First, we can choose ξ^0 such that the trace of $\tilde{h}_{\mu v}$ vanishes, i.e. $\tilde{h} = 0$. Note that such a choice implies that the trace-reversed perturbation $\tilde{h}_{\mu v}$ and the metric perturbation $h_{\mu v}$ coincide

$$h_{\mu v} = \tilde{h}_{\mu v} \,, \tag{12.26}$$

and in what follows we can omit the tilde for simplicity when we are in this case. Second, we can choose the three functions ξ^is such that $h^{0i} = 0$.

Imposing $h^{0i} = 0$, the harmonic gauge condition (12.14) becomes

$$\partial^0 h_{00} = 0 \,, \tag{12.27}$$

namely h_{00} is independent of time and therefore corresponds to the Newtonian potential of the source. Restricting the attention to gravitational waves (i.e. the time-dependent part of h_{00}), we can set $h_{00} = 0$. Eventually we have

$$h_{0\mu} = 0 \,, \quad h = 0 \,, \quad \partial^i h_{ij} = 0 \,, \tag{12.28}$$

which defines the *transverse-traceless gauge* (TT gauge). Quantities in the TT gauge are often indicated with TT, e.g. $h_{\mu\nu}^{TT}$.

Note that the TT gauge is only possible in vacuum, namely when Eq. (12.15) reads

$$\Box_\eta \tilde{h}_{\mu\nu} = 0 \,. \tag{12.29}$$

Inside a source we can choose the harmonic gauge and we still have the freedom to choose the four functions ξ_μs satisfying the equation $\Box_\eta \xi_\mu = 0$. However, we cannot set to zero any further component of $\tilde{h}_{\mu\nu}$ by choosing suitable ξ_μs because $\Box_\eta \tilde{h}_{\mu\nu} \neq 0$.

The vacuum equation $\Box_\eta h_{\mu\nu}^{TT} = 0$ has plane wave solutions ($h_{0\mu}^{TT} = 0$ because we are in the TT gauge)

$$h_{ij}^{TT} = \varepsilon_{ij} e^{ik^\mu x_\mu} \,, \tag{12.30}$$

where $k^\mu = (\omega/c, \mathbf{k})$, $\omega = |\mathbf{k}|c$ is the angular frequency of the gravitational wave, and ε_{ij} is the polarization tensor. For a gravitational wave propagating along the z direction, we have (ignoring the imaginary part and imposing that h_{ij} is symmetric and traceless)

$$h_{\mu\nu}^{TT} = \begin{pmatrix} 0 & 0 & 0 & 0 \\ 0 & h_+ & h_\times & 0 \\ 0 & h_\times & -h_+ & 0 \\ 0 & 0 & 0 & 0 \end{pmatrix} \cos\left[\omega\left(t - z/c\right)\right] \,, \tag{12.31}$$

where h_+ and $h_\times$ are the amplitudes of the gravitational wave in the two polarizations. The ij components of $h_{\mu\nu}^{TT}$ can be written as

$$h_{ij}^{TT} = h_+ \varepsilon_{ij}^+ \cos\left[\omega\left(t - z/c\right)\right] + h_\times \varepsilon_{ij}^\times \cos\left[\omega\left(t - z/c\right)\right] \tag{12.32}$$

where

$$\varepsilon_{ij}^+ = \begin{pmatrix} 1 & 0 & 0 \\ 0 & -1 & 0 \\ 0 & 0 & 0 \end{pmatrix} \quad \text{(plus mode)} \,,$$

$$\varepsilon_{ij}^\times = \begin{pmatrix} 0 & 1 & 0 \\ 1 & 0 & 0 \\ 0 & 0 & 0 \end{pmatrix} \quad \text{(cross mode)} \,. \tag{12.33}$$

With the metric perturbation (12.31), the line element reads

$$ds^2 = -c^2 dt^2 + (1 + h_+ \cos\phi)\, dx^2 + (1 - h_+ \cos\phi)\, dy^2 \\ + 2h_\times \cos\phi\, dxdy + dz^2 \,, \tag{12.34}$$

where $\phi = \omega\,(t - z/c)$.

Let us now check the effect of the gravitational wave in (12.34) on a free point-like particle. The particle is at rest at $\tau = 0$. The geodesic equations of the particle at $\tau = 0$ read

$$\left(\ddot{x}^i + \Gamma_{00}^i \dot{x}^0 \dot{x}^0\right)_{\tau=0} = 0\,, \tag{12.35}$$

because $\dot{x}^i = 0$ at $\tau = 0$. Γ_{00}^i is

$$\Gamma_{00}^i = \frac{1}{2}\eta^{i\mu}\left(\partial_0 h_{\mu 0} + \partial_0 h_{0\mu} - \partial_\mu h_{00}\right)$$

$$= \partial_0 h_0^i - \frac{1}{2}\partial^i h_{00}\,, \tag{12.36}$$

and in the TT gauge vanishes, $\Gamma_{00}^i = 0$. This means that $\ddot{x}^i = 0$, and therefore $\dot{x}^i = 0$ at all times and the particle remains at rest. This result should not be interpreted as the passage of gravitational waves having no physical effects, because in general relativity the choice of the coordinate system is arbitrary. In fact, the opposite is true, namely that the coordinates do not have a direct physical meaning.

Let us now consider two free point-like particles at rest, respectively with space coordinates $(x_0, 0, 0)$ and $(-x_0, 0, 0)$. The proper distance between the two particles is

$$L(t) = \int_{-x_0}^{x_0} \sqrt{g_{xx}}dx' = \int_{-x_0}^{x_0} \sqrt{1 + h_{xx}}dx'$$

$$\approx L_0\left[1 + \frac{1}{2}h_+\cos\left(\omega t\right)\right]\,, \tag{12.37}$$

where $L_0 = 2x_0$ is their proper distance in the absence of gravitational waves. If the two particles have instead, respectively, coordinates $(0, x_0, 0)$ and $(0, -x_0, 0)$, we find

$$L(t) \approx L_0\left[1 - \frac{1}{2}h_+\cos\left(\omega t\right)\right]\,. \tag{12.38}$$

We thus see that the proper distance between the two particles changes with time periodically and the variation is proportional to the amplitude of the gravitational wave. If we consider a rotation of $45°$ in the xy plane, we find the expressions in Eqs. (12.37) and (12.38) with h_+ replaced by $h_\times$. In the end, it is easy to see that the passage of a gravitational wave propagating along the z direction on a ring of particles in the xy-plane is that illustrated in Fig. 12.2.

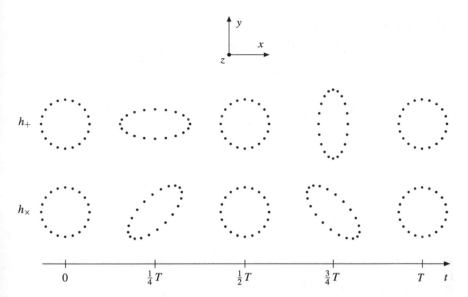

Fig. 12.2 Impact of a gravitational wave traveling along the z-axis on a ring of test-particles in the xy-plane. The effect of the polarization modes h_+ and $h_\times$ is the same modulo a rotation of $45°$ on the xy-plane. $T = 2\pi/\omega$ is the period of the gravitational wave

12.3 Quadrupole Formula

If we can assume that the region where the source is confined (i.e. $T^{\mu\nu}$ is non-vanishing) is much smaller than the wavelength of the emitted radiation, Eq. (12.24) can be approximated as

$$\tilde{h}^{\mu\nu}(t, \mathbf{x}) = \frac{4G_N}{c^4 r} \int d^3\mathbf{x}' T^{\mu\nu}\left(t - r/c, \mathbf{x}'\right), \qquad (12.39)$$

where $r = |\mathbf{x}|$.

In the linearized theory, $h_{\mu\nu}$ and $T_{\mu\nu}$ are of the same order and therefore we have $\partial_\mu T^{\mu\nu} = 0$ with the partial derivative. We write $\partial_\mu T^{\mu\nu} = 0$ as $\partial_0 T^{0\nu} = -\partial_k T^{k\nu}$ and then we integrate over the volume V containing all of the region with $T_{\mu\nu} \neq 0$

$$\frac{1}{c}\frac{\partial}{\partial t} \int_V T^{0\nu} d^3\mathbf{x} = -\int_V \frac{\partial T^{k\nu}}{\partial x^k} d^3\mathbf{x} = -\int_\Sigma T^{k\nu} d\Sigma_k = 0, \qquad (12.40)$$

where Σ is the surface of the volume V and $T^{\mu\nu} = 0$ on Σ. We have thus

$$\int_V T^{0\nu} d^3\mathbf{x} = \text{constant}, \qquad (12.41)$$

which implies $\tilde{h}^{0\nu}$ is constant too. Since here we are interested in gravitational waves, namely in the time-dependent part of the gravitational field, we can put $\tilde{h}^{0\nu} = 0$.

We write $\partial_\mu T^{\mu i} = 0$ as $\partial_0 T^{0i} = -\partial_k T^{ki}$, we multiply both sides by x^j, and we integrate over the volume V. We obtain

$$
\frac{1}{c}\frac{\partial}{\partial t}\int_V T^{i0}x^j\,d^3\mathbf{x} = -\int_V \frac{\partial T^{ik}}{\partial x^k}x^j\,d^3\mathbf{x}
$$

$$
= -\int_V \frac{\partial}{\partial x^k}\left(T^{ik}x^j\right)d^3\mathbf{x} + \int_V T^{ik}\frac{\partial x^j}{\partial x^k}d^3\mathbf{x}
$$

$$
= -\int_\Sigma T^{ik}x^j\,d\Sigma_k + \int_V T^{ij}d^3\mathbf{x}
$$

$$
= \int_V T^{ij}d^3\mathbf{x}. \tag{12.42}
$$

Since $T^{\mu\nu}$ is a symmetric tensor, we can also write Eq. (12.42) by exchanging i and j and

$$
\frac{1}{c}\frac{\partial}{\partial t}\int_V \left(T^{i0}x^j + T^{j0}x^i\right)d^3\mathbf{x} = 2\int_V T^{ij}d^3\mathbf{x}. \tag{12.43}
$$

Let us now write $\partial_\mu T^{\mu 0} = 0$ as $\partial_0 T^{00} = -\partial_k T^{k0}$. This time we multiply both sides by $x^i x^j$. We integrate over the volume V and we find

$$
\frac{1}{c}\frac{\partial}{\partial t}\int_V T^{00}x^i x^j\,d^3\mathbf{x} = -\int_V \frac{\partial T^{k0}}{\partial x^k}x^i x^j\,d^3\mathbf{x}
$$

$$
= -\int_V \frac{\partial}{\partial x^k}\left(T^{k0}x^i x^j\right)d^3\mathbf{x} + \int_V \left(T^{k0}\frac{\partial x^i}{\partial x^k}x^j + T^{k0}x^i\frac{\partial x^j}{\partial x^k}\right)d^3\mathbf{x}
$$

$$
= -\int_\Sigma T^{k0}x^i x^j\,d\Sigma_k + \int_V \left(T^{i0}x^j + T^{j0}x^i\right)d^3\mathbf{x}
$$

$$
= \int_V \left(T^{i0}x^j + T^{j0}x^i\right)d^3\mathbf{x}. \tag{12.44}
$$

We take a derivative with respect to the time t and we use Eq. (12.43)

$$
\frac{1}{c^2}\frac{\partial^2}{\partial^2 t}\int_V T^{00}x^i x^j\,d^3\mathbf{x} = \frac{1}{c}\frac{\partial}{\partial t}\int_V \left(T^{i0}x^j + T^{j0}x^i\right)d^3\mathbf{x}
$$

$$
= 2\int_V T^{ij}d^3\mathbf{x}. \tag{12.45}
$$

We define the *quadrupole moment* of the source as

$$
Q^{ij}(t) = \frac{1}{c^2}\int_V T^{00}(t,\mathbf{x})\,x^i x^j\,d^3\mathbf{x}. \tag{12.46}
$$

Equation (12.39) becomes

$$\tilde{h}_{\mu 0} = 0 \,,$$

$$\tilde{h}_{ij} = \frac{2G_N}{c^4 r} \ddot{Q}_{ij} \left(t - r/c \right) \,, \tag{12.47}$$

where the double dot stands for the double derivative with respect to t and $\ddot{Q}_{ij}$ is evaluated at the time $t - r/c$. It is worth noting that a spherical or axisymmetric distribution of matter has a constant quadrupole moment, even if the body is rotating. This implies, for instance, that there is no emission of gravitational waves in a perfectly spherically symmetric collapse, in a perfectly axisymmetric rotating body, etc. Gravitational waves are emitted when there is a certain "degree of asymmetry", e.g. the coalescence of two objects, non-radial pulsation of a body, etc.

If we want Eq. (12.47) in the TT gauge, we need a coordinate transformation that preserves the harmonic gauge and switches to the TT gauge. This can be done with a projector operator. We already have $\tilde{h}_{\mu 0} = 0$. The traceless and transverse wave conditions read, respectively,

$$\delta^{ij} h_{ij}^{\text{TT}} = 0 \,, \quad n^i h_{ij}^{\text{TT}} = 0 \,, \tag{12.48}$$

where $\mathbf{n} = \mathbf{x}/r$ is the unit vector normal to the wavefront.

The operator to project a vector onto the plane orthogonal to the direction of $\mathbf{n}$ is

$$P_{ij} = \delta_{ij} - n_i n_j \,. \tag{12.49}$$

P_{ij} is symmetric and projects out any component parallel to $\mathbf{n}$:

$$P_{ij} n^i = \delta_{ij} n^i - n_i n_j n^i = n_j - n_j = 0 \,. \tag{12.50}$$

The transverse-traceless projector is

$$P_{ijkl} = P_{ik} P_{jl} - \frac{1}{2} P_{ij} P_{kl} \,, \tag{12.51}$$

and extracts the transverse-traceless part of any tensor of type $(0, 2)$. h_{ij}^{TT} is given by

$$h_{ij}^{\text{TT}} = P_{ijkl} \tilde{h}_{kl} \,. \tag{12.52}$$

It is easy to check that the new tensor is traceless

$$\delta^{ij} h_{ij}^{\text{TT}} = \delta^{ij} \left(P_{ik} P_{jl} - \frac{1}{2} P_{ij} P_{kl} \right) \tilde{h}_{kl}$$

$$= \delta^{ij} \left[(\delta_{ik} - n_i n_k)(\delta_{jl} - n_j n_l) - \frac{1}{2} (\delta_{ij} - n_i n_j)(\delta_{kl} - n_k n_l) \right] \tilde{h}_{kl}$$

$$= \delta^{ij} \left(\delta_{ik}\delta_{jl} - \delta_{ik}n_j n_l - \delta_{jl}n_i n_k + n_i n_k n_j n_l \right) \tilde{h}_{kl}$$

$$- \frac{1}{2}\delta^{ij} \left(\delta_{ij}\delta_{kl} - \delta_{ij}n_k n_l - \delta_{kl}n_i n_j + n_i n_j n_k n_l \right) \tilde{h}_{kl}$$

$$= \tilde{h} - n_k n_l \tilde{h}_{kl} - \frac{1}{2} \left(3\tilde{h} - 3n_k n_l \tilde{h}_{kl} - \tilde{h} + n_k n_l \tilde{h}_{kl} \right) = 0 \,. \qquad (12.53)$$

Note that

$$h_{ij}^{\mathrm{TT}} = P_{ijkl}h_{kl} = P_{ijkl}\tilde{h}_{kl} \,, \qquad (12.54)$$

because h_{ij} and $\tilde{h}_{ij}$ only differ by the trace, which is projected out by P_{ijkl}.

Applying the projector P_{ijkl} to $\tilde{h}_{kl}$, Eq. (12.47) becomes

$$h_{\mu 0}^{\mathrm{TT}} = 0 \,,$$

$$h_{ij}^{\mathrm{TT}} = \frac{2G_{\mathrm{N}}}{c^4 r} \ddot{Q}_{ij}^{\mathrm{TT}} \left(t - r/c \right) \,, \qquad (12.55)$$

where

$$Q_{ij}^{\mathrm{TT}} = P_{ijkl}Q_{kl} \,. \qquad (12.56)$$

It is sometimes convenient to introduce the *reduced quadrupole moment*, which is defined as

$$\tilde{Q}_{ij} = Q_{ij} - \frac{1}{3}\delta_{ij}Q \,, \qquad (12.57)$$

where $Q = Q_i^i$ is the trace of the quadrupole moment. Note that

$$Q_{ij}^{\mathrm{TT}} = P_{ijkl}\tilde{Q}_{kl} = P_{ijkl}Q_{kl} \,, \qquad (12.58)$$

because Q_{ij} and $\tilde{Q}_{ij}$ only differ by the trace.

12.4   Energy of Gravitational Waves

Let us now evaluate the energy carried by a gravitational wave with the help of the pseudo-tensor of Landau–Lifshitz met in Sect. 7.5.

We consider a Cartesian coordinate system (x, y, z) in which the astrophysical source emitting gravitational waves is at the origin and the observer is far from the source with coordinates $(0, 0, z)$. The observer detects a gravitational wave traveling along the z axis. For simplicity, let us assume that the wave has only the plus polarization. The metric perturbation in the TT gauge reads

$$||h_{\mu\nu}^{TT}|| = \begin{pmatrix} 0 & 0 & 0 & 0 \\ 0 & h_+(t,z) & 0 & 0 \\ 0 & 0 & -h_+(t,z) & 0 \\ 0 & 0 & 0 & 0 \end{pmatrix},$$ (12.59)

where z is the z-coordinate of the observer. The amplitude of the gravitational wave h_+ has the form

$$h_+(t,z) = \frac{C f(t-z/c)}{z}.$$ (12.60)

where C and f are, respectively, a constant and a function that depends on the parameters and the nature of the emitting source. This is the form of the amplitudes that we can expect from Eq. (12.39) and that we will find in the examples in the next section.

Since we want to compute the pseudo-tensor of Landau–Lifshitz, we need to compute the Christoffel symbols of the metric of the spacetime $g_{\mu\nu} = \eta_{\mu\nu} + h_{\mu\nu}^{TT}$. The only time varying metric coefficients are g_{xx} and g_{yy} and they only depend on the coordinates t and z. The derivatives of h_+ with respect to t and z are, respectively,

$$\frac{\partial h_+}{\partial t} = \frac{C\dot{f}}{z},$$

$$\frac{\partial h_+}{\partial z} = -\frac{Cf}{z^2} - \frac{C\dot{f}}{cz} = -\frac{Cf}{z^2} - \frac{1}{c}\frac{\partial h_+}{\partial t}.$$ (12.61)

In what follows, we will only consider the leading order term proportional to $1/z$ and we will ignore the contribution proportional to $1/z^2$. The non-vanishing Christoffel symbols are

$$\Gamma_{xx}^0 = -\Gamma_{yy}^0 = \Gamma_{0x}^x = \Gamma_{x0}^x = -\Gamma_{0y}^y = -\Gamma_{y0}^y = \frac{1}{2}\dot{h}_+,$$

$$\Gamma_{xx}^z = -\Gamma_{yy}^z = \frac{1}{2c}\dot{h}_+,$$

$$\Gamma_{xz}^x = \Gamma_{zx}^x = -\Gamma_{yz}^y = -\Gamma_{zy}^y = -\frac{1}{2c}\dot{h}_+.$$ (12.62)

After tedious but straightforward calculations, we find the expression of t^{0z}

$$t^{0z} = \frac{c^2}{16\pi G_N}\dot{h}_+^2.$$ (12.63)

ct^{0z} is the energy flux measured by the distant observer, namely the energy per unit time and unit surface flowing orthogonal to the z axis

$$ct^{0z} = \frac{d E_{\text{GW}}}{dt\,dS} = \frac{c^3}{16\pi\,G_{\text{N}}}\dot{h}_+^2\,. \tag{12.64}$$

If we assume that both polarizations are present and we repeat the calculations, we would find that the energy flux is given by

$$ct^{0z} = \frac{c^3}{16\pi\,G_{\text{N}}}\left(\dot{h}_+^2 + \dot{h}_\times^2\right) = \frac{c^3}{32\pi\,G_{\text{N}}}\sum_{ij}\left(\dot{h}_{ij}^{\text{TT}}\right)^2\,. \tag{12.65}$$

Since in general relativity we cannot provide a local definition of the energy of the gravitational field, it is more correct to rewrite Eq. (12.65) averaging over several wavelengths

$$\frac{d E_{\text{GW}}}{dt\,dS} = \langle ct^{0z}\rangle = \frac{c^3}{32\pi\,G_{\text{N}}}\langle\sum_{ij}\left(\dot{h}_{ij}^{\text{TT}}\right)^2\rangle\,. \tag{12.66}$$

Let us now write the gravitational wave luminosity in terms of the quadrupole moment of the emitting source. From Eq. (12.55) we can write

$$\frac{d E_{\text{GW}}}{dt\,dS} = \frac{G_{\text{N}}}{8\pi\,c^5 r^2}\langle\sum_{ij}\left[\dddot{Q}_{ij}^{\text{TT}}\left(t - r/c\right)\right]^2\rangle$$

$$= \frac{G_{\text{N}}}{8\pi\,c^5 r^2}\langle\sum_{ij}\left[P_{ijkl}\dddot{Q}_{kl}\left(t - r/c\right)\right]^2\rangle \tag{12.67}$$

The luminosity of the source can be obtained by integrating over the whole solid angle

$$L_{\text{GW}} = \frac{d E_{\text{GW}}}{dt} = \int \frac{d E_{\text{GW}}}{dt\,dS}\,dS = \int \frac{d E_{\text{GW}}}{dt\,dS}r^2\,d\Omega$$

$$= \frac{G_{\text{N}}}{8\pi\,c^5}\int d\Omega\,\langle\sum_{ij}\left[P_{ijkl}\dddot{Q}_{kl}\left(t - r/c\right)\right]^2\rangle$$

$$= \frac{G_{\text{N}}}{8\pi\,c^5}\int d\Omega\,\langle\sum_{ij}\left[P_{ijkl}\dddot{\tilde{Q}}_{kl}\left(t - r/c\right)\right]^2\rangle\,, \tag{12.68}$$

where $d\Omega$ is the infinitesimal solid angle and in the last passage we have replaced the quadrupole moment with the reduced quadrupole moment because it is more convenient for the next calculations. Remember that Q_{ij} and $\tilde{Q}_{ij}$ only differ by the trace, which is projected out by P_{ijkl}.

To compute the integral over the solid angle, first we rewrite the expression inside the integral as follows

$$\sum_{ij}\left(P_{ijkl}\dddot{\tilde{Q}}_{kl}\right)^2 = \sum_{ij}\left(P_{ijkl}\dddot{\tilde{Q}}_{kl}\right)\left(P_{ijmn}\dddot{\tilde{Q}}_{mn}\right)$$

$$= \sum_{ij}\left(P_{ijkl}P_{ijmn}\dddot{\tilde{Q}}_{kl}\dddot{\tilde{Q}}_{mn}\right) = P_{klmn}\dddot{\tilde{Q}}_{kl}\dddot{\tilde{Q}}_{mn}$$

$$= \Big[(\delta_{km}-n_k n_m)(\delta_{ln}-n_l n_n)$$

$$-\frac{1}{2}(\delta_{kl}-n_k n_l)(\delta_{mn}-n_m n_n)\Big]\dddot{\tilde{Q}}_{kl}\dddot{\tilde{Q}}_{mn} \,. \quad (12.69)$$

Since $\tilde{Q}_{ij}$ is traceless and symmetric, we have

$$\delta_{kl}\dddot{\tilde{Q}}_{kl} = \delta_{mn}\dddot{\tilde{Q}}_{mn} = 0\,,$$

$$n_k n_m \delta_{ln}\dddot{\tilde{Q}}_{kl}\dddot{\tilde{Q}}_{mn} = n_l n_n \delta_{km}\dddot{\tilde{Q}}_{kl}\dddot{\tilde{Q}}_{mn}\,, \quad (12.70)$$

and Eq. (12.69) becomes

$$\sum_{ij}\left(P_{ijkl}\dddot{\tilde{Q}}_{kl}\right)^2 = \dddot{\tilde{Q}}_{lm}\dddot{\tilde{Q}}_{lm} - 2n_l n_n \dddot{\tilde{Q}}_{lm}\dddot{\tilde{Q}}_{mn} + \frac{1}{2}n_k n_l n_m n_n \dddot{\tilde{Q}}_{kl}\dddot{\tilde{Q}}_{mn}\,. \quad (12.71)$$

The expression for the luminosity of the source now reads

$$L_{\rm GW} = \frac{G_{\rm N}}{8\pi c^5}\int d\Omega \left(\dddot{\tilde{Q}}_{lm}\dddot{\tilde{Q}}_{lm} - 2n_l n_n \dddot{\tilde{Q}}_{lm}\dddot{\tilde{Q}}_{mn} + \frac{1}{2}n_k n_l n_m n_n \dddot{\tilde{Q}}_{kl}\dddot{\tilde{Q}}_{mn}\right)\,, \quad (12.72)$$

and we have to evaluate the following integrals

$$\int d\Omega\, n_i n_j\,, \quad \int d\Omega\, n_i n_j n_k n_l\,. \quad (12.73)$$

The unit vector **n** is

$$\mathbf{n} = (\sin\theta\cos\phi, \sin\theta\sin\phi, \cos\theta)\,. \quad (12.74)$$

It is easy to see that, for $i \ne j$, we have

$$\int d\Omega\, n_i n_j = \int_0^\pi d\theta\,\sin\theta\int_0^{2\pi} d\phi\, n_i n_j = 0\,, \quad (12.75)$$

while, for $i = j$, we find

$$\int d\Omega\, n_x^2 = \int d\Omega\, n_y^2 = \int d\Omega\, n_z^2 = \frac{4\pi}{3}\,. \quad (12.76)$$

The first integral in Eq. (12.73) can be written in the following compact form

$$\frac{1}{4\pi} \int d\Omega \, n_i n_j = \frac{1}{3}\delta_{ij} \,.$$
(12.77)

We can proceed in a similar way and evaluate the second integral in Eq. (12.73). It turns out that

$$\frac{1}{4\pi} \int d\Omega \, n_i n_j n_k n_l = \frac{1}{15} \left(\delta_{ij}\delta_{kl} + \delta_{ik}\delta_{jl} + \delta_{il}\delta_{jk} \right) \,.$$
(12.78)

We put together all the results found in this section and we obtain the total luminosity (or power) of the source in the form of gravitational radiation

$$L_{\mathrm{GW}} = \frac{G_{\mathrm{N}}}{5c^5} \langle \dddot{\tilde{Q}}_{ij} \left(t - r/c \right) \cdot \dddot{\tilde{Q}}_{ij} \left(t - r/c \right) \rangle \,.$$
(12.79)

12.5   Examples

Let us now calculate the gravitational waves emitted by two simple systems: (i) a rotating compact object (we can think about a neutron star) which is not perfectly axisymmetric, and (ii) a binary system in circular orbit far from coalescence.

12.5.1   Gravitational Waves from a Rotating Neutron Star

Let us approximate a non-rotating neutron star with an ellipsoid of uniform mass density ρ. The quadrupole moment of the neutron star is

$$Q_{ij} = \int_V \rho x_i x_j \, d^3x \,,$$
(12.80)

where V is the volume of the body. The inertia tensor of the neutron star is

$$I_{ij} = \int_V \rho \left(r^2 \delta_{ij} - x_i x_j \right) d^3x \,,$$
(12.81)

and is related to the quadrupole moment by the equation

$$I_{ij} = \delta_{ij} Q - Q_{ij} \,.$$
(12.82)

The reduced quadrupole moment introduced in Eq. (12.57) is

$$\tilde{Q}_{ij} = Q_{ij} - \frac{1}{3}\delta_{ij}Q = -I_{ij} + \frac{1}{3}\delta_{ij}I, \tag{12.83}$$

where I is the trace of the inertia tensor.

The inertia tensor of a non-rotating ellipsoid of constant mass density and with semi-axes α, β, and γ (respectively along the x, y, and z axes) is given by

$$||I_{ij}|| = \frac{M}{5}\begin{pmatrix} \beta^2 + \gamma^2 & 0 & 0 \\ 0 & \alpha^2 + \gamma^2 & 0 \\ 0 & 0 & \alpha^2 + \beta^2 \end{pmatrix} = \begin{pmatrix} I_{xx} & 0 & 0 \\ 0 & I_{yy} & 0 \\ 0 & 0 & I_{zz} \end{pmatrix}, \tag{12.84}$$

where M is the mass of the ellipsoid and I_{xx}, I_{yy}, and I_{zz} are the principal moments of inertia.

Let us now assume that the neutron star is rotating about the z axis with angular velocity Ω. To compute the inertia tensor of the rotating ellipsoid, we consider the rotation matrix connecting the co-rotating and the inertial reference frames

$$||R_{ij}|| = \begin{pmatrix} \cos\Omega t & -\sin\Omega t & 0 \\ \sin\Omega t & \cos\Omega t & 0 \\ 0 & 0 & 1 \end{pmatrix}. \tag{12.85}$$

If the inertia tensor in the co-rotating frame is given by Eq. (12.84), that in the inertial reference frame can be obtained with a rotation

$$I_{ij} \rightarrow I'_{ij} = R_{ik}R_{jl}I_{kl} = \begin{pmatrix} I'_{xx} & I'_{xy} & 0 \\ I'_{yx} & I'_{yy} & 0 \\ 0 & 0 & I'_{zz} \end{pmatrix}, \tag{12.86}$$

where

$$\begin{aligned}
I'_{xx} &= I_{xx}\cos^2\Omega t + I_{yy}\sin^2\Omega t, \\
I'_{xy} &= -\left(I_{yy} - I_{xx}\right)\sin\Omega t\cos\Omega t, \\
I'_{yx} &= I'_{xy}, \\
I'_{yy} &= I_{xx}\sin^2\Omega t + I_{yy}\cos^2\Omega t, \\
I'_{zz} &= I_{zz}.
\end{aligned} \tag{12.87}$$

We use the trigonometric identity $\cos 2\Omega t = 2\cos^2\Omega t - 1$, and we write the reduced quadrupole moment of the rotating neutron star as

$$||\tilde{Q}_{ij}|| = \frac{I_{yy} - I_{xx}}{2}\begin{pmatrix} \cos 2\Omega t & \sin 2\Omega t & 0 \\ \sin 2\Omega t & -\cos 2\Omega t & 0 \\ 0 & 0 & 0 \end{pmatrix} + \text{constant}, \tag{12.88}$$

where the constant part can be ignored because the gravitational wave emission involves the time derivative of the quadrupole moment. Note that

$$I_{yy} - I_{xx} = \frac{M}{5} \left(\alpha^2 + \gamma^2\right) - \frac{M}{5} \left(\beta^2 + \gamma^2\right) = \frac{M}{5} \left(\alpha^2 - \beta^2\right) . \quad (12.89)$$

and therefore, if $\alpha = \beta$, the reduced quadrupole moment is constant and there is no emission of gravitational waves. A perfectly axisymmetric object rigidly rotating about its symmetry axis does not emit gravitational waves.

We define the *oblateness* of the ellipsoid, ε, as

$$\varepsilon = 2\frac{\alpha - \beta}{\alpha + \beta} , \quad (12.90)$$

and we have

$$\frac{I_{yy} - I_{xx}}{I_{zz}} = \varepsilon + O(\varepsilon^3) . \quad (12.91)$$

The reduced quadrupole moment now reads

$$||\tilde{Q}_{ij}|| = \frac{\varepsilon I_{zz}}{2} \begin{pmatrix} \cos 2\Omega t & \sin 2\Omega t & 0 \\ \sin 2\Omega t & -\cos 2\Omega t & 0 \\ 0 & 0 & 0 \end{pmatrix} + \text{constant} . \quad (12.92)$$

The trace-reversed perturbation is

$$\tilde{h}_{ij} = \frac{2G_N}{c^4 r} \ddot{\tilde{Q}}_{ij} \left(t - r/c\right) . \quad (12.93)$$

h_{ij}^{TT} can be obtained with the use of the projector operator, as done in Eq. (12.52)

$$h_{ij}^{TT} = \frac{4G_N}{c^4 r} \varepsilon \Omega^2 I_{zz} P \begin{pmatrix} -\cos \varphi & -\sin \varphi & 0 \\ -\sin \varphi & \cos \varphi & 0 \\ 0 & 0 & 0 \end{pmatrix} , \quad (12.94)$$

where P indicates the transverse-traceless projector and $\varphi = 2\Omega \left(t - r/c\right)$. Note that the frequency of the gravitational wave is twice the rotational frequency of the neutron star. An estimate of the amplitude of the gravitational wave can be obtained by plugging in some reasonable numbers for a neutron star in our Galaxy

$$\frac{4G_N}{c^4 r} \varepsilon \Omega^2 I_{zz} \sim 10^{-25} \left(\frac{10 \text{ kpc}}{r}\right) \left(\frac{\varepsilon}{10^{-7}}\right) \left(\frac{\Omega}{1 \text{ kHz}}\right)^2 \left(\frac{I_{zz}}{10^{38} \text{ kg} \cdot \text{m}^2}\right) . \quad (12.95)$$

Upper bounds on ε have been obtained from the observed slowing down of the period of pulsars under the assumption that it is entirely due to the emission of gravitational waves [8].

The energy released can be evaluated from Eq. (12.79) and turns out to be

$$L_{GW} = \frac{32G_N}{5c^5}\varepsilon^2\Omega^6 I_{zz}^2 . \tag{12.96}$$

12.5.2 Gravitational Waves from a Binary System

Let us consider a binary system in circular orbit in Newtonian gravity. We have body 1 with mass m_1, body 2 with mass m_2, and the total mass is $M = m_1 + m_2$. We choose the coordinate system such that the motion is in the xy-plane and that the origin of the coordinate system coincides with the center of mass

$$r_1 m_1 + r_2 m_2 = 0 , \tag{12.97}$$

where r_1 and r_2 are the distances from of origin of, respectively, body 1 and body 2 and are given by

$$r_1 = \frac{m_2 R}{M} , \quad r_2 = \frac{m_1 R}{M} , \tag{12.98}$$

and $R = r_1 + r_2$ is the orbital separation. The orbital frequency Ω follows from Newton's Universal Law of Gravitation

$$\frac{G_N m_1 m_2}{R^2} = m_1\Omega^2 r_1 = m_1\Omega^2\frac{m_2 R}{M} \Rightarrow \Omega = \sqrt{\frac{G_N M}{R^3}} . \tag{12.99}$$

The trajectories of body 1 and of body 2 are, respectively,

$$\mathbf{x}_1 = \begin{pmatrix} r_1\cos\Omega t \\ r_1\sin\Omega t \\ 0 \end{pmatrix} , \quad \mathbf{x}_2 = \begin{pmatrix} -r_2\cos\Omega t \\ -r_2\sin\Omega t \\ 0 \end{pmatrix} . \tag{12.100}$$

Let us now compute the quadrupole moment of the binary system. The 00-component of the energy-momentum tensor of the system is

$$T^{00} = \sum_{i=1}^{2} m_i c^2 \delta\,(x - x_i)\,\delta\,(y - y_i)\,\delta\,(z) . \tag{12.101}$$

The xx-component of the quadrupole moment is

$$Q_{xx} = \sum_{i=1}^{2} \int_V m_i x^2 \delta(x - x_i)\, \delta(y - y_i)\, \delta(z)\, dx\, dy\, dz$$

$$= m_1 x_1^2(t) + m_2 x_2^2(t) = m_1 r_1^2 \cos^2 \Omega t + m_2 r_2^2 \cos^2 \Omega t$$

$$= \frac{\mu R^2}{2} \cos 2\Omega t + \text{constant}, \tag{12.102}$$

where $\mu = m_1 m_2 / M$ is the reduced mass of the binary system, in the last passage we have employed the identity $\cos 2\Omega t = 2\cos^2 \Omega t - 1$, and, as before, we can ignore the constant part because we are interested in the emission of gravitational waves. The other components can be computed in a similar way. Eventually we find the following quadrupole moment

$$\|Q_{ij}\| = \frac{\mu R^2}{2} \begin{pmatrix} \cos 2\Omega t & \sin 2\Omega t & 0 \\ \sin 2\Omega t & -\cos 2\Omega t & 0 \\ 0 & 0 & 0 \end{pmatrix} + \text{constant}. \tag{12.103}$$

Let us assume that the observer is along the z axis far from the source. The unit vector normal to the wavefront is $\mathbf{n} = (0, 0, 1)$. The projector P_{ij} is

$$\|P_{ij}\| = \|\delta_{ij} - n_i n_j\| = \begin{pmatrix} 1 & 0 & 0 \\ 0 & 1 & 0 \\ 0 & 0 & 0 \end{pmatrix}. \tag{12.104}$$

With the projector P_{ij}, we construct the transverse-traceless projector P_{ijkl} and we apply it to the quadrupole moment of the system to obtain the quadrupole moment in the TT gauge. The xx-component of the quadrupole moment in the TT gauge is thus

$$Q_{xx}^{TT} = P_{xxij} Q_{ij} = \left(P_{xi} P_{xj} - \frac{1}{2} P_{xx} P_{ij} \right) Q_{ij}$$

$$= \left(P_{xx}^2 - \frac{1}{2} P_{xx}^2 \right) Q_{xx} - \frac{1}{2} P_{xx} P_{yy} Q_{yy}$$

$$= \frac{1}{2} \left(Q_{xx} - Q_{yy} \right), \tag{12.105}$$

and we can evaluate the other components in a similar way. Eventually, we find

$$\|Q_{ij}^{TT}\| = \begin{pmatrix} \frac{1}{2}\left(Q_{xx} - Q_{yy}\right) & Q_{xy} & 0 \\ Q_{xy} & -\frac{1}{2}\left(Q_{xx} - Q_{yy}\right) & 0 \\ 0 & 0 & 0 \end{pmatrix}. \tag{12.106}$$

The metric perturbation in the TT gauge can be written as

$$||h_{\mu\nu}^{\text{TT}}|| = \begin{pmatrix} 0 & 0 & 0 & 0 \\ 0 & h_+(t) & h_\times(t) & 0 \\ 0 & h_\times(t) & -h_+(t) & 0 \\ 0 & 0 & 0 & 0 \end{pmatrix} , \tag{12.107}$$

where

$$h_+(t) = \frac{G_N}{c^4 z} \frac{d^2}{dt^2} \left(Q_{xx} - Q_{yy} \right) = -\frac{4G_N \mu R^2 \Omega^2}{c^4 z} \cos\left[2\Omega\left(t - z/c\right) \right] ,$$

$$h_\times(t) = \frac{2G_N}{c^4 z} \frac{d^2}{dt^2} Q_{xy} = -\frac{4G_N \mu R^2 \Omega^2}{c^4 z} \sin\left[2\Omega\left(t - z/c\right) \right] . \tag{12.108}$$

As in the case of the rotating ellipsoid in the previous subsection, the frequency of the gravitational wave is twice the frequency of the system. Note that in the case of elliptic orbits, we do not have the emission of a monochromatic gravitational wave. We have instead a discrete spectrum in which the frequencies of the gravitational waves are multiple of the orbital frequency Ω.

12.6 Astrophysical Sources

Generally speaking, potential astrophysical sources of gravitational waves are all the astrophysical systems in which the motion of a large amount of matter alters the background metric. Before listing some detected/expected astrophysical sources in this section and discussing how the generated gravitational waves can be observed in the next section, it is convenient to point out a few important differences between electromagnetic radiation and gravitational waves.

1. Unlike electromagnetic radiation, gravitational waves interact only very weakly with matter, which means that they can travel for very long distances almost unaltered.
2. Electromagnetic radiation is typically generated by moving charged particles in the astrophysical source. The photon wavelength is usually much smaller than the size of the source and is determined by the microphysics. Gravitational waves are instead generated by the motion of the astrophysical source itself, and therefore the emitted radiation has a wavelength comparable to (or larger than) the size of the source.
3. The amplitude of a gravitational wave scales with distance as $1/r$, while electromagnetic signals scale as $1/r^2$. For instance, this implies that, if we double the sensitivity of a gravitational wave detector, we double the distance to which sources can be detected, and we thus increase the number of detectable sources by a factor of 8.

As we have already pointed out, the wavelength (and therefore the frequency) of a gravitational wave depends on the size of the source. For a compact source of mass

M and size L, the characteristic frequency is

$$\nu \sim \frac{1}{2\pi} \sqrt{\frac{G_N M}{L^3}} \, . \tag{12.109}$$

Since the size of the source cannot be smaller than its gravitational radius, i.e. $L > G_N M/c^2$, we find the following upper bound for the frequency of a compact source

$$\nu < \frac{1}{2\pi} \frac{c^3}{G_N M} \sim 10 \left(\frac{M_\odot}{M} \right) \text{ kHz} \, . \tag{12.110}$$

Compact sources can generate high frequency gravitational waves only if they have a small mass. Very heavy systems inevitably produce low frequency gravitational waves.

12.6.1 Coalescing Black Holes

Coalescing black holes are among the leading candidate sources for detection by present and future gravitational wave observatories. Ground-based laser interferometers can detect gravitational waves in the frequency range 10 Hz–10 kHz and consequently can observe the last stage of the coalescence of stellar-mass black holes. Gravitational wave experiments sensitive at lower frequencies can detect signals from the coalescence of two supermassive black holes or from a system of a supermassive black hole and a stellar-mass compact object.

The first direct detection of gravitational waves was in September 2015 [1]. The event was called GW150914 and was the coalescence of two black holes with a mass, respectively, of $36 \pm 5 \, M_\odot$ and $29 \pm 4 \, M_\odot$. The merger produced a black hole with a mass $62 \pm 4 \, M_\odot$, while $3.0 \pm 0.5 \, M_\odot$ was radiated in the form of gravitational waves.

The coalescence of a system of two black holes is characterized by three stages (see Fig. 12.3):

1. *Inspiral.* The two objects rotate around each other. This causes the emission of gravitational waves. As the system loses energy and angular momentum, the separation between the two objects decreases and their relative velocity increases. The frequency and the amplitude of the gravitational waves increase (leading to the so-called "chirping" characteristic in the waveform) until the moment of merger.
2. *Merger.* The two black holes merge into a single black hole.
3. *Ringdown.* The newly born black hole emits gravitational waves to settle down to an equilibrium configuration.

Because of the complexity of the Einstein equations, it is necessary to employ certain approximation methods to compute the gravitational wave signal. In the case

Fig. 12.3 Temporal evolution of the strain, of the black hole separation, and of the black hole relative velocity in the event GW150914 (see Sect. 12.7 for the definition of strain). From [1] under the terms of the Creative Commons Attribution 3.0 License

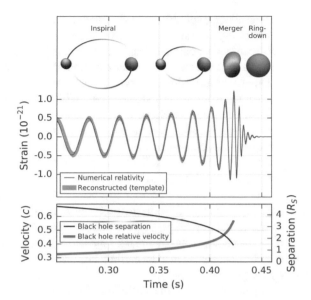

of (roughly) equal mass black holes, the three stages above are treated with the following methods:

1. *Post-Newtonian (PN) methods.* They are based on an expansion in $\varepsilon \sim U/c^2 \sim v^2/c^2$, where $U \sim G_N M/R$ is the Newtonian gravitational potential and v is the black hole relative velocity. The 0PN term is the Newtonian solution of the binary system. The nPN term is the $O(\varepsilon^n)$ correction to the Newtonian solution. In Einstein's gravity, radiation backreaction shows up at 3.5PN order in g_{00}, at 3PN order in g_{0i}, and at 2.5PN order in g_{ij}.
2. *Numerical relativity.* When the PN approach breaks down because ε is not a small parameter any longer, one has to solve numerically the field equations of the complete theory. Since the spacetime is not resolved to infinite precision, even this approach is an approximate method. For non-spinning black holes, the stage of merger smoothly connects the stages of inspiral and of ringdown. For spinning black holes, merger may be a more violent event, depending on the black hole spins and their alignments with respect to the orbital angular momentum.
3. *Black hole perturbation theory.* It is based on the study of small perturbations over a background metric. The method is used to describe the ringdown stage.

12.6.2 Extreme-Mass Ratio Inspirals

An extreme-mass ratio inspiral is a system of a stellar-mass compact object (black hole, neutron star, or white dwarf with a mass $\mu \sim 1$–$10\,M_\odot$) orbiting a supermassive black hole ($M \sim 10^6$–$10^{10}\,M_\odot$). Since the system emits gravitational waves, the stellar-mass compact object slowly inspirals into the supermassive black hole until the final plunge. A similar system can easily form by multi-body interactions in galactic centers. Initially, the captured object is in a "generic" orbit, namely the orbit will have a high eccentricity and any inclination angle with respect to the black hole spin. Due to the emission of gravitational waves, the eccentricity tends to decreases (if we are far from the last stable orbit), while the inclination of the orbit remains approximately constant [4, 7].

Extreme-mass ratio inspirals are an important class of detection candidates for future space-based gravitational wave interferometers. Detection will be dominated by sources in which the stellar-mass compact object is a black hole rather than a neutron star or a white dwarf. There are two reasons. First, the heaviest bodies tend to concentrate at the center. Second, the signal produced by a 10 $M_\odot$ black hole is stronger than the signal from a neutron star or a white dwarf with a mass $\mu \approx 1 \ M_\odot$.

In the frequency range 1–100 mHz, it is possible to detect the last few years of inspiral into a supermassive black hole of $\sim 10^6 \ M_\odot$. If the supermassive object is heavier or the inspiral is at an earlier stage, the emission of gravitational waves is at lower frequencies.

Since $\mu/M < 10^{-5}$, the evolution of the system is adiabatic; that is, the orbital parameters evolve on a timescale much longer than the orbital period of the stellar-mass compact object. A rough estimate can be obtained as follows [5]. If we are far from the last stable orbit, the orbital period is

$$T \sim 8 \left(1 - e^2\right)^{-3/2} \left(\frac{M}{10^6 \ M_\odot}\right) \left(\frac{p}{6M}\right)^{3/2} \text{min}, \tag{12.111}$$

where p is the semi-latus rectum of the orbit and e is the eccentricity. The timescale of the radiation back-reaction can be estimated as $T_R \sim -p/\dot{p}$ and we find

$$T_R \sim 100 \left(1 - e^2\right)^{-3/2} \left(\frac{M}{\mu}\right) \left(\frac{M}{10^6 \ M_\odot}\right) \left(\frac{p}{6M}\right)^4 \text{min}. \tag{12.112}$$

As long as $M/\mu \gg 1$, we have $T_R \gg T$ and the evolution of the system is adiabatic. This simplifies the description, as we can neglect the radiation back-reaction and assume that the small object follows the geodesics of the spacetime. Extreme-mass ratio inspirals are thus relatively simple systems and offer a unique opportunity to map the metric around supermassive black holes [3, 6, 12]. Since the value of μ/M is so low, the inspiral process is slow, and it is possible to observe the signal for many ($> 10^5$) cycles. In such a case, the signal-to-noise ratio can be high and it is possible to accurately measure the parameters of the system. In particular, extreme-mass ratio inspiral detections promise to provide unprecedented accurate measurements of the mass and the spin of supermassive black holes [2].

12.6.3 Neutron Stars

Even compact binary systems with neutron stars (i.e. neutron star-neutron star or neutron star-black hole) are promising candidate sources for ground-based laser interferometers. Since the mass of a neutron star cannot be more than 2–3 $M_\odot$, such binaries generate gravitational waves with an amplitude smaller than those emitted by binaries in which both bodies are black holes. This, in turn, makes their detection more difficult and explains why the first detections of gravitational waves are associated with binary systems in which both objects are black holes.

The coalescence of a system in which at least one of the bodies is a neutron star is still characterized by the three stages discussed in Sect. 12.6.1; that is, inspiral, merger, and ringdown. However, in the presence of a neutron star there are some additional complications because we do not have a purely gravitational system, and also the equation of state of the matter the neutron star is made of can have an impact on the gravitational wave signal.

Isolated neutron stars can also be sources of gravitational waves, for instance after the formation of the neutron star (as in the case of black holes, we could observe the stage of ringdown, and the body emits gravitational waves to settle down to an equilibrium configuration), because of the neutron star rotation and in the presence of deviations from a perfect axisymmetry of the body (see the example discussed in Sect. 12.5.1), or for various stellar oscillation modes and core superfluid turbulences. In principle, the detection of gravitational waves from these systems can be also exploited to study the matter equation of state at super-nuclear densities, which is not possible to do in laboratories on Earth.

12.7 Gravitational Wave Detectors

As we have seen in Sect. 12.2.2 and as illustrated in Fig. 12.2, the passage of a gravitational wave has the effect of altering the proper distances among the particles of a certain system. Gravitational wave detectors can reveal the passage of a gravitational wave by monitoring, in different ways, the proper distance of test-bodies. If the proper distance of two test-bodies is L in flat spacetime and the passage of a gravitational wave causes a variation ΔL, we call the *strain* $h = \Delta L / L$. The strain is the quantity measured by the detector, is related to the amplitude of the gravitational wave, and depends on the orientation of the detector with respect to the propagation direction and on the polarization of the gravitational wave.

Direct detection of gravitational waves is very challenging. This is because the expected amplitude of gravitational waves passing through Earth is extremely small, with h of order 10^{-20}. If r is the distance of the source from the detection point, $h \propto 1/r$. To have a simple idea of the technological difficulties to detect gravitational waves, we can consider that the Earth's radius $R \approx 6,000$ km would change by

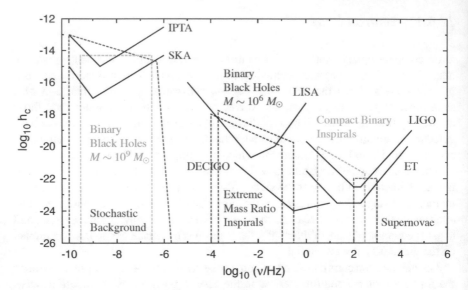

Fig. 12.4 Sketch of the sensitivity curves of a selection of present and proposed gravitational wave detectors in terms of the characteristic strain h_c and of the gravitational wave frequency v. Resonance spikes in the detector noise curves have been removed for clarity. The figure also shows the expected characteristic strain and frequency of a number of possible astrophysical and cosmological sources

$\Delta R = 60$ fm (1 fm $= 10^{-15}$ m) because of the passage of a gravitational wave with $h = 10^{-20}$. Such a value of ΔR is much smaller than the radius of an atom ($\sim 10^5$ fm).

Table 12.1 lists a selection of recent, present, and future/proposed gravitational wave detectors. There are three main types of detectors, which will be briefly reviewed in the next sections: resonant detectors, interferometers (either ground-based or space-based), and pulsar timing arrays.

Figure 12.4 illustrates the sensitivity curves of a selection of present and proposed gravitational wave detectors together with the expected strength of some gravitational wave sources. In the x-axis, v is the gravitational wave frequency. The y-axis is for the *characteristic strain* h_c, which is defined as [11]

$$|h_c(v)|^2 = 4v^2|\tilde{h}(v)|^2, \qquad (12.113)$$

where $\tilde{h}(v)$ is the Fourier transform of the strain $h(t)$. h_c is not directly related to the amplitude of the gravitational wave, as it includes the effect of integrating an inspiralling signal.

interferometer. Notes: [1] While there may not be an official end of the experiment, resonant detectors are not operative any longer. [2] Expected/proposed

Project	Activity period	Location	Detector type	Frequency range
EXPLORER http://www.roma1.infn.it/rog/explorer/	1990–2002	Geneva (Switzerland)	RD	~900 Hz
ALLEGRO	1991–2008	Baton Rouge (Louisiana)	RD	~900 Hz
NAUTILUS http://www.roma1.infn.it/rog/nautilus/	1995–2002	Frascati (Italy)	RD	~900 Hz
TAMA 300 http://tamago.mtk.nao.ac.jp/spacetime/tama300_e.html	1995–	Mitaka (Japan)	GBLI	10 Hz–10 kHz
GEO 600 http://www.geo600.org	2001–	Sarstedt (Germany)	GBLI	50 Hz–1.5 kHz
AURIGA http://www.auriga.lnl.infn.it	1997–x[1]	Padua (Italy)	RD	~900 Hz
Gravitational Radiation Antenna in Leiden (MiniGRAIL) http://www.minigrail.nl	2001–x[1]	Leiden (Netherland)	RD	2–4 kHz
Laser Interferometer Gravitational Wave Observatory (LIGO) http://www.ligo.org	2004–	Hanford (Washington) Livingston (Louisiana)	GBLI	30 Hz–7 kHz
Mario Schenberg	2006–x[1]	Sao Paulo (Brazil)	RD	3.0–3.4 kHz
Virgo Interferometer http://www.virgo-gw.eu	2007–	Cascina (Italy)	GBLI	10 Hz–10 kHz
Kamioka Gravitational Wave Detector (KAGRA) http://gwcenter.icrr.u-tokyo.ac.jp/en/	From 2018[2]	Kamioka mine (Japan)	GBLI	10 Hz–10 kHz

(continued)

Table 12.1 (continued)

Project	Activity period	Location	Detector type	Frequency range
Indian Initiative in Gravitational Wave Observations (IndiGO) http://gw-indigo.org/tiki-index.php	From 2023[2]	India	GBLI	30 Hz–7 kHz
Deci-hertz Interferometer Gravitational Wave Observatory (DECIGO) http://tamago.mtk.nao.ac.jp/decigo/index_E.html	From 2027[2]	Space	SBLI	1 mHz–10 Hz
Einstein Telescope (ET) http://www.et-gw.eu	From ~2030[2]	To be decided	GBLI	1 Hz–10 kHz
Laser Interferometer Space Antenna (LISA) https://www.lisamission.org	From 2034[2]	Space	SBLI	0.1 mHz–1 Hz
TianQin	From ~2040[2]	Space	SBLI	0.1–100 mHz

12.7.1 Resonant Detectors

In the case of resonant detectors, we have a large resonant body (bar) which is stretched and squeezed by the passage of a gravitational wave. The sensitivity of the detector is peaked at its mechanical resonance, which corresponds to the first longitudinal mode of the bar and in most detectors is around 1 kHz. Experiments with resonant detectors quit in the early 2010s because they were not competitive any longer with respect to interferometer detectors.

The first gravitational wave detector was the resonant detector constructed by Joseph Weber in the 1960s. While the claim of a detection of gravitational waves was reported in [14], the sensitivity of that detector was not good enough to detect gravitational waves from astrophysical sources and there is a common consensus on the fact that the claim in [14] was not a real detection.

EXPLORER, ALLEGRO, NAUTILUS, and AURIGA were all cylindrical bar detectors and worked in a similar way. The detector was a heavy bar cooled down to very low temperatures to reduce the thermal noise, namely the motion of the atoms of the bar. The bar was inside vacuum chambers to reduce the acoustic noise of the laboratory. The vibration of the bar was read out by a smaller mass (resonant transducer) of about 1 kg. The transducer had the same resonant frequency as the bar, so it could resonantly pick up the bar vibrations. Since it was much lighter, the amplitude of its vibrations could be much larger.

MiniGRAIL and Mario Schenberg were instead spherical detectors. Spherical antennas are technologically more challenging, but they present some advantages in the possibility of detecting gravitational waves. In particular, a spherical detector can detect gravitational waves arriving from any direction.

12.7.2 Interferometers

Gravitational wave laser interferometers are based on a Michelson interferometer. The set-up of the advanced LIGO detectors is sketched in Fig. 12.5. There are two arms, which are orthogonal to each other. A beamsplitter splits the original laser beam into two beams, which are reflected by the two mirrors at the end of the two arms and eventually recombine and produce an interference pattern. In general, the passage of a gravitational wave would change the travel time in the two arms in a different way: depending on the propagation direction of the gravitational wave with respect to the orientation of the interferometer, one of the arms can be stretched, while the other can be squeezed. The photodetector at the location of the interference pattern can measure a change in the proper length of the arms. To increase the effective path length of the laser light in the arms, there are partially reflecting mirrors, which make the laser light run along the arms many (typically hundreds) times.

TAMA 300 was the first laser interferometer to work, but its sensitivity is limited by its small size. The length of its arms is 300 m. GEO 600 is a laser interferometer

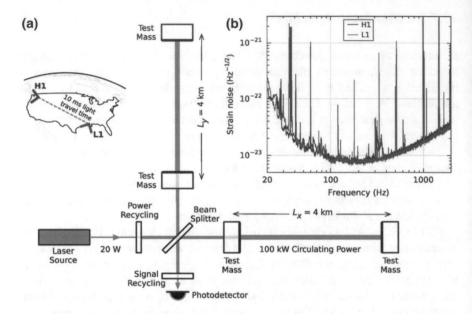

Fig. 12.5 Sketch of the advanced LIGO detectors, location and orientation of the LIGO detectors at Hanford (H1) and Livingston (L1) (**a**), and sensitivity curves in terms of equivalent gravitational wave strain amplitude (**b**). From [1] under the terms of the Creative Commons Attribution 3.0 License

with arms of 600 m. In the case of LIGO and Virgo, the length of the arms is, respectively, 4 and 3 km. Figure 12.6 shows the aerial view of the Virgo detector, near Pisa, in Italy. KAGRA and the Einstein Telescope are underground detectors, to reduce the seismic noise. The Einstein Telescope will have an equilateral triangle geometry, with three arms of 10 km and two detectors at each corner.

Space-based laser interferometers work with a constellation of satellites (e.g. three satellites in LISA and four clusters of three satellites in the case of DECIGO). The working principle is the same as the ground-based laser interferometers, and one wants to monitor the proper distance among mirrors located in different satellites. The distance between satellites is larger than the length of the arms of ground-based laser interferometers, so these experiments are sensitive to gravitational waves at lower frequencies (see Fig. 12.4). This is also possible because the limitation at low frequencies in ground-based experiments is due to seismic noise, but there is no seismic noise in space. In the case of DECIGO, the distance among the satellites should be $\sim 10^3$ km, while the arms of LISA should be $\sim 10^6$ km.

The sensitivity achievable by a laser interferometer can be understood as follows. We can consider the case of LIGO, where the laser wavelength is $\lambda \sim 1\,\mu$m and the interferometer arms have $L = 4$ km. If we could measure ΔL only with a precision of order the size of a fringe, i.e. $\Delta L \sim \lambda$, the minimum detectable strain would be $h \sim \lambda/L \sim 3 \cdot 10^{-10}$. Gravitational waves with a strain of order 10^{-20} can be

Fig. 12.6 Aerial view of the interferometer detector Virgo (Cascina, Pisa, Italy). From the Virgo collaboration under the Creative Commons CC0 1.0 Universal Public Domain Dedication

detected if we can measure changes in the arm length much smaller than λ. The photodetector of the experiment can indeed monitor changes in the photon flux and reach a sensitivity $\Delta L \sim \lambda / \sqrt{N}$, where N is the number of photons arriving at the detector and $\sqrt{N}$ is its fluctuation, being a Poisson process. If P is the laser power, $N \sim P/(\nu E_\gamma)$, where ν is the frequency of the gravitational wave (we can collect photons for a time $t \sim 1/\nu$) and E_γ is the photon energy. If we consider $P \sim 1$ W, $\nu = 100$ Hz, and $\lambda \sim 1$ μm, we find $N \sim 10^{16}$ and $h \sim 10^{-18}$. Moreover, the two arms of the interferometer are two Fabry–Perot optical cavities, and they can store the light for many round trips. For $\nu = 100$ Hz and $L = 4$ km, the light can make about a thousand round trips during the passage of a gravitational wave, which increases the effective arm length by a factor $\sim 10^3$ and the interferometer sensitivity becomes $h \sim 10^{-21}$.

12.7.3 Pulsar Timing Arrays

In a pulsar timing array experiment, the test-bodies are represented by 20–50 ms pulsars. Like in the interferometers of the previous section, the passage of gravitational waves contracts the space in one direction and expands the space in the other direction, thus changing the arrival time of the pulsar signals on Earth. Since millisecond pulsars can be used as very precise clocks, it is possible to infer variations in the time arrival of the signal of order of some ns.

The distances between Earth and these pulsars in the Galaxy are of the order of 1–10 kpc, which are definitively larger than the distances that we can monitor in interferometric detectors. Pulsar timing array experiments can thus detect gravitational waves of very low frequency, in the range 1–100 nHz. There are two main possible sources for such low frequency gravitational waves. (i) Binary systems of two supermassive black holes with an orbital period ranging from a few months to a few years. Even if they are far from us, the power emitted in gravitational waves is huge, and the signal may be strong enough to be detected. (ii) Gravitational waves produced in the early Universe. There are a number of different scenarios predicting a background of low frequency gravitational waves, like decay of cosmic strings, inflationary models, and first order phase transitions. In all these cases, the frequency of the gravitational waves would be very low because of the cosmological redshift.

At the moment, there are a few operative experiments: the European Pulsar Timing Array (EPTA), the Parkes Pulsar Timing Array (PPTA), the North American Nanohertz Observatory for Gravitational Waves (NANOGrav), and the International Pulsar Timing Array (IPTA). The Square Kilometre Array (SKA) is expected to start in 2020. With the available pulsar data, it is only possible to get some upper bounds on the amplitude of low frequency gravitational waves. These bounds can be improved with time, because the precision is determined by the observational time of the pulsars. We may also discover new millisecond pulsars suitable for these measurements. This would increase the number of sources monitored, which is also helpful to improve the sensitivity.

More details on pulsar timing arrays can be found in [9, 10] and references therein.

Problem

12.1 Derive the estimate of the maximum gravitational wave frequency for black holes with mass $M = 10^6 \, M_\odot$ and $10^9 \, M_\odot$ from Eq. (12.110) and compare the result with Fig. 12.4.

References

1. B.P. Abbott et al., LIGO scientific and virgo collaborations. Phys. Rev. Lett. **116**, 061102 (2016), arXiv:1602.03837 [gr-qc]
2. L. Barack, C. Cutler, Phys. Rev. D **69**, 082005 (2004). [gr-qc/0310125]
3. L. Barack, C. Cutler, Phys. Rev. D **75**, 042003 (2007). [gr-qc/0612029]
4. J.R. Gair, K. Glampedakis, Phys. Rev. D **73**, 064037 (2006). [gr-qc/0510129]
5. K. Glampedakis, Class. Quant. Grav. **22**, S605 (2005). [gr-qc/0509024]
6. K. Glampedakis, S. Babak, Class. Quant. Grav. **23**, 4167 (2006). [gr-qc/0510057]

7. K. Glampedakis, S.A. Hughes, D. Kennefick, Phys. Rev. D **66**, 064005 (2002). [gr-qc/0205033]
8. E. Gourgoulhon, S. Bonazzola, Gravitational waves from isolated neutron stars, in *Gravitational Waves: Sources and Detectors*, pp. 51–60 (World Scientific, Singapore, 1997). [astro-ph/9605150]
9. G. Hobbs et al., Class. Quant. Grav. **27**, 084013 (2010), arXiv:0911.5206 [astro-ph.SR]
10. A.N. Lommen, Rept. Prog. Phys. **78**, 124901 (2015)
11. C.J. Moore, R.H. Cole, C.P.L. Berry, Class. Quant. Grav. **32**, 015014 (2015), arXiv:1408.0740 [gr-qc]
12. F.D. Ryan, Phys. Rev. D **52**, 5707 (1995)
13. J.M. Weisberg, Y. Huang, Astrophys. J. **829**, 55 (2016), arXiv:1606.02744 [astro-ph.HE]
14. J. Weber, Phys. Rev. Lett. **22**, 1320 (1969)

Chapter 13
Beyond Einstein's Gravity

To conclude our introductory course on general relativity, in this final chapter we will briefly present the main theoretical problems that plague Einstein's gravity as well as some attempts to solve them. The discussion will be necessarily at a very qualitative level, without details and far from being complete. A more accurate and complete study of these topics would be beyond the purposes of the present textbook.

13.1 Spacetime Singularities

Generally speaking, a spacetime singularity is a "region" of the spacetime with some pathological properties. For example, there are several physically relevant solutions of Einstein's equations in which the spacetime is *geodesically incomplete*; that is, there are geodesics that cannot be extended beyond a certain point. At the end of the 1960s, Roger Penrose and Stephen Hawking discussed in a number of theorems the conditions that make the formation of this kind of spacetime singularities unavoidable [8, 9]. The singularities at $r = 0$ in the Schwarzschild and in the Reissner–Nordström solutions are singularities in the sense that the spacetime is geodetically incomplete there. In the case of the Kerr metric, $r = 0$ is geodetically incomplete only for the geodesics in the equatorial plane, while the spacetime can be extended beyond $r = 0$ for off-equatorial trajectories (see Sect. 10.6). The cosmological models discussed in Sects. 11.3.2 and 11.3.3 are singular at $t = 0$ as geodesics cannot be extended into the past.

Note that the fact that a spacetime is geodetically incomplete has profound physical implications. If we cannot extend a geodesic beyond a certain point, predictability is lost. We simply do not know what is going on at the singularity, where standard calculation methods clearly break down. With the terminology of Appendix C, at the

© Springer Nature Singapore Pte Ltd. 2018   257
C. Bambi, *Introduction to General Relativity*, Undergraduate Lecture Notes
in Physics, https://doi.org/10.1007/978-981-13-1090-4_13

singularity we do not have a differentiable manifold, and therefore we cannot do any calculation. This is not a minor issue. For instance, in the case of the Schwarzschild spacetime, every particle crossing the event horizon reaches the central singularity in a finite time. So we do not know what happens to the matter swallowed by the black hole!

Some spacetime singularities are curvature singularities, i.e. some curvature invariants, like the scalar curvature R and/or the Kretschmann scalar $\mathscr{K}$, diverge there. Physical quantities that diverge are often a symptom of the breakdown of a theory.

In Einstein's gravity, there is only one dimensional coupling constant, Newton's constant of gravitation G_N. If we combine G_N with the speed of light c and Dirac's constant $\hbar$, we obtain the *Planck length* L_{Pl}, the *Planck time* T_{Pl}, the *Planck mass* M_{Pl}, and the *Planck energy* E_{Pl}

$$L_{Pl} = \sqrt{\frac{G_N \hbar}{c^3}} = 1.616 \cdot 10^{-33} \text{ cm},$$

$$T_{Pl} = \sqrt{\frac{G_N \hbar}{c^5}} = 5.391 \cdot 10^{-44} \text{ s},$$

$$M_{Pl} = \sqrt{\frac{\hbar c}{G_N}} = 2.176 \cdot 10^{-5} \text{ g},$$

$$E_{Pl} = \sqrt{\frac{\hbar c^5}{G_N}} = 1.221 \cdot 10^{19} \text{ GeV}. \tag{13.1}$$

In units in which $c = \hbar = 1$, $L_{Pl} = T_{Pl} = 1/M_{Pl} = 1/E_{Pl}$, and therefore we can use just one of them. We can generically talk about *Planck scale* and often we use the Planck mass M_{Pl} in energy units: $M_{Pl} = 10^{19}$ GeV.

In Einstein's gravity, the Planck scale looks like the natural UV cut-off and therefore we should expect new physics beyond it (see the next section for more details). New physics may thus show up when curvature invariants approach the Planck scale, e.g. $R \to M_{Pl}^2$ and $\mathscr{K} \to M_{Pl}^4$, before they diverge to infinity.

It is often claimed that the problem of spacetime singularities should be fixed by a yet unknown theory of quantum gravity. As we will briefly discuss in the next section, Einstein's gravity can be quantized, but the result is an effective theory valid at scales much smaller than the Planck one, so it is not the theoretical framework to address the problem of spacetime singularities. In extensions of Einstein's gravity, at least some spacetime singularities can be avoided, but this is not an easy job in general. Moreover, there are many attempts to extend Einstein's gravity. Every model has its own predictions, which are difficult or impossible to test, and therefore it is extremely challenging to make progress in the field.

13.2 Quantization of Einstein's Gravity

Contrary to what is sometimes claimed, Einstein's gravity can be quantized, and the result is a self-consistent theory [5, 6, 10] (see Ref. [4] for a pedagogical introduction). However, it is an *effective field theory*[1] valid for energies $E \ll M_{\text{Pl}}$. At low energies, quantum corrections are extremely small, and experimental tests seem to be very unlikely, even in the future. If we consider processes approaching the Planck scale, the theory breaks down, and therefore it cannot address the most interesting questions concerning spacetime singularities at the center of black holes and in cosmology.

The procedure to quantize Einstein's gravity can be summarized as follows. We write the spacetime metric $g_{\mu\nu}$ as a background field $\bar{g}_{\mu\nu}$ plus a perturbation $h_{\mu\nu}$

$$g_{\mu\nu} = \bar{g}_{\mu\nu} + \tilde{\kappa} h_{\mu\nu} . \tag{13.2}$$

In general, $\bar{g}_{\mu\nu}$ is a solution of Einstein's equations, not necessarily the Minkowski metric $\eta_{\mu\nu}$. $h_{\mu\nu}$ is the field to quantize. $\tilde{\kappa}^2 = 32\pi/M_{\text{Pl}}^2$ is introduced to provide the right dimensions to $h_{\mu\nu}$. The inverse metric reads

$$g^{\mu\nu} = \bar{g}^{\mu\nu} - \tilde{\kappa} h^{\mu\nu} + \tilde{\kappa}^2 h_\rho^\mu h^{\rho\nu} + \cdots . \tag{13.3}$$

We expand the action around the background metric $\bar{g}_{\mu\nu}$

$$S = \int d^4x \sqrt{-\bar{g}} \left(\mathscr{L}^{(0)} + \mathscr{L}^{(1)} + \mathscr{L}^{(2)} + \cdots \right) , \tag{13.4}$$

where

$$\mathscr{L}^{(0)} = \frac{2}{\tilde{\kappa}^2} \bar{R} ,$$

$$\mathscr{L}^{(1)} = \frac{1}{\tilde{\kappa}} h_{\mu\nu} \left(\bar{g}^{\mu\nu} \bar{R} - 2\bar{R}^{\mu\nu} \right) ,$$

$$\mathscr{L}^{(2)} = \frac{1}{2} \left(\bar{\nabla}_\rho h_{\mu\nu} \right) \left(\bar{\nabla}^\rho h^{\mu\nu} \right) - \frac{1}{2} \left(\bar{\nabla}_\mu h \right) \left(\bar{\nabla}^\mu h \right)$$
$$+ \left(\bar{\nabla}_\mu h \right) \left(\bar{\nabla}_\nu h^{\mu\nu} \right) - \left(\bar{\nabla}_\rho h_{\mu\nu} \right) \left(\bar{\nabla}^\nu h^{\mu\rho} \right)$$
$$+ \frac{1}{2} \bar{R} \left(\frac{1}{2} h^2 - h_{\mu\nu} h^{\mu\nu} \right) + \left(2h_\mu^\rho h_{\nu\rho} - h h_{\mu\nu} \right) \bar{R}^{\mu\nu} . \tag{13.5}$$

---

[1] An effective field theory is a field theory that provides reliable predictions in its realm of validity but it breaks down beyond it. The crucial point of the known effective field theories is the possibility of separating the physics at low energies from that at much higher energies. This permits us to make predictions at low energies without making unwarranted assumptions about what is going on at high energies.

The quantities with a bar are evaluated with the background metric $\bar{g}_{\mu\nu}$. h is the trace of $h_{\mu\nu}$, i.e. $h = h^{\mu}_{\mu}$.

At this point, there are some technical problems to solve. It is necessary to fix the gauge, but this causes the appearance of some non-physical degrees of freedom. The latter can be removed with some tricks. After that, we can infer the Feynman rules of the theory in the same way as when we quantize other interactions. The result is a non-renormalizable theory: all amplitudes are divergent at a sufficiently high order in the perturbation, and therefore it is not possible to absorb the divergent terms into a finite number of observables. The theory is predictable at low energies ($E \ll M_{\text{Pl}}$), when low order terms are dominant and higher order terms can be neglected, and breaks down when we approach the Planck scale. We have a situation similar to the Fermi theory of the weak interactions, which is a viable effective theory for energies $E \ll M_W$, where $M_W = 80$ GeV is the mass of the W-boson, while it breaks down when $E \to M_W$.

The quantum theory obtained from this procedure is predictable at low energies. As an example, we can consider the Schwarzschild spacetime. The classical Schwarzschild metric in the harmonic gauge reads (see e.g. Ref. [11])

$$g_{00} = -\frac{1 - \frac{G_N M}{c^2 r}}{1 + \frac{G_N M}{c^2 r}} = -\left(1 - \frac{2G_N M}{c^2 r} + \frac{2G_N^2 M^2}{c^4 r^2} + \cdots\right)$$

$$g_{0i} = 0$$

$$g_{ij} = \left(1 + \frac{G_N M}{c^2 r}\right)^2 \delta_{ij} + \frac{G_N^2 M^2}{c^4 r^2} \left(\frac{1 + \frac{G_N M}{c^2 r}}{1 - \frac{G_N M}{c^2 r}}\right) \frac{x_i x_j}{r^2}$$

$$= \left(1 + \frac{2G_N M}{c^2 r} + \frac{G_N^2 M^2}{c^4 r^2}\right) \delta_{ij} + \frac{G_N^2 M^2}{c^4 r^2} \frac{x_i x_j}{r^2} + \cdots. \qquad (13.6)$$

In Ref. [3], the authors employed a particular set of Feynman diagrams and found long distance quantum corrections to the solution in (13.6). The quantum-corrected metric reads

$$g_{00} = -\left(1 - \frac{2G_N M}{c^2 r} + \frac{2G_N^2 M^2}{c^4 r^2} + \frac{62G_N^2 M \hbar}{15\pi c^5 r^3} + \cdots\right)$$

$$g_{0i} = 0$$

$$g_{ij} = \left(1 + \frac{2G_N M}{c^2 r} + \frac{G_N^2 M^2}{c^4 r^2} + \frac{14G_N^2 M \hbar}{15\pi c^5 r^3}\right) \delta_{ij}$$

$$+ \left(\frac{G_N^2 M^2}{c^4 r^2} + \frac{76G_N^2 M \hbar}{15\pi c^5 r^3}\right) \frac{x_i x_j}{r^2} + \cdots. \qquad (13.7)$$

If we compare the metric in Eq. (13.6) with that in Eq. (13.7), we see that the leading order quantum corrections are proportional to $\hbar$, as it was to be expected. However, the contribution of these corrections is very small and therefore any observational test turns out to be extremely challenging.

The "UV completion", namely how to extend the theory in order to have a good model even at high energies, is a completely open problem. There are many attempts in the literature, but none is satisfactory at the moment. The lack of experimental tests because of the huge value of the Planck energy with respect to particle physics energies is an additional limitation of this line of research.

13.3 Black Hole Thermodynamics and Information Paradox

As we have pointed out in Sect. 10.4, the black holes of Einstein's gravity are very simple objects, in the sense that they are characterized by a small number of parameters. In the case of a Kerr–Newman black hole, there are three parameters: the mass M, the spin angular momentum J, and the electric charge Q. If the black hole swallows a number of particles, the black hole only changes the values of M, J, and Q, and the particles disappear. At first, one may naively argue that such a process could permit the reduction of the entropy of the whole system, violating the Second Law of Thermodynamics. Entropy is a measurement of the number of microscopic configurations that a thermodynamical system can have for certain macroscopic variables. When the system is made of a black hole and many particles, we have a large number of possible configurations, because every particle has its own position and velocity. After the black hole has swallowed all particles, we have just a black hole characterized by a few parameters and nothing more.

Such a conclusion is not correct. It turns out that the entropy of a black hole is proportional to the area of its event horizon [1]

$$S_{\mathrm{BH}} = \frac{k_{\mathrm{B}} A_{\mathrm{H}}}{4 L_{\mathrm{Pl}}^2},$$ (13.8)

where k_{B} is Boltzmann's constant and A_{H} is the area of the event horizon of the black hole. Equation (13.8) is usually referred to as the *Bekenstein-Hawking formula*. However, the exact origin of the black hole entropy is not yet known.

This is not the end of the story. A black hole is not completely "black", in the sense that it can only swallow matter. It has instead a finite temperature and, therefore, it emits radiation (*Hawking radiation*) [7]. In the case of a Schwarzschild black hole, the temperature is

$$T_{\mathrm{BH}} = \frac{\hbar c^3}{8\pi G_{\mathrm{N}} M k_{\mathrm{B}}}$$ (13.9)

The black hole temperature is proportional to $\hbar$, which means that classically ($\hbar \rightarrow 0$) the temperature vanishes. Since $\hbar \neq 0$, the temperature is finite, but in the case of astrophysical black holes it is extremely low and can be ignored. For a Solar mass Schwarzschild black hole, we have

Fig. 13.1 Penrose diagram
for the formation and the
evaporation of a black hole.
The letter S indicates the
interior region of the
collapsing star. The black arc
extending from i^- to the
singularity (the horizontal
line with wiggles) is the
surface of the star. See the
text for the details

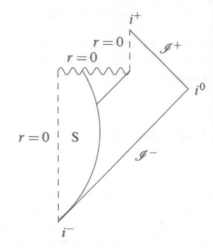

$$T_{BH} = 6 \cdot 10^{-8} \left(\frac{M_\odot}{M} \right) \, \text{K} . \tag{13.10}$$

A black hole emits radiation (almost) as a black body and its luminosity is given by

$$L_{BH} \sim \sigma_{BH} A_H T_{BH}^4 \sim 10^{-21} \left(\frac{M_\odot}{M} \right)^2 \, \text{erg/s} . \tag{13.11}$$

where σ_{BH} is a Stefan–Bolzmann-like constant whose numerical value depends on
the particles that can be emitted, so it depends on the black hole mass and the particle
content of the theory. It is clear that a luminosity like that predicted by Eq. (13.11)
will not likely be detected, even in the future.

Since a black hole emits radiation, it loses mass and eventually can "evaporate". It
turns out that the Penrose diagram of the gravitational collapse of a massive body that
creates a black hole, illustrated in Fig. 10.8, changes when the evaporation process is
taken into account. The (possible) new Penrose diagram is shown in Fig. 13.1. As an
artifact of the coordinates of the Penrose diagram, it seems like the evaporation is an
instantaneous process, but this is not true. The evaporation process is actually very
slow and can be estimated from the luminosity in (13.11), since $L_{BH} = dM/dt$. At
the end we have a Minkowski spacetime.

The evaporation of a black hole presents the following open problem. Let us
assume that the initial collapsing body is a well-defined quantum state (a "pure"
state described by a single ket vector). In quantum mechanics, the evolution operator
is unitary and a pure state evolves into a pure state. However, in the formation and
evaporation of a black hole, it seems that a pure state could be transformed into
thermal radiation, which is a mixed state. If so, the laws of quantum mechanics
are not consistent with the black hole evaporation process. This is the *black hole
information paradox*. There are several proposals to solve the problem, but we do

not know which one, if any, is correct. For example, it is possible that there are small but non-negligible deviations from the standard semiclassical predictions at the event horizon and that there is no information loss once these corrections are properly taken into account.

13.4 Cosmological Constant Problem

When we construct the action of a theory, we have to add to the Lagrangian any term that is not forbidden by the symmetries of the theory. Even if we did not do so, such terms would show up after quantum corrections are taken into account, as nothing can prevent their presence. This is well established in particle physics, where it is possible to test quantum field theory.

From this point of view, when we write the action of the gravitational sector, we should have

$$S = \int d^4x \sqrt{-g} \left(-\frac{\Lambda}{\kappa c} + \frac{R}{2\kappa c} + a_1 R^2 + a_2 R_{\mu\nu} R^{\mu\nu} + \cdots \right) \quad (13.12)$$

instead of the simple Einstein-Hilbert action of Eq. (7.24). The terms proportional to R^2 and $R_{\mu\nu} R^{\mu\nu}$, as well as those of higher order that are omitted in Eq. (13.12), become important only when we approach the Planck scale, which is not the case in astrophysical or cosmological environments that can be observed (at least for the moment). So they can be safely neglected for most purposes. On the contrary, the cosmological constant Λ cannot be ignored and a number of theoretical arguments would suggest a value much higher than the one compatible with that in the Universe. This is the *cosmological constant problem*.

If we write the action of the gravitational sector with a cosmological constant term and we do not want to introduce a new scale, we should expect that the effective energy density associated to the cosmological constant is

$$\rho_\Lambda = \frac{\Lambda}{\kappa} \sim M_{\text{Pl}}^4 \sim 10^{76} \text{ GeV}^4, \quad (13.13)$$

because M_{Pl} is the only scale in the gravitational sector. However, if Λ introduces a new scale in the system, we cannot make any prediction about its value, that should thus be obtained from observations.

The cosmological constant problem arises because the effective cosmological constant, namely the one appearing in the Friedmann equations and contributing to the expansion of the Universe, should be the sum of a number of different contributions. While we cannot make exact predictions of the final value resulting from this sum, we can estimate the order of magnitude of the single contributions. An almost perfect cancellation from so many different terms of different magnitude would be very unnatural.

For example, if we have the scalar field ϕ with the action (in this discussion, we use natural units in which $\hbar = c = 1$ for convenience)

$$S = -\int \left[\frac{1}{2} g^{\mu\nu} \left(\partial_\mu \phi\right) \left(\partial_\nu \phi\right) + V(\phi)\right] \sqrt{-g}\, d^4 x\,, \tag{13.14}$$

its energy-momentum tensor is

$$T^{\mu\nu} = \left(\partial^\mu \phi\right)\left(\partial^\nu \phi\right) - \frac{1}{2} g^{\mu\nu} g^{\rho\sigma} \left(\partial_\rho \phi\right)\left(\partial_\sigma \phi\right) - V(\phi) g^{\mu\nu}\,. \tag{13.15}$$

In the state of minimum energy, the kinetic energy of the scalar field vanishes, and ϕ sits at the minimum of the potential. Equation (13.15) becomes

$$T^{\mu\nu} = -V(\phi_{\min})\, g^{\mu\nu}\,. \tag{13.16}$$

This generates the vacuum energy $\rho_\Lambda = V(\phi_{\min})$.

From particle physics, we would expect a number of terms similar to this example. In the Standard Model of Particle Physics, we have the Higgs boson. The absolute value of its potential cannot be measured in particle colliders, because with the exception of gravity only energy differences matter. However, the Higgs sector is at the electroweak scale $M_{\mathrm{EW}} \sim 100$ GeV, and therefore it is natural to expect $V(\phi_{\min}) \sim (100\text{ GeV})^4$. If we believe in the Grand Unification Theories, we have similar situations but involving the GUT scale $M_{\mathrm{GUT}} \sim 10^{16}$ GeV, and therefore $V(\phi_{\min}) \sim (10^{16}\text{ GeV})^4$. From quantum chromodynamics (QCD), we expect that the world around us is a condensate of quarks and gluons. The QCD scale is of order 100 MeV, and the corresponding contribution to the effective cosmological constant should be roughly $(100\text{ MeV})^4$. Note that the value of the quark and gluon condensates can be measured by observations, even if we cannot measure the absolute value contributing to the cosmological constant.

Lastly, a non-vanishing cosmological constant should be expected even from the quantum fluctuation of any field of the theory. Within a semiclassical approach, we treat gravity as a classical field and we quantize matter. The result is that the Einstein equations read

$$G_{\mu\nu} = \frac{8\pi G_{\mathrm{N}}}{c^4} \langle T_{\mu\nu} \rangle\,, \tag{13.17}$$

where $\langle T_{\mu\nu} \rangle$ is the expectation value of the matter energy-momentum tensor $T_{\mu\nu}$. An effective cosmological constant arises from the fluctuations of the matter field. After standard renormalization procedures, we obtain a renormalized cosmological constant [2], whose value should be measured in experiments. While it is not possible to make theoretical predictions on it, it would still be natural to expect $\rho_\Lambda \sim M_{\mathrm{Pl}}^4$, because there are no other scales in the gravity sector.

In the end, from theoretical arguments we cannot make any clear prediction about the value of the effective cosmological constant in the Universe, but it would be quite natural to have $\rho_\Lambda \sim M_{\text{Pl}}^4$, or at least $\rho_\Lambda \sim M_{\text{EW}}^4$. However, this is not what we see in the Universe. Such a high value of ρ_Λ would lead to a very fast expansion of the Universe, making impossible the formation of any structure, including galaxies and galaxy clusters. Current astronomical data are consistent with a non-vanishing vacuum energy density of order

$$\rho_\Lambda \sim 10^{-47} \ \text{GeV}^4 . \tag{13.18}$$

This is ~ 120 orders of magnitude smaller than the expectation $\rho_\Lambda \sim M_{\text{Pl}}^4$!

There are several proposals on how to solve the cosmological constant problem, but none seems to be satisfactory. Note that there is an implicit but strong assumption on this issue: we assume that we can discuss this problem within an effective field theory, in which low energy physics is decoupled from that at much higher energies and therefore we can get reliable predictions at low energies without knowing what happens at much higher energies. Since we know a number of expected contributions to an effective cosmological constant from the known physics below the electroweak scale, we argue that at least those contributions should be there, and other contributions coming from the physics at higher energies may be added. Such a conclusion is not guaranteed. It is instead possible that the explanation to this puzzle comes from unknown physics at higher energies.

Problem

13.1 Get a rough estimate of the evaporation time of a Schwarzschild black hole of one Solar mass.

References

1. J.D. Bekenstein, Phys. Rev. D **7**, 2333 (1973)
2. N.D. Birrell, P.C.W. Davies, *Quantum Fields in Curved Space* (Cambridge University Press, Cambridge, 1984)
3. N.E.J. Bjerrum-Bohr, J.F. Donoghue, B.R. Holstein, Phys. Rev. D **68**, 084005 (2003). [Erratum: ibidem **71**, 069904 (2005)], arXiv:hep-th/0211071
4. J.F. Donoghue, arXiv:gr-qc/9512024
5. J.F. Donoghue, Phys. Rev. D **50**, 3874 (1994), arXiv:gr-qc/9405057
6. R.P. Feynman, Acta Phys. Pol. **24**, 697 (1963)
7. S.W. Hawking, Commun. Math. Phys. **43**, 199 (1975). [erratum: Commun. Math. Phys. **46**, 206 (1976)]
8. S.W. Hawking, R. Penrose, Proc. R. Soc. Lond. A **314**, 529 (1970)
9. R. Penrose, Phys. Rev. Lett. **14**, 57 (1965)
10. G.'t Hooft, M.J.G. Veltman, Ann. Inst. H. Poincare Phys. Theor. A **20**, 69 (1974)
11. S. Weinberg, *Gravitation and Cosmology* (Wiley, New York, 1972)

Correction to: Introduction to General Relativity

Correction to:
C. Bambi, *Introduction to General Relativity*,
Undergraduate Lecture Notes in Physics,
https://doi.org/10.1007/978-981-13-1090-4

In the original version of this book, equations and paragraphs in the following sections were corrected. Sections 1.2, 4.6.1, 7.3, 8.5, 8.6.2, 9.5, 12.5.2, I.2 and I.4.
The erratum book has been updated with the changes.

The updated version of the book can be found at https://doi.org/10.1007/978-981-13-1090-4

© Springer Nature Singapore Pte Ltd. 2020
C. Bambi, *Introduction to General Relativity*, Undergraduate Lecture
Notes in Physics, https://doi.org/10.1007/978-981-13-1090-4_14

Appendix A
Algebraic Structures

An algebraic structure is a set X with one or more operations satisfying certain axioms. In this appendix, we will briefly review groups and vector spaces.

A.1 Groups

Group. A *group* is a set G equipped with an operation $m : G \times G \rightarrow G$ such that:

1. $\forall x, y, z \in G$, $m(x, m(y, z)) = m(m(x, y), z)$ (associativity);
2. $\forall x \in G$, there exists an element $u \in G$ such that $m(u, x) = m(x, u) = x$ (existence of the identity element);
3. $\forall x \in G$, there exists x^{-1} such that $m(x, x^{-1}) = m(x^{-1}, x) = u$ (existence of the inverse element).

If $m(x, y) = m(y, x) \ \forall x, y \in G$ then G is an *abelian group*.

Note that in every group the identity element is unique. If we assume there are two identity elements, say u and u', we should have $m(u, u') = u$ and $m(u, u') = u'$, and therefore $u = u'$. Even the inverse element must be unique. If x had two inverse elements, say x^{-1} and x'^{-1}, then

$$
\begin{aligned}
x^{-1} &= m(u, x^{-1}) = m(m(x'^{-1}, x), x^{-1}) \\
&= m(x'^{-1}, m(x, x^{-1})) = m(x'^{-1}, u) = x'^{-1} .
\end{aligned} \tag{A.1}
$$

The set of real numbers $\mathbb{R}$ with the common operation of sum is an abelian group with the identity element 0; the inverse of the element x is denoted $-x$. The set $\mathbb{R}/\{0\}$ with the common operation of product is an abelian group with the identity element 1; the inverse of the element x is denoted $1/x$.

© Springer Nature Singapore Pte Ltd. 2018, corrected publication 2020
C. Bambi, *Introduction to General Relativity*, Undergraduate Lecture Notes in Physics, https://doi.org/10.1007/978-981-13-1090-4

Let $M(n, \mathbb{R})$ be the set of real matrices $n \times n$. Then

$$GL(n, \mathbb{R}) = \{A \in M(n, \mathbb{R}) \mid \det A \neq 0\}, \tag{A.2}$$

is the subset of invertible matrices in $M(n, \mathbb{R})$. If we introduce the common product of matrices, $GL(n, \mathbb{R})$ is a non-abelian group.

The Galilei group, the Lorentz group, and the Poincaré group introduced in Chaps. 1 and 2 are other examples of groups.

A.2   Vector Spaces

Vector space. A *vector space* over a field K is a set V equipped with two operations, $m : V \times V \to V$ (addition) and $p : K \times V \to V$ (multiplication by an element of K), such that:
1. $\forall \mathbf{x}, \mathbf{y} \in V$, $m(\mathbf{x}, \mathbf{y}) = m(\mathbf{y}, \mathbf{x})$ (commutativity);
2. $\forall \mathbf{x}, \mathbf{y}, \mathbf{z} \in V$, $m(\mathbf{x}, m(\mathbf{y}, \mathbf{z})) = m(m(\mathbf{x}, \mathbf{y}), \mathbf{z})$ (associativity);
3. $\forall \mathbf{x} \in V$, there exists an element $\mathbf{0} \in V$ such that $m(\mathbf{x}, \mathbf{0}) = \mathbf{x}$ (existence of the identity element);
4. $\forall \mathbf{x} \in V$, there exists $-\mathbf{x}$ such that $m(\mathbf{x}, -\mathbf{x}) = \mathbf{0}$ (existence of the inverse element);
5. $\forall \mathbf{x}, \mathbf{y} \in V$ and $\forall a \in K$, $p(a, m(\mathbf{x}, \mathbf{y})) = m(p(a, \mathbf{x}), p(a, \mathbf{y}))$;
6. $\forall \mathbf{x} \in V$ and $\forall a, b \in K$, $p(a + b, \mathbf{x}) = m(p(a, \mathbf{x}), p(b, \mathbf{x}))$;
7. $\forall \mathbf{x} \in V$ and $\forall a, b \in K$, $p(ab, \mathbf{x}) = p(a, p(b, \mathbf{x}))$,
8. $p(1, \mathbf{x}) = \mathbf{x}$.

The elements of V are called *vectors* and the elements of K are called *scalars*.

Note that the definition above of vector space is equivalent to saying that V is an abelian group with the operation $p : K \times V \to V$ satisfying the points 5–8. It is common to employ the symbol $+$ to denote the operation m, i.e. $m(\mathbf{x}, \mathbf{y}) = \mathbf{x} + \mathbf{y}$, and no symbol for the operation p, i.e. $p(a, \mathbf{x}) = a\mathbf{x}$. Vector spaces over $\mathbb{R}$ are called real vector spaces, while those over $\mathbb{C}$ are called complex vector spaces.

Linear operator. Let V and W be two vector spaces over K. The function $f : V \to W$ is a *linear operator* if:

$$f(a\mathbf{v} + b\mathbf{w}) = af(\mathbf{v}) + bf(\mathbf{w}), \tag{A.3}$$

$\forall \mathbf{v}, \mathbf{w} \in V$ and $\forall a, b \in K$.

Let us denote with $L(V; W)$ the set of all linear operators from V into W. We define the operator

$$\tilde{m} : L(V; W) \times L(V; W) \to L(V; W) \qquad (A.4)$$

as the sum at every point of two elements of $L(V; W)$, and the operator

$$\tilde{p} : K \times L(V; W) \to L(V; W) \qquad (A.5)$$

as the product at every point of an element of $L(V; W)$ with an element of K. The operators $\tilde{m}$ and $\tilde{p}$ provide $L(V; W)$ the structure of a vector space. The concept of linear operators can be extended to define bilinear operators and, more in general, n-linear operators. The function $f : V \times W \to Z$ is called a bilinear operator if $f(\mathbf{v}, \cdot) : W \to Z$ and $f(\cdot, \mathbf{w}) : V \to Z$ are linear operators $\forall \mathbf{v} \in V$ and $\forall \mathbf{w} \in W$. Let us indicate with $L(V, W; Z)$ the set of all bilinear operators from $V \times W$ into Z. If we define the operation of sum between two bilinear operators and the operation of product between a bilinear operator and an element of K as done in the case of linear operators, then $L(V, W; Z)$ becomes a vector space. The generalization to n-linear operators from $V_1 \times V_2 \times \cdots \times V_n$ into Z and to the vector space $L(V_1, V_2, \ldots, V_n; Z)$ is straightforward.

Isomorphism. Let V and W be two vector spaces. The function $f : V \to W$ is an *isomorphism* if it is bijective, linear, and the application inverse is also linear. The vector spaces V and W are said to be *isomorphic* if there exists an isomorphism between V and W.

Dual space. Let V be a vector space. The *dual space* of V is the vector space $L(V; \mathbb{R})$ and is denoted V^*.

Subspace. A *subspace* of a vector space V over K is a subset W of V such that, $\forall \mathbf{w}_1, \mathbf{w}_2 \in W$ and $\forall a, b \in K, a\mathbf{w}_1 + b\mathbf{w}_2 \in W$.

Let U be a subset of a vector space V over K. A linear combination of elements of U is an element $\mathbf{v} \in V$ of the form $\mathbf{v} = a_1\mathbf{u}_1 + a_2\mathbf{u}_2 + \cdots + a_n\mathbf{u}_n$, where $\{\mathbf{u}_i\}$ is a finite subset of U and $\{a_i\}$ are elements of K. The subspace generated by U is the set of linear combinations that can be obtained from the elements of U and is denoted $\langle U \rangle$. The elements $\{u_i\}$ are linearly independent if

$$a_1\mathbf{u}_1 + a_2\mathbf{u}_2 + \cdots, +a_n\mathbf{u}_n = \mathbf{0} \qquad (A.6)$$

if and only if (note the difference between the element $\mathbf{0} \in V$ and the element $0 \in K$)

$$a_1 = a_2 = \cdots = a_n = 0. \qquad (A.7)$$

Basis. A *basis* of a vector space V is a subset B of V consisting of linearly independent elements and such that $\langle B \rangle = V$. If B has n elements, then we say that the vector space V has dimension n.

Every vector space admits a basis and the number of elements of the basis is independent of the choice of the basis (for the proof, see e.g. Ref. [1]). Let us now consider a vector space V of dimension n and with the basis $B = \{\mathbf{e}_i\}$. If n is a finite number, then V^* has the same dimension as V [1]. We indicate with $\mathbf{e}^i$ the elements of V^* such that $\mathbf{e}^i(\mathbf{e}_j) = \delta^i_j$, where δ^i_j is the Kronecker delta. The set $B^* = \{\mathbf{e}^i\}$ is a basis of V^* and is called the dual basis of B.

Let us consider another basis of V, say $B' = \{\mathbf{e}'_i\}$, and the transformation of the change of basis M defined as (note that we use the Einstein convention of summation over repeated indices)

$$\mathbf{e}'_i = M_i^j \mathbf{e}_j. \qquad (A.8)$$

Every element $\mathbf{v} \in V$ can be written in terms of the basis B, $\mathbf{v} = v^i\mathbf{e}_i$, as well as in terms of the basis B', $\mathbf{v} = v'^i\mathbf{e}'_i$. v^i and v'^i are the components of the vector $\mathbf{v}$ with respect to the bases B and B', respectively. It is easy to see that the components of a vector must change with the inverse transformation with respect to the basis vectors

$$v'^i = (M^{-1})^i_j v^j. \qquad (A.9)$$

We say that the basis vectors *transform covariantly* under a change of basis and we employ lower indices. The components of a vector *transform contravariantly* and are written with upper indices. In a similar way, we can see that the vectors of the dual basis transform contravariantly, while the components of an element of V^* transform covariantly. If $B'^* = \{\mathbf{e}'^i\}$ is the dual basis of B' and N is the transformation of the change of basis from B^* and B'^* defined as

$$\mathbf{e}'^i = N^i_j \mathbf{e}^j, \qquad (A.10)$$

then

$$\delta^i_j = \mathbf{e}'^i(\mathbf{e}'_j) = N^i_k \mathbf{e}^k(M_j^m \mathbf{e}_m) = N^i_k M_j^m \mathbf{e}^k(\mathbf{e}_m) = N^i_k M_j^m \delta^k_m = N^i_k M_j^k, \quad (A.11)$$

and therefore $N_k^i = (M^{-1})_k^i$. Since these transformations map quantities that transform covariantly (contravariantly) into quantities that transform in the same way, indices under summation are placed as upper (lower) indices, while free indices are placed as lower (upper) indices. A deeper investigation can show that M and N are matrices in which the indices of row and column transform covariantly or contravariantly, depending on the cases.

If V is of finite dimension, V and V^* are isomorphic. However, in general there is no preferred isomorphism between the two vector spaces. The situation is different if V is a real vector space and is provided with a non-degenerate symmetric bilinear form.

Bilinear form. A *bilinear form* on a real vector space V is an operator $g \in L(V, V; \mathbb{R})$. The bilinear form g is called *symmetric* if $g(\mathbf{v}, \mathbf{w}) = g(\mathbf{w}, \mathbf{v})$ $\forall \mathbf{v}, \mathbf{w} \in V$.

If g is a symmetric bilinear form on a real vector space V, then an element $v \in V$ is called (with respect to g):

1. *Time-like* if $g(\mathbf{v}, \mathbf{v}) < 0$;
2. *Light-like* if $g(\mathbf{v}, \mathbf{v}) = 0$;
3. *Space-like* if $g(\mathbf{v}, \mathbf{v}) > 0$.

As already pointed out in Sect. 2.2, there are two opposite conventions in the literature. The definition above is more popular in the gravity community. In the particle physics community, it is more common to say that $\mathbf{v}$ is a time-like vector if $g(\mathbf{v}, \mathbf{v}) > 0$ and a space-like vector if $g(\mathbf{v}, \mathbf{v}) < 0$.

If g is a symmetric bilinear form on a real vector space V and $\mathbf{v}$ and $\mathbf{w}$ are two elements of V, $\mathbf{v}$ and $\mathbf{w}$ are said to be *orthogonal* if $g(\mathbf{v}, \mathbf{w}) = 0$.

A symmetric bilinear form g on a real vector space V is called *non-degenerate* if the function $\tilde{g} : V \to V^*$ that transforms an element $\mathbf{v} \in V$ into the element $g(\mathbf{v}, \cdot) \in V^*$ is injective.

If the vector space V is of finite dimension, then the form g is represented, with respect to a basis $B = \{\mathbf{e}_i\}$, by an $n \times n$ matrix in which the element ij is

$$g_{ij} = g(\mathbf{e}_i, \mathbf{e}_j). \tag{A.12}$$

If g is a non-degenerate symmetric bilinear form, then $\tilde{g}$ is the natural isomorphism of V onto V^*, while its inverse $\tilde{g}^{-1}$ is the natural isomorphism of V^* onto V: an element $\mathbf{v} \in V$ is associated to the element $\tilde{\mathbf{v}} \in V^*$ given by $\tilde{\mathbf{v}} = \tilde{g}(\mathbf{v}) = g(\mathbf{v}, \cdot)$. If $B = \{\mathbf{e}_i\}$ is a basis of V and $B^* = \{\mathbf{e}^i\}$ is its dual basis, we write $\mathbf{v} = v^i \mathbf{e}_i$ and $\tilde{\mathbf{v}} = v_i \mathbf{e}^i$. From Eq. (A.12), we find that the components of $\tilde{\mathbf{v}}$ are obtained by *lowering the indices* with g

$$v_i = g_{ij} v^j. \tag{A.13}$$

Denoting g^{ij} the components of the inverse matrix associated to the bilinear form g, the components of the vector **v** are obtained by *raising the indices* of the components of the vector $\tilde{\mathbf{v}}$

$$v^i = g^{ij} v_j \, .$$

(A.14)

Note that, by definition, $g_{ij} g^{jk} = \delta_i^k$.

Reference

1. M. Nakahara, *Geometry, Topology and Physics* (IOP, Bristol, 1990).

Appendix B
Vector Calculus

This appendix briefly reviews some basic operators and identities of vector calculus in a 3-dimensional Euclidean space with a Cartesian coordinate system (x, y, z). The metric is the Kronecker delta δ_{ij}

$$||\delta_{ij}|| - \begin{pmatrix} 1 & 0 & 0 \\ 0 & 1 & 0 \\ 0 & 0 & 1 \end{pmatrix} . \tag{B.1}$$

Indices are (trivially) raised and lowered with δ^{ij} and δ_{ij}, respectively. The inner product is the algebraic operation that transforms two vectors, say $\mathbf{V}$ and $\mathbf{W}$, into the number

$$\mathbf{V} \cdot \mathbf{W} = \delta_{ij} V^i W^j , \tag{B.2}$$

where the Einstein convention of summation over repeated indices is used.

In what follows, $\phi = \phi(x, y, z)$ denotes a generic scalar function and $\mathbf{V} = (V^x, V^y, V^z)$, where $V^i = V^i(x, y, z)$, is a generic vector field.

B.1 Operators

The *del* operator ∇ is defined as

$$\nabla = \frac{\partial}{\partial x}\hat{\mathbf{x}} + \frac{\partial}{\partial y}\hat{\mathbf{y}} + \frac{\partial}{\partial z}\hat{\mathbf{z}} , \tag{B.3}$$

where $(\hat{\mathbf{x}}, \hat{\mathbf{y}}, \hat{\mathbf{z}})$ is the natural basis, namely $\hat{\mathbf{x}}$, $\hat{\mathbf{y}}$, and $\hat{\mathbf{z}}$ are the unit vectors pointing in the direction of the axes of the Cartesian coordinate system.

The *gradient* of the scalar function ϕ is the vector field

© Springer Nature Singapore Pte Ltd. 2018, corrected publication 2020
C. Bambi, *Introduction to General Relativity*, Undergraduate Lecture Notes
in Physics, https://doi.org/10.1007/978-981-13-1090-4

$$\nabla\phi = \frac{\partial\phi}{\partial x}\hat{\mathbf{x}} + \frac{\partial\phi}{\partial y}\hat{\mathbf{y}} + \frac{\partial\phi}{\partial z}\hat{\mathbf{z}}.$$ (B.4)

The components of the resulting vector field are

$$(\nabla\phi)^i = \partial^i\phi.$$ (B.5)

The *divergence* of the vector field $\mathbf{V}$ is the scalar function

$$\nabla\cdot\mathbf{V} = \frac{\partial V^x}{\partial x} + \frac{\partial V^y}{\partial y} + \frac{\partial V^z}{\partial z},$$ (B.6)

so we can write

$$\nabla\cdot\mathbf{V} = \partial_i V^i.$$ (B.7)

The *curl* or *rotor* of the vector field $\mathbf{V}$ is the vector field

$$\nabla\times\mathbf{V} = \left(\frac{\partial V^z}{\partial y} - \frac{\partial V^y}{\partial z}\right)\hat{\mathbf{x}} + \left(\frac{\partial V^x}{\partial z} - \frac{\partial V^z}{\partial x}\right)\hat{\mathbf{y}} + \left(\frac{\partial V^y}{\partial x} - \frac{\partial V^x}{\partial y}\right)\hat{\mathbf{z}},$$ (B.8)

The components of the resulting vector field can be written as

$$(\nabla\times\mathbf{V})^i = \varepsilon^{ijk}\partial_j V_k,$$ (B.9)

where ε^{ijk} is the Levi-Civita symbol (see next section).

The *Laplacian* of the scalar function ϕ is the scalar function

$$\Delta\phi = \nabla^2\phi = (\nabla\cdot\nabla)\phi = \frac{\partial^2\phi}{\partial x^2} + \frac{\partial^2\phi}{\partial y^2} + \frac{\partial^2\phi}{\partial z^2}.$$ (B.10)

We can also write

$$\Delta\phi = \delta^{ij}\partial_i\partial_j\phi.$$ (B.11)

The *d'Alembertian* of the scalar function ϕ is the scalar function

$$\Box\phi = \left(-\frac{1}{c^2}\frac{\partial^2}{\partial t^2} + \nabla^2\right)\phi = -\frac{1}{c^2}\frac{\partial^2\phi}{\partial t^2} + \frac{\partial^2\phi}{\partial x^2} + \frac{\partial^2\phi}{\partial y^2} + \frac{\partial^2\phi}{\partial z^2}.$$ (B.12)

The d'Alembertian can be seen as the Laplacian in Minkowski spacetime

$$\Box\phi = \eta^{\mu\nu}\partial_\mu\partial_\nu\phi.$$ (B.13)

B.2 Levi-Civita Symbol

The *Levi-Civita symbol* ε_{ijk} is defined as

$$
\varepsilon_{ijk} = \begin{cases} +1 & \text{if } (i, j, k) \text{ is an } even \text{ permutation of } (1, 2, 3), \\ -1 & \text{if } (i, j, k) \text{ is an } odd \text{ permutation of } (1, 2, 3), \\ 0 & \text{if any index is repeated.} \end{cases} \tag{B.14}
$$

Indices can be raised and lowered with δ^{ij} and δ_{ij}, but often the position of the indices is ignored because δ^{ij} and δ_{ij} have a trivial effect. For this reason, sometimes only lower indices are used. The following formulas hold

$$
\varepsilon_{ijk}\varepsilon^{imn} = \delta_j^m \delta_k^n - \delta_j^n \delta_k^m , \tag{B.15}
$$

$$
\varepsilon_{imn}\varepsilon^{jmn} = 2\delta_i^j , \tag{B.16}
$$

$$
\varepsilon_{ijk}\varepsilon^{ijk} = 6 . \tag{B.17}
$$

B.3 Properties

The divergence of the curl of a vector field is identically zero

$$
\nabla \cdot (\nabla \times \mathbf{V}) = 0 . \tag{B.18}
$$

This can be easily proved by rewriting this expression with the Einstein convention

$$
\partial_i \left(\varepsilon^{ijk} \partial_j V_k \right) = 0 . \tag{B.19}
$$

The expression vanishes because we sum over i and j, and ε_{ijk} and $\partial_i \partial_j$ are, respectively, antisymmetric and symmetric with respect to i and j.

The curl of the gradient of a vector field is identically zero

$$
\nabla \times (\nabla \phi) = 0 . \tag{B.20}
$$

If we rewrite this expression as

$$
\varepsilon^{ijk} \partial_j (\partial_k \phi) = 0 , \tag{B.21}
$$

we see that, again, we sum over i and j, and ε_{ijk} and $\partial_i \partial_j$ are, respectively, antisymmetric and symmetric with respect to i and j.

The following identity holds

$$
\nabla \times (\nabla \times \mathbf{V}) = \nabla (\nabla \cdot \mathbf{V}) - \nabla^2 \mathbf{V} , \tag{B.22}
$$

As in the previous cases, we write the component i

$$\varepsilon^{ijk}\partial_j\left(\nabla\times\mathbf{V}\right)_k = \varepsilon^{ijk}\partial_j\varepsilon_{kmn}\partial^m V^n. \tag{B.23}$$

Let us exchange i and k, so we can directly apply Eq. (B.15). We find

$$\varepsilon^{kji}\partial_j\varepsilon_{imn}\partial^m V^n = -\varepsilon^{ijk}\partial_j\varepsilon_{imn}\partial^m V^n = \left(\delta_n^j\delta_m^k - \delta_m^j\delta_n^k\right)\partial_j\partial^m V^n$$
$$= \partial^k\partial_n V^n - \partial_m\partial^m V^k, \tag{B.24}$$

and we recover Eq. (B.22).

Appendix C
Differentiable Manifolds

Differentiable manifolds are the natural generalization of curves and surfaces in the case of spaces of arbitrary dimensions. Their key-property is that we can locally define a one-to-one correspondence between the points of a differentiable manifold of dimension n with the points of an open subset of $\mathbb{R}^n$. In this way we can "label" every point of the manifold with n numbers, representing the local coordinates of that point, and use the differential calculus developed in $\mathbb{R}^n$. If the differentiable manifold is globally different from $\mathbb{R}^n$, it is necessary to introduce more than one coordinate system to have all the points of the manifold under control. In such a case, we need that the transformations to change coordinate system are differentiable in order to have results independent of the choice of the coordinate system.

C.1 Local Coordinates

Differentiable manifold. A *differentiable manifold* of dimension n is a set M equipped with a family of bijective operators $\varphi_i : U_i \subset M \rightarrow \mathbb{R}^n$ such that:
1. $\{U_i\}$ is a family of open sets and $\bigcup_i U_i = M$;
2. $\forall i, j$ such that $U_i \cap U_j \neq \varnothing$, the operator $\lambda_{ij} = \varphi_i \varphi_j^{-1}$ from $\varphi_j(U_i \cap U_j)$ to $\varphi_i(U_i \cap U_j)$ is infinitely differentiable.

The pair (U_i, φ_i) is a local parametrization of M or a *map* of M. The family of maps $\{(U_i, \varphi_i)\}$ is called the *atlas* of M.

Figure C.1 illustrates the concepts of differentiable manifold and maps. The function φ_i can be thought of as an object with n components in which every component is a function $x^k(p)$. These n functions evaluated at a point p of the differentiable manifold are the *coordinates* of p with respect to the map (U_i, φ_i).

Let us consider an example. The spherical surface of dimension n is the subset of $\mathbb{R}^{n+1}$ defined as

© Springer Nature Singapore Pte Ltd. 2018, corrected publication 2020
C. Bambi, *Introduction to General Relativity*, Undergraduate Lecture Notes
in Physics, https://doi.org/10.1007/978-981-13-1090-4

Fig. C.1 Differentiable
manifold M with the maps
(U_i, φ_i) and (U_j, φ_j)

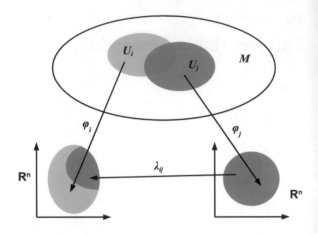

$$S^n = \left\{ (x^1, x^2, \ldots, x^{n+1}) \mid \sum_{i=1}^{n+1} (x^i)^2 = 1 \right\}. \tag{C.1}$$

We define the operator

$$\varphi_1 : U_1 = S^n / \{(0, \ldots, 0, 1)\} \to \mathbb{R}^n \tag{C.2}$$

as

$$\varphi_1(x^1, x^2, \ldots, x^{n+1}) = \left(\frac{x^1}{1 - x^{n+1}}, \frac{x^2}{1 - x^{n+1}}, \ldots, \frac{x^n}{1 - x^{n+1}} \right), \tag{C.3}$$

and the operator

$$\varphi_2 : U_2 = S^n / \{(0, \ldots, 0, -1)\} \to \mathbb{R}^n \tag{C.4}$$

as

$$\varphi_2(x^1, x^2, \ldots, x^{n+1}) = \left(\frac{x^1}{1 + x^{n+1}}, \frac{x^2}{1 + x^{n+1}}, \ldots, \frac{x^n}{1 + x^{n+1}} \right). \tag{C.5}$$

The differentiable manifold S^n is parametrized by the maps (U_1, φ_1) and (U_2, φ_2).
The two maps together are an atlas of S^n.

Let M and N be two differentiable manifolds, respectively of dimension m and
n. The function $f : M \to N$ is said to be *differentiable* at a point $p \in M$ if, for any
parametrization $\psi : V \subset N \to \mathbb{R}^n$ with $f(p) \in V$, there exists a parametrization
$\varphi : U \subset M \to \mathbb{R}^m$ with $p \in U$, such that the function

$$\pi = \psi f \varphi^{-1} : \mathbb{R}^m \to \mathbb{R}^n \tag{C.6}$$

is differentiable in $\varphi(p)$. From the definition of differentiable manifold, the differentiability of f is independent of the choice of the coordinate system.

C.2   Tangent Vectors

Since a differentiable manifold of dimension n can be identified only locally with an open set of $\mathbb{R}^n$, even all objects that we can construct on the manifold are just locally defined. Tangent vectors can be defined after introducing the concept of curve.

> **Curve.** A *curve* on a differentiable manifold M is a differentiable function $\gamma : [t_1, t_2] \subset \mathbb{R} \to M$.

With the concept of curve, we can define tangent vectors and their vector space.

> **Tangent vector.** Let $\gamma(t)$ be a curve on a differentiable manifold M and $\mathscr{M}$ be the set of differentiable functions $f : M \to \mathbb{R}$. We call the *tangent vector* to M at $p = \gamma(t_0)$ along the direction of the curve $\gamma(t)$ the operator $X : \mathscr{M} \to \mathbb{R}$ that provides the directional derivative of the function $f \in \mathscr{M}$ along the curve $\gamma(t)$ at p; that is,
>
> $$X[f] = \left. \frac{df[\gamma(t)]}{dt} \right|_{t=t_0} = X^\mu \frac{\partial f}{\partial x^\mu}. \tag{C.7}$$
>
> The *tangent space* to the differentiable manifold M at the point p is the space generated by the tangent vectors to M at p and is denoted $T_p M$.

The tangent space at a point on a differentiable manifold is a vector space whose elements are the tangent vectors at that point.

Let us consider the map (U, φ) of the differentiable manifold M around the point $p = \varphi^{-1}(\tilde{x}^1, \tilde{x}^2, \ldots, \tilde{x}^n)$. With this parametrization, the curve γ can be written as

$$\varphi[\gamma(t)] = \begin{pmatrix} x^1(t) \\ x^2(t) \\ \vdots \\ x^n(t) \end{pmatrix}, \tag{C.8}$$

and therefore

$$\gamma(t) = \varphi^{-1} \begin{pmatrix} x^1(t) \\ x^2(t) \\ \vdots \\ x^n(t) \end{pmatrix}. \tag{C.9}$$

The tangent vector X applied at the function f at $\gamma(t_0) = p$ is

$$X[f] = \frac{d}{dt} f \varphi^{-1} \begin{pmatrix} x^1(t) \\ x^2(t) \\ \vdots \\ x^n(t) \end{pmatrix}_{t=t_0} = \frac{\partial f \varphi^{-1}}{\partial x^k}\bigg|_{\mathbf{x}=\bar{\mathbf{x}}} \frac{dx^k}{dt}\bigg|_{t=t_0}. \tag{C.10}$$

The tangent vector X can be written as

$$X = X^k \mathbf{e}_k, \tag{C.11}$$

where

$$X^k = \frac{dx^k}{dt}\bigg|_{t=t_0} \tag{C.12}$$

are the components of the vector X, while

$$\mathbf{e}_k = \frac{\partial}{\partial x^k} \tag{C.13}$$

are the basis vectors of the tangent space at the point p of the parametrization (U, φ). The set $\{\mathbf{e}_k\}$ is called the *basis of the coordinates* and is usually denoted $\{\partial_k\}$.

We can now show that, given $p \in M$ and $X \in T_p M$, it is always possible to find a curve $\gamma(t)$ such that $\gamma(t_0) = p$ and $\dot{\gamma}(t_0) = X$. It is sufficient to write the vector X in terms of some local parametrization (U, φ)

$$X = X^k \frac{\partial}{\partial x^k} \tag{C.14}$$

and solve n differential equations

$$\frac{d}{dt} u^i(t)\bigg|_{t=t_0} = X^i. \tag{C.15}$$

The curve is

$$\gamma(t) = \varphi^{-1} \begin{pmatrix} u^1(t) \\ u^2(t) \\ \vdots \\ u^n(t) \end{pmatrix}. \tag{C.16}$$

A vector $X = X^k \mathbf{e}_k$ exists independently of the choice of the coordinate system on M and therefore it can also be written in terms of another local parametrization with coordinates $\{x'^k\}$

$$X' = X'^k \mathbf{e}'_k, \tag{C.17}$$

where

$$X'^k = \frac{dx'^k}{dt'} \bigg|_{t'=t'_0}. \tag{C.18}$$

The components of the vector X in the two systems of coordinates are related by

$$X'^k = \frac{\partial x'^k}{\partial x^j} X^j, \tag{C.19}$$

while the relation between the two bases is

$$\mathbf{e}'_k = \frac{\partial x^j}{\partial x'^k} \mathbf{e}_j. \tag{C.20}$$

As already seen in Sect. A.2, the basis vectors of a vector space transform with the opposite rule with respect to the one for the components of vectors.

C.3 Cotangent Vectors

Cotangent vector. Let M be a differentiable manifold. The *cotangent space* at the point p is the vector space $T_p^* M$, dual of $T_p M$. A *cotangent vector* is an element of the cotangent space.

A cotangent vector is thus a linear function $\omega : T_p M \to \mathbb{R}$. If $B^* = \{\mathbf{e}^i\}$ is a basis of $T_p^* M$, we can write

$$\omega = \omega_k \mathbf{e}^k. \tag{C.21}$$

If $X = X^k e_k$ is an element of $T_p M$ written in terms of the basis B, we have

$$\omega(X) = \omega_k X^k \, . \tag{C.22}$$

It is common to use the notation $\{dx^k\}$ to indicate the basis of $T_p^* M$ with respect to the coordinates $\{x^k\}$.

Let us now consider another local parametrization in the neighborhood of the point p with local coordinates $\{x'^k\}$ and that provides the basis $\{\mathbf{e}'_k\}$ to the vector space $T_p M$ and the basis $\{\mathbf{e}'^k\}$ to $T_p^* M$. Since the quantity $\omega(X) \in \mathbb{R}$ does not depend on the choice of the basis, we have $\omega_k X^k = \omega'_k X'^k$, where ω'_k and X'^k are, respectively, the components of ω and X with respect to the new bases. From Eqs. (C.19) and (C.20), we find

$$\omega'_k = \frac{\partial x^j}{\partial x'^k}\omega_j \, , \quad \mathbf{e}'^k = \frac{\partial x'^k}{\partial x^j}\mathbf{e}^j \, . \tag{C.23}$$

C.4   Tensors

Tensors are the generalization of tangent and cotangent vectors.

Tensor. Let M be a differentiable manifold and p an element of M. A *tensor* of type (r, s) and of order $r + s$ at p is an $(r + s)$-linear function

$$\tau : \underbrace{T_p^* M \times \cdots \times T_p^* M}_{r \text{ times}} \times \underbrace{T_p M \times \cdots \times T_p M}_{s \text{ times}} \rightarrow \mathbb{R} \, . \tag{C.24}$$

With such a definition of tensor, tangent vectors are tensors of type $(1, 0)$, while cotangent vectors are tensors of type $(0, 1)$. The set of tensors of type (r, s) at p forms a vector space that we indicate with $\chi_s^r(M, p)$. Let $\{\mathbf{e}_k\}$ be a basis of $T_p M$ and $\{\mathbf{e}^k\}$ its dual basis. A tensor $\tau \in \chi_s^r(M, p)$ can be written as

$$\tau = \tau_{j_1 j_2 \ldots j_s}^{i_1 i_2 \ldots i_r} \, \mathbf{e}_{i_1} \, \mathbf{e}_{i_2} \, \ldots \, \mathbf{e}_{i_r} \, \mathbf{e}^{j_1} \, \mathbf{e}^{j_2} \, \ldots \, \mathbf{e}^{j_s} \, . \tag{C.25}$$

Let $\{\mathbf{e}'_k\}$ be another basis of $T_p M$ with dual basis $\{\mathbf{e}'^k\}$. Since the rules of transformation for the bases and for the dual bases are the same encountered in the previous sections for tangent and cotangent vectors, the components of the tensor τ change as follows

$$\tau'^{i_1 i_2 \ldots i_r}_{j_1 j_2 \ldots j_s} = \frac{\partial x'^{i_1}}{\partial x^{p_1}} \frac{\partial x'^{i_2}}{\partial x^{p_2}} \cdots \frac{\partial x'^{i_r}}{\partial x^{p_r}} \frac{\partial x^{q_1}}{\partial x'^{j_1}} \frac{\partial x^{q_2}}{\partial x'^{j_2}} \cdots \frac{\partial x^{q_s}}{\partial x'^{j_s}} \tau^{p_1 p_2 \ldots p_r}_{q_1 q_2 \ldots q_s} \, . \tag{C.26}$$

> **Tensor field.** Let M be a differentiable manifold. A *tensor field* of type (r, s) on M is a function that at every point $p \in M$ associates with an element $\tau \in \chi_s^r(M, p)$ in a differentiable way.

Let τ be a tensor field of type (r, s) on a differentiable manifold M. If all the components of τ vanish in a particular coordinate system, then they vanish in any coordinate system. Such a conclusion directly follows from Eq. (C.26).

C.5 Example: Spherical Surface of Dimension 2

Let us consider the spherical surface of dimension 2. The definition in (C.1) can be rewritten as

$$S^2 = \left\{ (x, y, z) \mid x^2 + y^2 + z^2 = 1 \right\} . \tag{C.27}$$

The map $(\varphi, S^2/\{(0, 0, 1), (0, 0, -1)\})$ is defined as

$$\varphi \begin{pmatrix} \sin\theta \cos\phi \\ \sin\theta \sin\phi \\ \cos\theta \end{pmatrix} = (\theta, \phi) , \quad \varphi^{-1}(\theta, \phi) = \begin{pmatrix} \sin\theta \cos\phi \\ \sin\theta \sin\phi \\ \cos\theta \end{pmatrix} , \tag{C.28}$$

where $\theta \in (0, \pi)$ and $\phi \in [0, 2\pi)$. A curve γ assumes the form

$$\varphi[\gamma(t)] = \begin{pmatrix} \theta(t) \\ \phi(t) \end{pmatrix} . \tag{C.29}$$

A generic $(1, 0)$ tensor field can be written as

$$V = V^\theta \mathbf{e}_\theta + V^\phi \mathbf{e}_\phi , \tag{C.30}$$

where $V^\theta = V^\theta(\theta, \phi)$ and $V^\phi = V^\phi(\theta, \phi)$ are the components of the tensor field with respect to the basis vectors

$$\mathbf{e}_\theta = \frac{\partial}{\partial\theta} , \quad \mathbf{e}_\phi = \frac{\partial}{\partial\phi} . \tag{C.31}$$

Figure C.2 shows the tangent space at two different points of the differentiable manifold.

Fig. C.2 S^2 with the
parametrization φ and the
tangent space at two points.
$\mathbf{e}_1$ and $\mathbf{e}_2$ are the two basis
vectors of the
parametrization φ and can be
employed to write tensor
fields on the manifold

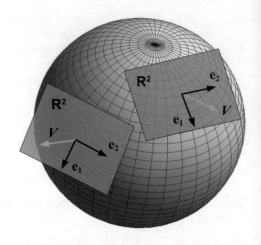

Appendix D
Ellipse Equation

An ellipse is a curve in a plane with two focal points. The sum of the distances of every point on the curve from the two focal points is constant. In Fig. D.1, the two focal points are F_1 and F_2. The distance between F_1 and F_2 is $2c$:

$$\overline{F_1 F_2} = 2c \,.$$ (D.1)

The ellipse with focal points F_1 and F_2 and eccentricity $e = c/a$ ($0 \le e < 1$) is the set of points P such that

$$\overline{F_1 P} + \overline{F_2 P} = 2a \,.$$ (D.2)

We adopt a polar coordinate system (r, ϕ) centered at the point F_1. r describes the distance of the point P from F_1 and ϕ the angle $\widehat{Q F_1 P}$. Equation (D.2) can be rewritten as

$$r + \sqrt{(2c + r \cos \phi)^2 + (r \sin \phi)^2} = 2a \,.$$ (D.3)

We consider the square of this equations and, after some simple calculations, we find

$$r \,(a + c \cos \phi) = a^2 - c^2 \,.$$ (D.4)

Since $e = c/a$, Eq. (D.4) can be written as

$$r \,(a + c \cos \phi) = a^2 \left(1 - e^2\right) \,,$$ (D.5)

and in the more compact form

$$\frac{1}{r} = A + B \cos \phi \,,$$ (D.6)

© Springer Nature Singapore Pte Ltd. 2018, corrected publication 2020
C. Bambi, *Introduction to General Relativity*, Undergraduate Lecture Notes in Physics, https://doi.org/10.1007/978-981-13-1090-4

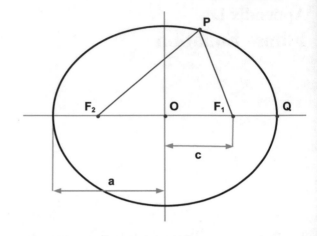

where

$$A = \frac{1}{a\left(1 - e^2\right)}, \quad B = \frac{e}{a\left(1 - e^2\right)}. \tag{D.7}$$

Appendix E
Mathematica Packages for Tensor Calculus

There are several Mathematica packages for tensor calculus available. Most of them can be downloaded from the web for free. It is also quite easy to write a Mathematica program to evaluate the main tensors and the Christoffel symbols for a given metric. In this appendix, we will introduce the package called RGTC (Riemannian Geometry and Tensor Calculus) for tensor calculus in Riemannian geometry. It is available for free at

http://www.inp.demokritos.gr/~sbonano/RGTC/

where there is also a manual with a number of examples.

With a few simple steps, RGTC computes the explicit expressions for some tensors (Riemann, Ricci, Einstein, Weyl) and checks if the spacetime belongs to any of the following categories: flat, conformally flat, Ricci flat, Einstein space or space of constant curvature.

First, it is necessary to initialize the code with the command

```
<< EDCRGTCcode.m
```

Then, we have to define the coordinates and the metric. As an example, we can consider the Schwarzschild metric in the usual Schwarzschild coordinates. For the coordinates, we can write

```
Coord = {t,r,θ,ϕ};
```

Then we write the non-vanishing metric coefficients

```
gtt = - (1 - 2M/r);
grr = 1/(1 - 2M/r);
gpp = r^2;
gvv = r^2 Sin[θ]^2;
```

and we define the metric

```
g = {{gtt, 0, 0, 0}, {0, grr, 0, 0}, {0, 0, gpp, 0},
    { 0, 0, 0, gvv }};
```

© Springer Nature Singapore Pte Ltd. 2018, corrected publication 2020
C. Bambi, *Introduction to General Relativity*, Undergraduate Lecture Notes
in Physics, https://doi.org/10.1007/978-981-13-1090-4

To launch the code, the command is

```
RGtensors[g, Coord, {1, 1, 1}]
```

and the output is something like

```
gdd = ...
LineElement = ...
gUU = ...
gUU computed in 0.007923 sec
Gamma computed in 0.003136 sec
Riemann(dddd) computed in 0.002824 sec
Riemann(Uddd) computed in 0.002247 sec
Ricci computed in 0.000167 sec
Weyl computed in 0.000015 sec
Ricci Flat
All tasks completed in 0.019439 seconds
```

where in the place of the ellipses . . . there are the expressions for the metric, the line element, and the inverse metric, respectively.

The option {1, 1, 1} can be manipulated to skip the evaluation of some tensors. If the first number is 0, i.e. we have {0, 1, 1}, the package does not compute the Riemann tensor $R^\mu{}_{\nu\rho\sigma}$ and instead of the line

```
Riemann(Uddd) computed in 0.002247 sec
```

the output produces the line

```
RUddd not computed
```

For {1, 0, 1}, the code skips the evaluation of the Weyl tensor. For {1, 1, 0}, it skips the evaluation of the Einstein tensor. It is just a command to save time if we do not need some tensors.

After RGTC has evaluated the tensors of the input metric, we can write their expression or work with them. For instance, if we want to visualize the Christoffel symbol Γ^t_{tr}, we write the command

```
GUdd[[1,1,2]]
```

The output will be the expression of Γ^t_{tr} of the input metric. To visualize the component R_{tt} of the Ricci tensor, we write

```
Rdd[[1,1]]
```

Note that indices run from 1 to n, where n is the number of the dimensions of the spacetime, according to the definition in Coord; that is, there is no index 0.

The package has some built-in functions to raise/lower indices, contract indices, evaluate the covariant derivative, etc. More details can be found in the manual of the package. We can also use standard Mathematica functions. For instance, the Kretschmann scalar can be evaluated with the command

```
Kretschmann = Simplify[ Sum[
   Rdddd[[i, j, k, l]]*RUddd[[i, m, n, o]]*gUU[[j, m]]
      *gUU[[k, n]]*gUU[[l, o]],
   {i, 1, 4}, {j, 1, 4}, {k, 1, 4}, {l, 1, 4},
      {m, 1, 4}, {n, 1, 4}, {o, 1, 4} ] ]
```

Appendix F
Interior Solution

The Schwarzschild metric found in Sect. 8.2 is the exterior vacuum solution for any spherically symmetric matter distribution in 4-dimensional Einstein's gravity; that is, it is the solution for the region $r > r_0$, where r_0 is the radius of the matter distribution. In this appendix, we want to find a simple solution for the interior region $r < r_0$.

First, we have to specify the matter energy-momentum tensor appearing on the right hand side of the Einstein equations. The simplest case is that of a perfect fluid and $T_{\mu\nu}$ reads

$$T_{\mu\nu} = (\rho + P)\frac{u_\mu u_\nu}{c^2} + Pg_{\mu\nu}, \qquad (F.1)$$

where ρ, P, and u^μ are, respectively, the energy density, the pressure, and the fluid 4-velocity. The most general line element for a spherically symmetric spacetime is given in (8.8). Here, we further simplify the problem, and we impose that the line element is also independent of time, namely we do not want possible radial inflows or outflows of matter. Our line element becomes

$$ds^2 = -f(r)c^2dt^2 + g(r)dr^2 + r^2\left(d\theta^2 + \sin^2\theta d\phi^2\right). \qquad (F.2)$$

Since we are employing the coordinate system in which matter is at rest, the only non-vanishing component of the fluid 4-velocity is the temporal one and u^μ is

$$||u^\mu|| = \left(\frac{c}{\sqrt{f}}, \mathbf{0}\right), \qquad (F.3)$$

because $g_{\mu\nu}u^\mu u^\nu = -c^2$. The matter energy-momentum tensor is

$$||T_{\mu\nu}|| = \begin{pmatrix} \rho f & 0 & 0 & 0 \\ 0 & Pg & 0 & 0 \\ 0 & 0 & Pr^2 & 0 \\ 0 & 0 & 0 & Pr^2\sin^2\theta \end{pmatrix}, \qquad (F.4)$$

© Springer Nature Singapore Pte Ltd. 2018, corrected publication 2020
C. Bambi, *Introduction to General Relativity*, Undergraduate Lecture Notes in Physics, https://doi.org/10.1007/978-981-13-1090-4

where ρ and P can be (at most) functions of the radial coordinate r.

At this point we have all the ingredients to write the Einstein equations. Unlike in Sect. 8.2, now we are not in vacuum, so the scalar curvature is non-vanishing in general and has to be calculated. Its expression is ($R_{\mu\nu}$s were calculated in Sect. 8.2)

$$
\begin{aligned}
R &= g^{tt} R_{tt} + g^{rr} R_{rr} + g^{\theta\theta} R_{\theta\theta} + g^{\phi\phi} R_{\phi\phi} \\
&= -\frac{f''}{2fg} + \frac{f'}{4fg}\left(\frac{f'}{f} + \frac{g'}{g}\right) - \frac{f'}{rfg} - \frac{f''}{2fg} + \frac{f'}{4fg}\left(\frac{f'}{f} + \frac{g'}{g}\right) + \frac{g'}{rg^2} \\
&\quad + \frac{2}{r^2} - \frac{2}{r^2 g} + \frac{1}{rg}\left(\frac{g'}{g} - \frac{f'}{f}\right) \\
&= -\frac{f''}{fg} + \frac{f'}{2fg}\left(\frac{f'}{f} + \frac{g'}{g}\right) + \frac{2}{r^2} - \frac{2}{r^2 g} - \frac{2}{rg}\left(\frac{f'}{f} - \frac{g'}{g}\right).
\end{aligned}
\tag{F.5}
$$

The non-vanishing components of the Einstein tensor $G_{\mu\nu}$ are

$$
\begin{aligned}
G_{tt} &= \frac{f''}{2g} - \frac{f'}{4g}\left(\frac{f'}{f} + \frac{g'}{g}\right) + \frac{f'}{rg} \\
&\quad -\frac{1}{2}\left[\frac{f''}{g} - \frac{f'}{2g}\left(\frac{f'}{f} + \frac{g'}{g}\right) - \frac{2f}{r^2} + \frac{2f}{r^2 g} + \frac{2f}{rg}\left(\frac{f'}{f} - \frac{g'}{g}\right)\right] \\
&= \frac{f}{r^2} - \frac{f}{r^2 g} + \frac{fg'}{rg^2},
\end{aligned}
\tag{F.6}
$$

$$
\begin{aligned}
G_{rr} &= -\frac{f''}{2f} + \frac{f'}{4f}\left(\frac{f'}{f} + \frac{g'}{g}\right) + \frac{g'}{rg} \\
&\quad +\frac{1}{2}\left[\frac{f''}{f} - \frac{f'}{2f}\left(\frac{f'}{f} + \frac{g'}{g}\right) - \frac{2g}{r^2} + \frac{2}{r^2} + \frac{2}{r}\left(\frac{f'}{f} - \frac{g'}{g}\right)\right] \\
&= -\frac{g}{r^2} + \frac{1}{r^2} + \frac{f'}{rf},
\end{aligned}
\tag{F.7}
$$

$$
\begin{aligned}
G_{\theta\theta} &= 1 - \frac{1}{g} + \frac{r}{2g}\left(\frac{g'}{g} - \frac{f'}{f}\right) \\
&\quad +\frac{1}{2}\left[\frac{r^2 f''}{fg} - \frac{r^2 f'}{2fg}\left(\frac{f'}{f} + \frac{g'}{g}\right) - 2 + \frac{2}{g} + \frac{2r}{g}\left(\frac{f'}{f} - \frac{g'}{g}\right)\right] \\
&= \frac{r}{2g}\left(\frac{f'}{f} - \frac{g'}{g}\right) + \frac{r^2 f''}{2fg} - \frac{r^2 f'}{4fg}\left(\frac{f'}{f} + \frac{g'}{g}\right),
\end{aligned}
\tag{F.8}
$$

$$
G_{\phi\phi} = G_{\theta\theta} \sin^2\theta. \tag{F.9}
$$

With the Einstein tensor $G_{\mu\nu}$ and the matter energy-momentum tensor $T_{\mu\nu}$, we can write the Einstein equations

$$G_{tt} = \frac{8\pi G_N}{c^4} T_{tt} \rightarrow \frac{1}{r^2} - \frac{1}{r^2 g} + \frac{g'}{rg^2} = \frac{8\pi G_N}{c^4} \rho \,, \tag{F.10}$$

$$G_{rr} = \frac{8\pi G_N}{c^4} T_{rr} \rightarrow \frac{1}{r^2 g} - \frac{1}{r^2} + \frac{f'}{rfg} = \frac{8\pi G_N}{c^4} P \,, \tag{F.11}$$

$$G_{\theta\theta} = \frac{8\pi G_N}{c^4} T_{\theta\theta} \rightarrow$$

$$\frac{1}{2rg}\left(\frac{f'}{f} - \frac{g'}{g}\right) + \frac{f''}{2fg} - \frac{f'}{4fg}\left(\frac{f'}{f} + \frac{g'}{g}\right) = \frac{8\pi G_N}{c^4} P \,. \tag{F.12}$$

The $\phi\phi$ component is as the $\theta\theta$ component with the factor $\sin^2\theta$ and therefore it is not an independent equation. The off-diagonal components are all trivial $0 = 0$. We have thus three equations and four unknown functions $(f, g, \rho, \text{ and } P)$. The system can be closed by specifying the matter equation of state. The simplest case is that in which the energy density is constant in the matter rest-frame, i.e.

$$\rho = \text{constant} \,. \tag{F.13}$$

More realistic equations of state typically require the equations to be solved numerically. Now we have three equations for three unknown functions $(f, g, \text{ and } P)$.

The covariant conservation of the matter energy-momentum tensor is a direct consequence of the Einstein equations

$$\nabla_\nu T^{\mu\nu} = \frac{\partial T^{\mu\nu}}{\partial x^\nu} + \Gamma^\mu_{\lambda\nu} T^{\lambda\nu} + \Gamma^\nu_{\lambda\nu} T^{\mu\lambda} = 0 \,. \tag{F.14}$$

Even if it does not provide an independent equation, it is sometimes more convenient to use. As it can be easily imagined from the symmetry of our system, only the equation for $\mu = r$ has a non-trivial solution. The equation reads (the Christoffel symbols were calculated in Sect. 8.2)

$$\frac{\partial T^{rr}}{\partial r} + \Gamma^r_{tt} T^{tt} + \Gamma^r_{rr} T^{rr} + \Gamma^r_{\theta\theta} T^{\theta\theta} + \Gamma^r_{\phi\phi} T^{\phi\phi}$$

$$+ \left(\Gamma^t_{rt} + \Gamma^r_{rr} + \Gamma^\theta_{r\theta} + \Gamma^\phi_{r\phi}\right) T^{rr} = 0 \,,$$

$$\frac{P'}{g} - \frac{Pg'}{g^2} + \frac{f'\rho}{2gf} + \frac{g'P}{2gg} - \frac{rP}{gr^2} - \frac{r\sin^2\theta}{g}\frac{P}{r^2\sin^2\theta}$$

$$+ \left(\frac{f'}{2f} + \frac{g'}{2g} + \frac{1}{r} + \frac{1}{r}\right)\frac{P}{g} = 0 \,,$$

$$\frac{P'}{g} + \frac{f'}{2fg}(\rho + P) = 0 \,,$$

$$P' = -\frac{f'}{2f}(\rho + P) \,. \tag{F.15}$$

From Eq. (F.10) we have

$$\frac{1}{g} - \frac{rg'}{g^2} = 1 - \frac{8\pi G_N}{c^4}\rho r^2 \,,$$

$$\frac{d}{dr}\frac{r}{g} = 1 - \frac{8\pi G_N}{c^4}\rho r^2 \,,$$

$$\frac{r}{g} = r - \frac{8\pi G_N}{c^4}\int_0^r \rho \tilde{r}^2 d\tilde{r} + C \,, \tag{F.16}$$

where C is an integration constant. For $r = 0$, we see that $C = 0$. If we employ Eq. (F.13), we find g

$$g = \frac{1}{1 - \frac{8\pi G_N}{3c^4}\rho r^2} \,. \tag{F.17}$$

We rewrite Eq. (F.15) as

$$\frac{d}{dr}(\rho + P) = \frac{dP}{dr} = -\frac{f'}{2f}(\rho + P) \,. \tag{F.18}$$

The solution is

$$\rho + P = \frac{C_1}{\sqrt{f}} \,, \tag{F.19}$$

where C_1 is a constant. We sum Eq. (F.10) with (F.11) and we get

$$\frac{g'}{rg^2} + \frac{f'}{rfg} = \frac{8\pi G_N}{c^4}(\rho + P) = \frac{8\pi G_N}{c^4}\frac{C_1}{\sqrt{f}} \,. \tag{F.20}$$

From Eq. (F.17) we can write

$$\frac{1}{g} = 1 - \frac{8\pi G_N}{3c^4}\rho r^2 \tag{F.21}$$

$$\frac{g'}{g^2} = -\frac{d}{dr}\frac{1}{g} = \frac{16\pi G_N}{3c^4}\rho r \,. \tag{F.22}$$

We combine Eqs. (F.20)–(F.22) to write

$$\frac{16\pi G_N}{3c^4}\rho + \frac{f'}{rf}\left(1 - \frac{8\pi G_N}{3c^4}\rho r^2\right) = \frac{8\pi G_N}{c^4}\frac{C_1}{\sqrt{f}} \,. \tag{F.23}$$

Let us define $h = \sqrt{f}$. Equation (F.23) can be rewritten as

$$\frac{16\pi G_N}{3c^4}\rho + \frac{2h'}{rh} - \frac{16\pi G_N}{3c^4}\rho r \frac{h'}{h} = \frac{8\pi G_N}{c^4}\frac{C_1}{h},$$

$$\frac{8\pi G_N}{3c^4}\rho r h + h' - \frac{8\pi G_N}{3c^4}\rho r^2 h' = \frac{4\pi G_N}{c^4} r C_1,$$

$$\frac{1 - \frac{8\pi G_N}{3c^4}\rho r^2}{\frac{8\pi G_N}{3c^4}\rho r}h' + h = \frac{3C_1}{2\rho}. \tag{F.24}$$

The homogeneous solution of this equation is

$$h = -C_2 \sqrt{1 - \frac{8\pi G_N}{3c^4}\rho r^2}, \tag{F.25}$$

where C_2 is a constant. An inhomogeneous solution is

$$h = \frac{3C_1}{2\rho}. \tag{F.26}$$

The function f is thus

$$f = \left(\frac{3C_1}{2\rho} - C_2\sqrt{1 - \frac{8\pi G_N}{3c^4}\rho r^2}\right)^2, \tag{F.27}$$

and the line element of the spacetime reads

$$ds^2 = \left(\frac{3C_1}{2\rho} - C_2\sqrt{1 - \frac{8\pi G_N}{3c^4}\rho r^2}\right)^2 dt^2$$

$$- \frac{dr^2}{1 - \frac{8\pi G_N}{3c^4}\rho r^2} + r^2\left(d\theta^2 + \sin^2\theta d\phi^2\right). \tag{F.28}$$

At this point we have two metrics. The Schwarzschild metric holds in the exterior region $r > r_0$ and is characterized by the parameter M. The matter solution holds in the interior region $r < r_0$ and has two parameters, C_1 and C_2. We can now link these constants by imposing physically reasonable conditions.

We require that at $r = r_0$, which is the surface separating the matter interior from the vacuum exterior, the metric is continuous and the pressure vanishes. The rr component of the metric tensor is continuous if

$$g_{\text{out}}(r_0) = g_{\text{in}}(r_0), \tag{F.29}$$

where g_{out} and g_{in} are, respectively, the g_{rr} coefficients of the exterior and of the interior solutions. Plugging in their explicit expression we have

$$1 - \frac{2G_N M}{c^2 r_0} = 1 - \frac{8\pi G_N}{3c^4} \rho r_0^2 , \tag{F.30}$$

and we find

$$M = \frac{4\pi}{3c^2} \rho r_0^3 . \tag{F.31}$$

M can thus be interpreted as the *effective mass* of the body generating the gravitational field. Note that $(4/3)\pi r_0^3$ is not the volume of the massive body. Indeed r_0 is only the value of the radial coordinate of its surface. If we had not imposed the equation of state in (F.13), $\rho = \rho(r)$, and Eq. (F.31) would read

$$M = \frac{4\pi}{c^2} \int_0^{r_0} \rho \tilde{r}^2 d\tilde{r} . \tag{F.32}$$

The total mass of the body in this spacetime should be given by

$$M' = \frac{4\pi}{c^2} \int_0^{r_0} \rho \frac{\tilde{r}^2 d\tilde{r}}{\sqrt{1 - \frac{8\pi G_N}{3c^4} \rho \tilde{r}^2}} , \tag{F.33}$$

and we can thus define as the *gravitational mass defect* the quantity

$$\Delta M = M' - M . \tag{F.34}$$

From Eqs. (F.19) and (F.27) we can write the pressure P as

$$P = \frac{C_1}{\sqrt{f}} - \rho = \frac{C_2 \rho \sqrt{1 - \frac{8\pi G_N}{3c^4} \rho r^2} - \frac{C_1}{2}}{\frac{3C_1}{2\rho} - C_2 \sqrt{1 - \frac{8\pi G_N}{3c^4} \rho r^2}} . \tag{F.35}$$

The condition $P(r_0) = 0$ reads

$$\frac{C_1}{2\rho} = C_2 \sqrt{1 - \frac{8\pi G_N}{3c^4} \rho r_0^2} \tag{F.36}$$

and links together the quantities C_1, C_2, and r_0.

Lastly, we impose that even the g_{tt} coefficient is continuous at the boundary $r = r_0$

$$f_{\text{out}}(r_0) = f_{\text{in}}(r_0) . \tag{F.37}$$

We find

$$1 - \frac{2G_{\mathrm{N}}M}{c^2 r_0} = \left(\frac{3C_1}{2\rho} - C_2 \sqrt{1 - \frac{8\pi G_{\mathrm{N}}}{3c^4} \rho r_0^2} \right)^2 . \tag{F.38}$$

Employing Eqs. (F.31) and (F.36), we find

$$1 - \frac{8\pi G_{\mathrm{N}}}{3c^4} \rho r_0^2 = 4C_2^2 \left(1 - \frac{8\pi G_{\mathrm{N}}}{3c^4} \rho r_0^2 \right) ,$$

$$C_2^2 = \frac{1}{4} . \tag{F.39}$$

The solution with the negative sign is not physical because it would imply $C_1 < 0$ and then $\rho + P < 0$. Eventually, the only solution is

$$M = \frac{4\pi}{3c^2} \rho r_0^3 ,$$

$$C_1 = \rho \sqrt{1 - \frac{8\pi G_{\mathrm{N}}}{3c^4} \rho r_0^2} ,$$

$$C_2 = \frac{1}{2} , \tag{F.40}$$

The three constants M, C_1, and C_2 are now completely determined by the energy density ρ and the radius of the body r_0.

The pressure P is given by

$$P = \frac{\sqrt{1 - \frac{8\pi G_{\mathrm{N}}}{3c^4} \rho r^2} - \sqrt{1 - \frac{8\pi G_{\mathrm{N}}}{3c^4} \rho r_0^2}}{3 \sqrt{1 - \frac{8\pi G_{\mathrm{N}}}{3c^4} \rho r_0^2} - \sqrt{1 - \frac{8\pi G_{\mathrm{N}}}{3c^4} \rho r^2}} \rho . \tag{F.41}$$

It remains finite at $r = 0$ if the denominator in (F.41) is larger than zero

$$3 \sqrt{1 - \frac{8\pi G_{\mathrm{N}}}{3c^4} \rho r_0^2} - 1 > 0$$

$$1 - \frac{8\pi G_{\mathrm{N}}}{3c^4} \rho r_0^2 > \frac{1}{9} ,$$

$$\frac{3\pi G_{\mathrm{N}}}{c^4} \rho r_0^2 < 1 . \tag{F.42}$$

We multiply both sides by r_0 and we find the condition

$$r_0 > \frac{3\pi G_{\mathrm{N}}}{c^4} \rho r_0^3 = \frac{9}{8} \left(\frac{2G_{\mathrm{N}}}{c^2} \frac{4\pi}{3c^2} \rho r_0^3 \right) = \frac{9}{8} \left(\frac{2G_{\mathrm{N}}M}{c^2} \right) = \frac{9}{8} r_{\mathrm{S}} , \tag{F.43}$$

where r_S is the Schwarzschild radius of the body. Our solution with constant energy density is only possible if the surface of the body satisfies Eq. (F.43). There is no solution for more compact objects.

Appendix G
Metric Around a Slow-Rotating Massive Body

In this appendix, we want to derive the metric around a slow-rotating and quasi-Newtonian (i.e. the gravitational field is weak) massive body. The result can be used to see that the parameter a in the Kerr metric is the specific spin of the black hole.

For simplicity, we consider a spherically symmetric and rigidly rotating homogeneous ball of dust. The matter energy-momentum tensor reduces to $T^{\mu\nu} = \rho u^\mu u^\nu$, where ρ is the mass density (not the energy density),

$$u^\mu = (\gamma c, \gamma \mathbf{v}) \qquad (G.1)$$

is the 4-velocity of every element of the ball of dust, and $\mathbf{v}$ is the 3-velocity. Let us assume that the object is rotating in the xy plane. The Lorentz factor of every element of the ball of dust is

$$\gamma = \frac{1}{\sqrt{1 - \mathbf{v}^2/c^2}} = 1 + O\left(\Omega^2\right), \qquad (G.2)$$

where $\mathbf{v}^2 = \Omega^2(x^2 + y^2)$ and $\Omega = $ constant is the angular velocity. The 3-velocity is $\mathbf{v} = (-\Omega y, \Omega x, 0)$. The matter energy-momentum tensor is

$$||T^{\mu\nu}|| = \begin{pmatrix} \rho c^2 & -\rho c \Omega y & \rho c \Omega x & 0 \\ -\rho c \Omega y & 0 & 0 & 0 \\ \rho c \Omega x & 0 & 0 & 0 \\ 0 & 0 & 0 & 0 \end{pmatrix} + O\left(\Omega^2\right), \qquad (G.3)$$

and, with lower indices,

$$||T_{\mu\nu}|| = \begin{pmatrix} \rho c^2 & \rho c \Omega y & -\rho c \Omega x & 0 \\ \rho c \Omega y & 0 & 0 & 0 \\ -\rho c \Omega x & 0 & 0 & 0 \\ 0 & 0 & 0 & 0 \end{pmatrix} + O\left(\Omega^2\right). \qquad (G.4)$$

© Springer Nature Singapore Pte Ltd. 2018, corrected publication 2020
C. Bambi, *Introduction to General Relativity*, Undergraduate Lecture Notes
in Physics, https://doi.org/10.1007/978-981-13-1090-4

We can now proceed as in Sect. 12.2. We write the metric $g_{\mu\nu}$ as the Minkowski metric plus a small perturbation

$$g_{\mu\nu} = \eta_{\mu\nu} + h_{\mu\nu}. \tag{G.5}$$

The Einstein equations provide the following solution for the trace-reversed perturbation $\tilde{h}_{\mu\nu}$

$$\tilde{h}_{\mu\nu} = \frac{4G_N}{c^4} \int d^3\mathbf{x}' \frac{T_{\mu\nu}(\mathbf{x}')}{|\mathbf{x} - \mathbf{x}'|}. \tag{G.6}$$

Note that here, unlike in Sect. 12.2, $T_{\mu\nu}$ is independent of time. If the ball of dust is at the origin of the coordinate system and we are interested in the metric at large radii, we can expand the term $1/|\mathbf{x} - \mathbf{x}'|$ inside the integral as

$$\frac{1}{|\mathbf{x} - \mathbf{x}'|} = \frac{1}{r} + \frac{x_i x'^i}{r^3} + \cdots. \tag{G.7}$$

For the tt-component we have

$$\tilde{h}_{tt} = \frac{4G_N}{c^2 r} \int d^3\mathbf{x}' \rho + \cdots = \frac{4G_N M}{c^2 r} + \cdots, \tag{G.8}$$

where M is the mass of the slow-rotating object. Since $\tilde{h} = \tilde{h}^\mu_\mu = -\tilde{h}_{tt}$, we find that the tt-component of the metric perturbation is

$$h_{tt} = \tilde{h}_{tt} - \frac{1}{2}\eta_{tt}\tilde{h} = \frac{2G_N M}{c^2 r}. \tag{G.9}$$

$\tilde{h}_{ij} = 0$ because $T_{ij} = 0$ (we ignore terms of order Ω^2 or higher because the rotation is slow). The ij-components of the metric perturbation are

$$h_{ij} = -\frac{1}{2}\eta_{ij}\tilde{h} = \begin{cases} 0 & \text{if } i \neq j, \\ \frac{2G_N M}{c^2 r} & \text{if } i = j. \end{cases} \tag{G.10}$$

Lastly, we have the terms h_{ti}s. $\tilde{h}_{ti} = h_{ti}$ because $\eta_{ti} = 0$. $h_{tz} = 0$ because $T_{tz} = 0$. For the tx-component we have

$$\begin{aligned} h_{tx} &= \frac{4G_N}{c^3 r} \int d^3\mathbf{x}' \rho\Omega y' + \frac{4G_N}{c^3 r^3} \int d^3\mathbf{x}' \rho\Omega y' (xx' + yy' + zz') + \cdots \\ &= \frac{4G_N}{c^3 r^3} \int d^3\mathbf{x}' \rho\Omega yy'^2 + \cdots, \end{aligned} \tag{G.11}$$

and for the ty component we have the same expression exchanging x and y and adding a minus sign

$$h_{ty} = -\frac{4G_N}{c^3 r^3} \int d^3x' \rho \Omega x x'^2 + \cdots .$$ (G.12)

We introduce the spin angular momentum of the object J as

$$J = 2 \int d^3x' \rho \Omega x'^2 = 2 \int d^3x' \rho \Omega y'^2 = \int d^3x' \rho \Omega \left(x'^2 + y'^2\right) , \quad \text{(G.13)}$$

and the h_{ti}s terms can be written as

$$h_{tx} = \frac{2G_N J y}{c^3 r^3} + \cdots , \quad h_{ty} = -\frac{2G_N J x}{c^3 r^3} + \cdots , \quad h_{tz} = 0 . \quad \text{(G.14)}$$

Considering only the leading order terms in $h_{\mu\nu}$, the line element of the spacetime at large radii reads

$$ds^2 = -\left(1 - \frac{2G_N M}{c^2 r}\right) c^2 dt^2 + \frac{4G_N J y}{c^2 r^3} dt dx - \frac{4G_N J x}{c^2 r^3} dt dy$$
$$+ \left(1 + \frac{2G_N M}{c^2 r}\right) \left(dx^2 + dy^2 + dz^2\right) . \quad \text{(G.15)}$$

Let us now rewrite the line element in spherical coordinates (ct, r, θ, ϕ). The relation between Cartesian and spherical coordinates is

$$\begin{aligned} t &= t , \\ x &= r \sin\theta \cos\phi , \\ y &= r \sin\theta \sin\phi , \\ z &= r \cos\theta . \end{aligned} \quad \text{(G.16)}$$

The metric tensor transforms as

$$g'_{\mu\nu} = \frac{\partial x^\alpha}{\partial x'^\mu} \frac{\partial x^\beta}{\partial x'^\nu} g_{\alpha\beta} . \quad \text{(G.17)}$$

g_{tt} does not change, because there is no mixing between the time and the space coordinates. g_{ij}s change as in the Euclidean space (see Sect. 1.2). g_{ti}s vanish except $g_{t\phi}$

$$\begin{aligned} g_{tr} &= \frac{\partial t}{\partial t} \frac{\partial x}{\partial r} g_{tx} + \frac{\partial t}{\partial t} \frac{\partial y}{\partial r} g_{ty} \\ &= \sin\theta \cos\phi \left(\frac{2G_N J r \sin\theta \sin\phi}{c^3 r^3}\right) + \sin\theta \sin\phi \left(-\frac{2G_N J r \sin\theta \cos\phi}{c^3 r^3}\right) = 0 , \\ g_{t\theta} &= \frac{\partial t}{\partial t} \frac{\partial x}{\partial \theta} g_{tx} + \frac{\partial t}{\partial t} \frac{\partial y}{\partial \theta} g_{ty} \\ &= r \cos\theta \cos\phi \left(\frac{2G_N J r \sin\theta \sin\phi}{c^3 r^3}\right) + r \cos\theta \sin\phi \left(-\frac{2G_N J r \sin\theta \cos\phi}{c^3 r^3}\right) = 0 , \end{aligned}$$

$$
\begin{aligned}
g_{t\phi} &= \frac{\partial t}{\partial t}\frac{\partial x}{\partial \phi}g_{tx} + \frac{\partial t}{\partial t}\frac{\partial y}{\partial \phi}g_{ty} \\
&= -r\sin\theta\sin\phi\left(\frac{2G_{N}Jr\sin\theta\sin\phi}{c^3 r^3}\right) + r\sin\theta\cos\phi\left(-\frac{2G_{N}Jr\sin\theta\cos\phi}{c^3 r^3}\right) \\
&= -\frac{2G_{N}J\sin^2\theta}{c^3 r}.
\end{aligned}
\tag{G.18}
$$

The line element in spherical coordinates thus reads

$$
\begin{aligned}
ds^2 = &-\left(1 - \frac{2G_{N}M}{c^2 r}\right)c^2 dt^2 - \frac{4G_{N}aM\sin^2\theta}{c^2 r}dtd\phi \\
&+ \left(1 + \frac{2G_{N}M}{c^2 r}\right)\left(dr^2 + r^2 d\theta^2 + r^2\sin^2\theta d\phi^2\right),
\end{aligned}
\tag{G.19}
$$

where we have introduced the specific spin $a = J/M$. These coordinates are still isotropic. If we want Boyer–Lindquist-like coordinates, we need another coordinate transformation, as done in Sect. 9.5.

Appendix H
Friedmann–Robertson–Walker Metric

Alexander Friedmann in the early 1920s and, independently, Georges Lemaitre in the late 1920s and early 1930s were the first to employ the Friedman–Robertson–Walker metric and study the corresponding cosmological models assuming the Einstein equations. In the mid 1930s, Howard Percy Robertson and Arthur Geoffrey Walker rigorously proved that the Friedman–Robertson–Walker metric is the only geometry compatible with the Cosmological Principle. Such a statement is independent of the field equations of the gravity theory, which can only determine the scale factor $a(t)$. In this appendix, we want to outline a possible derivation of the Friedman–Robertson–Walker metric.

We want to obtain the most general metric describing a spatially homogeneous and isotropic spacetime. Isotropy means that there are no preferred directions: the spacetime should thus look spherically symmetric and we can proceed as in Sect. 8 1, finding the line element in Eq. (8.5). We can then consider a coordinate transformation to remove the off-diagonal metric coefficient and we get the metric

$$ds^2 = -f(t, r)c^2dt^2 + g(t, r)\left(dr^2 + r^2d\theta^2 + r^2\sin^2\theta d\phi^2\right). \qquad \text{(H.1)}$$

Unlike the spherically symmetric spacetime in Chap. 8, here the spacetime is also spatially homogeneous; that is, there are no preferred points. This means, in particular, that the clock of any static observer should measure the same time and therefore $f = f(t)$ because it cannot depend on the radial coordinate r. If g_{tt} only depends on the time coordinate, we can always consider the transformation (which is a redefinition of time and is called *synchronization*)

$$dt \rightarrow dt' = \sqrt{f}dt, \qquad \text{(H.2)}$$

and set $g_{tt} = -1$. Because of isotropy, we can write $g(t, r) = a^2(t)h(r)$ and the line elements becomes

$$ds^2 = -c^2dt^2 + a^2(t)dl^2, \qquad \text{(H.3)}$$

© Springer Nature Singapore Pte Ltd. 2018, corrected publication 2020
C. Bambi, *Introduction to General Relativity*, Undergraduate Lecture Notes
in Physics, https://doi.org/10.1007/978-981-13-1090-4

where dl^2 is given by

$$
\begin{aligned}
dl^2 &= h(r)\left(dr^2 + r^2 d\theta^2 + r^2 \sin^2\theta d\phi^2\right) \\
&= h(r)\left(dx^2 + dy^2 + dz^2\right) .
\end{aligned}
\tag{H.4}
$$

It is straightforward to calculate the scalar curvature of the 3-metric g_{ij} with a Mathematica package for tensor calculus (see Appendix E). We find

$$
R = \frac{3h'^2 - h\left(8h'/r + 4h''\right)}{2h^3} .
\tag{H.5}
$$

Because of homogeneity, R must be spatially constant. Imposing this condition, the solution for the function h is

$$
h(r) = \frac{1}{\left(1 + kr^2/4\right)^2} ,
\tag{H.6}
$$

where $k = 0, \pm 1$. $R = 6k$ and therefore $R > 0$, < 0, and 0 for, respectively, $k = 1, -1$, and 0. The line element of the spacetime turns out to be

$$
\begin{aligned}
ds^2 &= -c^2 dt^2 + a^2(t)\frac{dx^2 + dy^2 + dz^2}{\left(1 + kr^2/4\right)^2} \\
&= -c^2 dt^2 + a^2(t)\frac{dr^2 + r^2 d\theta^2 + r^2 \sin^2\theta d\phi^2}{\left(1 + kr^2/4\right)^2} .
\end{aligned}
\tag{H.7}
$$

Note that we have obtained the most general expression for a metric describing a spatially homogeneous and isotropic spacetime. We can always rescale r to have $k = 0, \pm 1$, and therefore other values of k do not represent different metrics but just the same metric with a different radial coordinate.

With the following transformation for the radial coordinate

$$
\tilde{r} = \frac{r}{1 + kr^2/4} ,
\tag{H.8}
$$

we get the Friedman–Robertson–Walker metric in the coordinates employed in Chap. 11

$$
ds^2 = -c^2 dt^2 + a^2(t)\left[\frac{d\tilde{r}^2}{1 - k\tilde{r}^2} + \tilde{r}^2 d\theta^2 + \tilde{r}^2 \sin^2\theta d\phi^2\right] .
\tag{H.9}
$$

Let us check that the transformation in (H.8) transforms the line element in (H.7) into the line element in (H.9). For $g_{\theta\theta}$ and $g_{\phi\phi}$, it is easy to see that

$$\frac{r^2}{\left(1 + kr^2/4\right)^2} \left(d\theta^2 + \sin^2\theta \, d\phi^2\right) = \tilde{r}^2 \left(d\theta^2 + \sin^2\theta \, d\phi^2\right) . \qquad \text{(H.10)}$$

For $g_{\tilde{r}\tilde{r}}$, we can write

$$
\begin{aligned}
g_{rr} &= \frac{\partial \tilde{r}}{\partial r} \frac{\partial \tilde{r}}{\partial r} g_{\tilde{r}\tilde{r}} \\
&= \frac{\left(1 - kr^2/4\right)}{\left(1 + kr^2/4\right)^2} \frac{\left(1 - kr^2/4\right)}{\left(1 + kr^2/4\right)^2} \frac{1}{1 - k\tilde{r}^2} \\
&= \frac{\left(1 - kr^2/4\right)}{\left(1 + kr^2/4\right)^2} \frac{\left(1 - kr^2/4\right)}{\left(1 + kr^2/4\right)^2} \frac{\left(1 + kr^2/4\right)^2}{\left(1 + kr^2/4\right)^2 - kr^2} \\
&= \frac{1}{\left(1 + kr^2/4\right)^2} , \qquad\qquad\qquad\qquad\qquad\qquad \text{(H.11)}
\end{aligned}
$$

and we obtain the correct metric coefficient in (H.7).

Appendix I
Suggestions for Solving the Problems

I.1   Chapter 1

Problem 1.1: In Cartesian coordinates $\{x^i\} = (x, y, z)$, the metric tensor is δ_{ij}. In spherical coordinates $\{x'^i\} = (r, \theta, \phi)$, the metric tensor is given by

$$g'_{ij} = \frac{\partial x^m}{\partial x'^i} \frac{\partial x^n}{\partial x'^j} \delta_{mn} . \tag{I.1}$$

For $i = j = r$, we have

$$g_{rr} = \frac{\partial x^m}{\partial r} \frac{\partial x^n}{\partial r} \delta_{mn} = \frac{\partial x}{\partial r} \frac{\partial x}{\partial r} + \frac{\partial y}{\partial r} \frac{\partial y}{\partial r} + \frac{\partial z}{\partial r} \frac{\partial z}{\partial r}$$
$$= \sin^2 \theta \cos^2 \phi + \sin^2 \theta \sin^2 \phi + \cos^2 \theta = 1 . \tag{I.2}$$

For $i = r$ and $j = \theta$, we have

$$g_{r\theta} = \frac{\partial x^m}{\partial r} \frac{\partial x^n}{\partial \theta} \delta_{mn} = \frac{\partial x}{\partial r} \frac{\partial x}{\partial \theta} + \frac{\partial y}{\partial r} \frac{\partial y}{\partial \theta} + \frac{\partial z}{\partial r} \frac{\partial z}{\partial \theta}$$
$$= r \sin \theta \cos \theta \cos^2 \phi + r \sin \theta \cos \theta \sin^2 \phi - r \sin \theta \cos \theta = 0 . \tag{I.3}$$

For $i = r$ and $j = \phi$, we have

$$g_{r\phi} = \frac{\partial x^m}{\partial r} \frac{\partial x^n}{\partial \phi} \delta_{mn} = \frac{\partial x}{\partial r} \frac{\partial x}{\partial \phi} + \frac{\partial y}{\partial r} \frac{\partial y}{\partial \phi} + \frac{\partial z}{\partial r} \frac{\partial z}{\partial \phi}$$
$$= -r \sin^2 \theta \cos \phi \sin \phi + r \sin^2 \theta \cos \phi \sin \phi = 0 . \tag{I.4}$$

We calculate the other components of the metric tensor in the same way and eventually we find that the only non-vanishing components are $g_{rr} = 1$, $g_{\theta\theta} = r^2$, and $g_{\phi\phi} = r^2 \sin^2 \theta$. The line element in spherical coordinates is thus

© Springer Nature Singapore Pte Ltd. 2018, corrected publication 2020
C. Bambi, *Introduction to General Relativity*, Undergraduate Lecture Notes
in Physics, https://doi.org/10.1007/978-981-13-1090-4

$$dl^2 = dr^2 + r^2 d\theta^2 + r^2 \sin^2 \theta d\phi^2 , \tag{I.5}$$

and we have verified Eq. (1.14).

Problem 1.2: We proceed as in Problem 1.1. In spherical coordinates $\{x^i\} = (r, \theta, \phi)$, the metric tensor g_{ij} is given in Eq. (1.15). In cylindrical coordinates $\{x'^i\} = (\rho, z, \phi')$, we calculate the metric tensor from

$$g'_{ij} = \frac{\partial x^m}{\partial x'^i} \frac{\partial x^n}{\partial x'^j} g_{mn} . \tag{I.6}$$

For $i = j = \rho$, we have

$$\begin{aligned}
g_{\rho\rho} &= \frac{\partial x^m}{\partial \rho} \frac{\partial x^n}{\partial \rho} g_{mn} = \frac{\partial r}{\partial \rho} \frac{\partial r}{\partial \rho} + \frac{\partial \theta}{\partial \rho} \frac{\partial \theta}{\partial \rho} r^2 + \frac{\partial \phi}{\partial \rho} \frac{\partial \phi}{\partial \rho} r^2 \sin^2 \theta \\
&= \frac{\rho}{\sqrt{\rho^2 + z^2}} \frac{\rho}{\sqrt{\rho^2 + z^2}} + \frac{z}{\rho^2 + z^2} \frac{z}{\rho^2 + z^2} \left(\rho^2 + z^2 \right) = 1 .
\end{aligned} \tag{I.7}$$

For $i = \rho$ and $j = z$, we have

$$\begin{aligned}
g_{\rho z} &= \frac{\partial x^m}{\partial \rho} \frac{\partial x^n}{\partial z} g_{mn} = \frac{\partial r}{\partial \rho} \frac{\partial r}{\partial z} + \frac{\partial \theta}{\partial \rho} \frac{\partial \theta}{\partial z} r^2 + \frac{\partial \phi}{\partial \rho} \frac{\partial \phi}{\partial z} r^2 \sin^2 \theta \\
&= \frac{\rho}{\sqrt{\rho^2 + z^2}} \frac{z}{\sqrt{\rho^2 + z^2}} + \frac{z}{\rho^2 + z^2} \frac{-\rho}{\rho^2 + z^2} \left(\rho^2 + z^2 \right) = 0 .
\end{aligned} \tag{I.8}$$

For $i = \rho$ and $j = \phi'$, we have

$$g_{\rho\phi'} = \frac{\partial x^m}{\partial \rho} \frac{\partial x^n}{\partial \phi'} g_{mn} = \frac{\partial r}{\partial \rho} \frac{\partial r}{\partial \phi'} + \frac{\partial \theta}{\partial \rho} \frac{\partial \theta}{\partial \phi'} r^2 + \frac{\partial \phi}{\partial \rho} \frac{\partial \phi}{\partial \phi'} r^2 \sin^2 \theta = 0 . \tag{I.9}$$

We calculate the other components of the metric tensor in the same way and eventually we find that the only non-vanishing components are $g_{\rho\rho} = 1, g_{zz} = 1$, and $g_{\phi'\phi'} = \rho^2$. The line element in cylindrical coordinates is thus

$$dl^2 = d\rho^2 + dz^2 + \rho^2 d\phi'^2 . \tag{I.10}$$

Problem 1.3: The Jacobian of the inverse transformation of the transformation in Eq. (1.36) is

$$\left\| \frac{\partial x^m}{\partial x'^i} \right\| = ||\delta_i^m|| . \tag{I.11}$$

Hence the new metric is $g'_{ij} = \delta_i^m \delta_j^n \delta_{mn} = \delta_{ij}$.

Problem 1.4: The transformation from the Cartesian coordinates (x, y, z) to the Cartesian coordinates (x', y', z') is given by

$$x' = x \cos \theta + y \sin \theta ,$$
$$y' = -x \sin \theta + y \cos \theta ,$$
$$z' = z . \tag{I.12}$$

The inverse transformation is

$$x = x' \cos \theta - y' \sin \theta ,$$
$$y = x' \sin \theta + y' \cos \theta ,$$
$$z = z' . \tag{I.13}$$

We proceed as in the previous exercises to find the metric in the new coordinate system

$$g'_{ij} = \frac{\partial x^m}{\partial x'^i} \frac{\partial x^n}{\partial x'^j} \delta_{mn} = \frac{\partial x}{\partial x'^i} \frac{\partial x}{\partial x'^j} + \frac{\partial y}{\partial x'^i} \frac{\partial y}{\partial x'^j} + \frac{\partial z}{\partial x'^i} \frac{\partial z}{\partial x'^j} . \tag{I.14}$$

For $i = j = x'$, we have

$$g_{x'x'} = \cos^2 \theta + \sin^2 \theta = 1 . \tag{I.15}$$

For $i = x'$ and $j = y'$, we have

$$g_{x'y'} = -\cos \theta \sin \theta + \sin \theta \cos \theta = 0 . \tag{I.16}$$

For $i = x'$ and $j = z'$, we have

$$g_{x'y'} = 0 . \tag{I.17}$$

We can calculate all the metric components. The result is that $g'_{ij} = \delta_{ij}$ and the expression of the Euclidean metric does not change.

Problem 1.5: The Lagrangian of a free point-like particle is

$$L = \frac{1}{2} m g_{ij} \dot{x}^i \dot{x}^j . \tag{I.18}$$

In cylindrical coordinates, the metric tensor is (see Problem 1.2)

$$||g_{ij}|| = \begin{pmatrix} 1 & 0 & 0 \\ 0 & 1 & 0 \\ 0 & 0 & \rho^2 \end{pmatrix} . \tag{I.19}$$

Equation (1.18) thus becomes

$$L = \frac{1}{2}m \left(\dot{\rho}^2 + \dot{z}^2 + \rho^2 \dot{\phi}'^2 \right) . \tag{I.20}$$

The Euler–Lagrange equations are

$$\ddot{\rho} - \rho\dot{\phi}'^2 = 0, \quad \ddot{z} = 0, \quad \ddot{\phi}' + \frac{2}{\rho}\dot{\rho}\dot{\phi}' = 0. \tag{I.21}$$

Problem 1.6: We have just to match the geodesic equations

$$\ddot{x}^i + \Gamma^i_{jk}\dot{x}^j\dot{x}^k = 0, \tag{I.22}$$

with Eq. (I.21). It is straightforward to see that

$$\Gamma^\rho_{\phi'\phi'} = -\rho, \quad \Gamma^{\phi'}_{\rho\phi'} = \Gamma^{\phi'}_{\phi'\rho} = \frac{1}{\rho}, \tag{I.23}$$

and all other Christoffel symbols vanish.

Problem 1.7: Here the Lagrangian coordinates are (θ, ϕ). The Euler–Lagrange equations are

$$\frac{d}{dt}\frac{\partial L}{\partial \dot{\theta}} - \frac{\partial L}{\partial \theta} = 0, \quad \frac{d}{dt}\frac{\partial L}{\partial \dot{\phi}} - \frac{\partial L}{\partial \phi} = 0. \tag{I.24}$$

We find

$$\ddot{\theta} - \sin\theta\cos\theta\dot{\phi}^2 = 0, \quad \ddot{\phi} + 2\cot\theta\dot{\theta}\dot{\phi} = 0. \tag{I.25}$$

Problem 1.8: The Lagrangian does not explicitly depend on the time t, so we have the conservation of the energy E

$$E = \frac{\partial L}{\partial \dot{x}}\dot{x} + \frac{\partial L}{\partial \dot{y}}\dot{y} - L = \frac{1}{2}m \left(\dot{x}^2 + \dot{y}^2 \right) + \frac{1}{2}k \left(x^2 + y^2 \right) . \tag{I.26}$$

The Euler–Lagrange equations are

$$\ddot{x} + \frac{k}{m}x = 0, \quad \ddot{y} + \frac{k}{m}y = 0. \tag{I.27}$$

Problem 1.9: Let us consider the Galilean transformation

$$x' = x - vt, \quad y' = y, \quad z' = z, \quad t' = t, \tag{I.28}$$

We have

$$\frac{\partial}{\partial x^i} = \frac{\partial x'^m}{\partial x^i}\frac{\partial}{\partial x'^m} = \frac{\partial}{\partial x'^i} \tag{I.29}$$

and therefore $\nabla = \nabla'$. For the derivative of the temporal coordinate, we have

$$\frac{\partial}{\partial t} = -v\frac{\partial}{\partial x'} + \frac{\partial}{\partial t'}. \tag{I.30}$$

Maxwell's third and fourth equations would change to

$$\nabla \times \mathbf{E}' = -\frac{1}{c'+v}\left(\frac{\partial}{\partial t'} - v\frac{\partial}{\partial x'}\right)\mathbf{B}', \tag{I.31}$$

$$\nabla \times \mathbf{B}' = \frac{1}{c'+v}\left(\frac{\partial}{\partial t'} - v\frac{\partial}{\partial x'}\right)\mathbf{E}', \tag{I.32}$$

since $c' = c - v$ in Galilean relativity. Independently of the transformation rule of the electric and magnetic fields, in general these equations are not invariant under a Galilean transformation.

I.2   Chapter 2

Problem 2.1: The relation between Cartesian coordinates $\{x^i\} = (ct, x, y, z)$ and spherical coordinates $\{x''^i\} = (ct', r, \theta, \phi)$ is

$$t = t', \quad x = r\sin\theta\cos\phi, \quad y = r\sin\theta\sin\phi, \quad z = r\cos\theta, \tag{I.33}$$

with inverse

$$t' = t, \quad r = \sqrt{x^2 + y^2 + z^2}, \quad \theta = \arccos\left(\frac{z}{\sqrt{x^2 + y^2 + z^2}}\right),$$

$$\phi = \arctan\left(\frac{y}{x}\right). \tag{I.34}$$

As a tensor, $T^{\mu\nu}$ transforms according to the rule in Eq. (1.30). From the expression in Eq. (2.60), we find

$$\begin{aligned}
T'^{\mu\nu} &= \frac{\partial x'^\mu}{\partial x^\alpha}\frac{\partial x'^\nu}{\partial x^\beta}T^{\alpha\beta} \\
&= \frac{1}{c^2}\frac{\partial x'^\mu}{\partial t}\frac{\partial x'^\nu}{\partial t}\varepsilon + \frac{\partial x'^\mu}{\partial x}\frac{\partial x'^\nu}{\partial x}P + \frac{\partial x'^\mu}{\partial y}\frac{\partial x'^\nu}{\partial y}P + \frac{\partial x'^\mu}{\partial z}\frac{\partial x'^\nu}{\partial z}P
\end{aligned} \tag{I.35}$$

For instance, for $\mu = \nu = t'$ we have

$$T^{t't'} = \frac{\partial t'}{\partial t}\frac{\partial t'}{\partial t}\varepsilon = \varepsilon .$$ (I.36)

After calculating all the components, we find that the energy-momentum tensor of a perfect fluid in spherical coordinates and in the rest-frame of the fluid has the following form

$$||T'^{\mu\nu}|| = \begin{pmatrix} \varepsilon & 0 & 0 & 0 \\ 0 & P & 0 & 0 \\ 0 & 0 & \frac{P}{r^2} & 0 \\ 0 & 0 & 0 & \frac{P}{r^2 \sin^2 \theta} \end{pmatrix} .$$ (I.37)

$T'^{\mu}{}_{\nu}$ is obtained by lowering the second index with $g_{\mu\nu}$

$$T'^{\mu}{}_{\nu} = g_{\nu\rho} T'^{\mu\rho} .$$ (I.38)

The result is

$$||T'^{\mu}{}_{\nu}|| = \begin{pmatrix} -\varepsilon & 0 & 0 & 0 \\ 0 & P & 0 & 0 \\ 0 & 0 & P & 0 \\ 0 & 0 & 0 & P \end{pmatrix} .$$ (I.39)

Similarly, $T'_{\mu\nu} = g_{\mu\rho} T'^{\rho}{}_{\nu}$, and we find

$$||T'_{\mu\nu}|| = \begin{pmatrix} \varepsilon & 0 & 0 & 0 \\ 0 & P & 0 & 0 \\ 0 & 0 & Pr^2 & 0 \\ 0 & 0 & 0 & Pr^2 \sin^2 \theta \end{pmatrix} .$$ (I.40)

Problem 2.2: We have to move from the Cartesian coordinates $\{x^{\mu}\}$ to the Cartesian coordinates $\{x'^{\mu}\}$, where

$$\frac{\partial x'^{\mu}}{\partial x^{\alpha}} = \Lambda^{\mu}{}_{\alpha}$$ (I.41)

and $\Lambda^{\mu}{}_{\alpha}$ is the transformation in Eq. (2.8). The energy-momentum tensor in the coordinates $\{x'^{\mu}\}$ can be calculated from

$$T'^{\mu\nu} = \Lambda^{\mu}{}_{\alpha} \Lambda^{\nu}{}_{\beta} T^{\alpha\beta} .$$ (I.42)

For instance, for $\mu = \nu = t'$ we have

$$T^{t't'} = \Lambda^{t'}{}_{\alpha} \Lambda^{t'}{}_{\beta} T^{\alpha\beta} = \Lambda^{t'}{}_{t} \Lambda^{t'}{}_{t} \varepsilon + \Lambda^{t'}{}_{x} \Lambda^{t'}{}_{x} P = \gamma^2 \left(\varepsilon + \beta^2 P \right) .$$ (I.43)

The other components can be computed with the same procedure.

Problem 2.3: The Lorentz boost connecting the references frames (ct, x, y, z) and (ct', x', y', z') is

$$||\Lambda_1{}^\mu{}_\alpha|| = \begin{pmatrix} \gamma & -\gamma\beta & 0 & 0 \\ -\gamma\beta & \gamma & 0 & 0 \\ 0 & 0 & 1 & 0 \\ 0 & 0 & 0 & 1 \end{pmatrix}, \tag{I.44}$$

where $\beta = v/c$, $\gamma = 1/\sqrt{1 - \beta^2}$, and we have $x'^\mu = \Lambda_1{}^\mu{}_\alpha x^\alpha$. The Lorentz boost connecting the references frames (ct', x', y', z') and (ct'', x'', y'', z'') is

$$||\Lambda_2{}^\mu{}_\alpha|| = \begin{pmatrix} \gamma' & -\gamma'\beta' & 0 & 0 \\ -\gamma'\beta' & \gamma' & 0 & 0 \\ 0 & 0 & 1 & 0 \\ 0 & 0 & 0 & 1 \end{pmatrix}, \tag{I.45}$$

where $\beta' = v'/c$, $\gamma' = 1/\sqrt{1 - \beta'^2}$, and we have $x''^\mu = \Lambda_2{}^\mu{}_\alpha x'^\alpha$. The Lorentz boost connecting the references frames (ct, x, y, z) and (ct'', x'', y'', z'') can be found from

$$\Lambda_3{}^\mu{}_\alpha = \Lambda_2{}^\mu{}_\sigma \Lambda_1{}^\sigma{}_\alpha. \tag{I.46}$$

Problem 2.4: We can just show that the matrices in Eqs. (2.28) and (2.29) do not commute. For instance

$$\Lambda_x{}^\mu{}_\sigma \Lambda_y{}^\sigma{}_\alpha \neq \Lambda_y{}^\mu{}_\sigma \Lambda_x{}^\sigma{}_\alpha. \tag{I.47}$$

Problem 2.5: Let us indicate with Δt a time interval measured by a clock on Earth and with $\Delta\tau$ the same time interval measured by a clock on one of these satellites. Considering only the effect of the orbital motion of the satellite, we have

$$\Delta t = \frac{\Delta\tau}{\sqrt{1 - \beta^2}} \approx \left(1 + \frac{\beta^2}{2}\right) \Delta\tau = \left(1 + 8.4 \cdot 10^{-11}\right) \Delta\tau, \tag{I.48}$$

where $\beta = v/c = 1.3 \cdot 10^{-5}$ is the satellite speed in units of the speed of light.

I.3   Chapter 3

Problem 3.1: We follow the 4-dimensional formalism. From the action in Eq. (3.22), in spherical coordinates the Lagrangian is

$$L = -\frac{1}{2}m \left(c^2 \dot{t}^2 - \dot{r}^2 - r^2\dot{\theta}^2 - r^2 \sin^2\theta\dot{\phi}^2 \right) . \tag{I.49}$$

We employ Eq. (3.23) to calculate the components of the conjugate momentum. For $\mu = t$, we have

$$p_t = \frac{\partial L}{\partial \dot{x}^0} = \frac{1}{c}\frac{\partial L}{\partial \dot{t}} = -mc\dot{t} . \tag{I.50}$$

For $\mu = r, \theta$, and ϕ, we find

$$p_r = m\dot{r} , \quad p_\theta = mr^2\dot{\theta} , \quad p_\phi = mr^2 \sin^2\theta\dot{\phi} . \tag{I.51}$$

The components of the 4-momentum can be obtained raising the index in p_μ with the inverse of the metric tensor. For $\mu = t$, we have

$$p^t = g^{t\mu} p_\mu = mc\dot{t} . \tag{I.52}$$

Similarly, for the spatial components we get

$$p^r = m\dot{r} , \quad p^\theta = m\dot{\theta} , \quad p^\phi = m\dot{\phi} . \tag{I.53}$$

Problem 3.2: The Lagrangian in Eq. (I.49) does not depend on the coordinates t and ϕ, and therefore we have the conservation of the energy E and of the axial component of the angular momentum L_z

$$E = -p_t = mc\dot{t} , \quad L_z = p_\phi = mr^2 \sin^2\theta\dot{\phi} . \tag{I.54}$$

Note that the energy E is defined as $-p_t$. If we adopt a metric with signature $(+ - --)$, as it is common in particle physics, we would define the energy $E = p_t$ and the axial component of the angular momentum $L_z = -p_\phi$. The system has also a third constant of motion, which is associated to the conservation of the norm of the 4-velocity and follows from Eq. (3.24).

Problem 3.3: In cylindrical coordinates, the Lagrangian in (I.49) becomes

$$L = -\frac{1}{2}m \left(c^2 \dot{t}^2 - \dot{\rho}^2 - \dot{z}^2 - \rho^2\dot{\phi}^2 \right) . \tag{I.55}$$

The components of the conjugate momentum are

$$p_t = -mc\dot{t} , \quad p_\rho = m\dot{\rho} , \quad p_z = m\dot{z} , \quad p_\phi = m\rho^2\dot{\phi} . \tag{I.56}$$

The components of the 4-momentum are obtained raising the index of p_μ with the inverse of the metric tensor

$$p^t = mc\dot{t}, \quad p^\rho = m\dot{\rho}, \quad p^z = m\dot{z}, \quad p^\phi = m\dot{\phi}. \tag{I.57}$$

The Lagrangian does not depend on the coordinates t, z, and ϕ, so we have the conservation of the energy $-p_t$, of the momentum along the z axis p_z ($=p^z$), and of the axial component of the angular momentum p_ϕ.

Problem 3.4: We assume that the high-energy photon moves in the xy plane and the CMB photon moves along the x axis. Their 4-momenta are, respectively,

$$||p_\gamma^\mu|| = (p, p\cos\theta, p\sin\theta, 0), \quad ||p_{CMB}^\mu|| = (q, q, 0, 0), \tag{I.58}$$

The reaction is energetically allowed when

$$- p_i^\mu p_\mu^i \geq 4m_e^2 c^2, \tag{I.59}$$

where $p_i^\mu = p_\gamma^\mu + p_{CMB}^\mu$. We find

$$p^2 + q^2 + 2pq - p^2\cos^2\theta - q^2 - 2pq\cos\theta - p^2\sin^2\theta \geq 4m_e^2 c^2,$$
$$2pq\,(1 - \cos\theta) \geq 4m_e^2 c^2. \tag{I.60}$$

The average energy of CMB photons is $\langle qc \rangle = 2 \cdot 10^{-4}$ eV. Ignoring the term $pq\cos\theta$ in Eq. (I.60), we find that the threshold energy for the high-energy photon is $E_\gamma \sim 10^{15}$ eV.

Problem 3.5: The binding energy of iron-56 is

$$
\begin{aligned}
E_B &= (26 \cdot m_p c^2 + 30 \cdot m_n c^2 - Mc^2) \\
&= (26 \cdot 0.938 + 30 \cdot 0.940 - 52.103)\,\text{GeV} = 485\,\text{MeV}. \tag{I.61}
\end{aligned}
$$

The binding energy per nucleon is $\varepsilon_B = E_B/56 = 8.7$ MeV.

Problem 3.6: The first part in the Euler–Lagrange equation is

$$
\begin{aligned}
& \frac{1}{\sqrt{-g}} \frac{\partial}{\partial x^\sigma} \left[\sqrt{-g} \frac{\partial \mathscr{L}}{\partial (\partial_\sigma \phi)} \right] \\
&= -\frac{1}{\sqrt{-g}} \frac{\partial}{\partial x^\sigma} \left[\sqrt{-g} \frac{\hbar}{2} g^{\mu\nu} \frac{\partial (\partial_\mu \phi)}{\partial (\partial_\sigma \phi)} (\partial_\nu \phi) + \sqrt{-g} \frac{\hbar}{2} g^{\mu\nu} (\partial_\mu \phi) \frac{\partial (\partial_\nu \phi)}{\partial (\partial_\sigma \phi)} \right] \\
&= -\frac{1}{\sqrt{-g}} \frac{\partial}{\partial x^\sigma} \left[\sqrt{-g} \frac{\hbar}{2} g^{\mu\nu} \delta_\mu^\sigma (\partial_\nu \phi) + \sqrt{-g} \frac{\hbar}{2} g^{\mu\nu} (\partial_\mu \phi) \delta_\nu^\sigma \right] \\
&= -\frac{\hbar}{\sqrt{-g}} \frac{\partial}{\partial x^\sigma} \left[\sqrt{-g}\, g^{\nu\sigma} (\partial_\nu \phi) \right] = -\hbar \Box \phi, \tag{I.62}
\end{aligned}
$$

where we have introduced the operator $\Box$

$$\Box = \frac{1}{\sqrt{-g}} \frac{\partial}{\partial x^\mu} \sqrt{-g} \, g^{\mu\nu} \frac{\partial}{\partial x^\nu} \tag{I.63}$$

The second part in the Euler–Lagrange equation reads

$$\frac{\partial \mathcal{L}}{\partial \phi} = -\frac{m^2 c^2}{\hbar} \phi \,. \tag{I.64}$$

In the end, the Euler–Lagrange equation can be written in the following form

$$\left(\Box - \frac{m^2 c^2}{\hbar^2}\right)\phi = 0 \,. \tag{I.65}$$

This is the Klein–Gordon equation.

Problem 3.7: The problem reduces to writing the operator $\Box$ in Cartesian and spherical coordinates. In Cartesian coordinates, this is trivial, as $g^{\mu\nu} = \mathrm{diag}(-1, 1, 1, 1)$ and therefore

$$\Box = \eta^{\mu\nu} \frac{\partial^2}{\partial x^\mu \partial x^\nu} = -\frac{1}{c^2}\frac{\partial^2}{\partial t^2} + \frac{\partial^2}{\partial x^2} + \frac{\partial^2}{\partial y^2} + \frac{\partial^2}{\partial z^2} \,. \tag{I.66}$$

In spherical coordinates, we have $\sqrt{-g} = r^2 \sin\theta$ and therefore

$$\Box = -\frac{1}{c^2}\frac{\partial^2}{\partial t^2} + \frac{\partial^2}{\partial r^2} + \frac{2}{r}\frac{\partial}{\partial r} + \frac{1}{r^2}\frac{\partial^2}{\partial \theta^2} + \frac{\cot\theta}{r^2}\frac{\partial}{\partial \theta} + \frac{1}{r^2 \sin^2\theta}\frac{\partial^2}{\partial \phi^2} \,. \tag{I.67}$$

Problem 3.8: From Eq. (I.62) we know that

$$\frac{\partial \mathcal{L}}{\partial \left(\partial_\mu \phi\right)} = -\hbar \, g^{\mu\sigma} \left(\partial_\sigma \phi\right) = -\hbar \left(\partial^\mu \phi\right) \,. \tag{I.68}$$

The energy-momentum tensor is thus

$$T^\mu_\nu = \hbar \left(\partial^\mu \phi\right)\left(\partial_\nu \phi\right) - \delta^\mu_\nu \left[\frac{\hbar}{2}\eta^{\sigma\rho} \left(\partial_\sigma \phi\right)\left(\partial_\rho \phi\right) + \frac{1}{2}\frac{m^2 c^2}{\hbar}\phi^2\right] \,, \tag{I.69}$$

or

$$T^{\mu\nu} = \hbar \left(\partial^\mu \phi\right)\left(\partial^\nu \phi\right) - \frac{\hbar}{2}\eta^{\mu\nu}\left[\eta^{\sigma\rho} \left(\partial_\sigma \phi\right)\left(\partial_\rho \phi\right) + \frac{m^2 c^2}{\hbar^2}\phi^2\right] \,. \tag{I.70}$$

Problem 3.9: In Eq. (3.104) we see $\eta^{\mu\nu}$, which is the metric tensor in Cartesian coordinates. If we replace $\eta^{\mu\nu}$ with $g^{\mu\nu}$, we have

$$T^{\mu\nu} = (\varepsilon + P)\frac{U^\mu U^\nu}{c^2} + P g^{\mu\nu} \,. \tag{I.71}$$

This is the energy-momentum tensor of a perfect fluid in a general coordinate system. For spherical coordinates $g_{\mu\nu} = \text{diag}\left(-1, 1, r^2, r^2 \sin^2\theta\right)$, and in the fluid rest-frame we have

$$||T^{\mu\nu}|| = \begin{pmatrix} \varepsilon & 0 & 0 & 0 \\ 0 & P & 0 & 0 \\ 0 & 0 & \frac{P}{r^2} & 0 \\ 0 & 0 & 0 & \frac{P}{r^2 \sin^2\theta} \end{pmatrix}, \quad ||T_{\mu\nu}|| = \begin{pmatrix} \varepsilon & 0 & 0 & 0 \\ 0 & P & 0 & 0 \\ 0 & 0 & Pr^2 & 0 \\ 0 & 0 & 0 & Pr^2 \sin^2\theta \end{pmatrix}. \quad (I.72)$$

I.4   Chapter 4

Problem 4.1: We can write

$$F^{\mu\nu} F_{\mu\nu} = F^{\mu t} F_{\mu t} + F^{\mu x} F_{\mu x} + F^{\mu y} F_{\mu y} + F^{\mu z} F_{\mu z}. \quad (I.73)$$

The first term on the right hand side is

$$F^{\mu t} F_{\mu t} = -E_x^2 - E_y^2 - E_z^2. \quad (I.74)$$

Similarly, we calculate the other terms

$$F^{\mu x} F_{\mu x} = -E_x^2 + B_z^2 + B_y^2,$$
$$F^{\mu y} F_{\mu y} = -E_y^2 + B_z^2 + B_x^2,$$
$$F^{\mu z} F_{\mu z} = -E_z^2 + B_y^2 + B_x^2. \quad (I.75)$$

So we find

$$F^{\mu\nu} F_{\mu\nu} = 2\left(\mathbf{B}^2 - \mathbf{E}^2\right), \quad (I.76)$$

where we have defined $\mathbf{B}^2 = B_x^2 + B_y^2 + B_z^2$ and $\mathbf{E}^2 = E_x^2 + E_y^2 + E_z^2$.

$\varepsilon^{\mu\nu\rho\sigma} F_{\mu\nu} F_{\rho\sigma}$ can be calculated with the same approach and the result is

$$\varepsilon^{\mu\nu\rho\sigma} F_{\mu\nu} F_{\rho\sigma} = -8\mathbf{E} \cdot \mathbf{B}, \quad (I.77)$$

where $\mathbf{E} \cdot \mathbf{B} = E_x B_x + E_y B_y + E_z B_z$.

Problem 4.2: Let us write the component i of the left hand side. We have

$$[\nabla (\mathbf{V} \cdot \mathbf{W})]^i = \partial^i \left(V^j W_j\right) = \left(\partial^i V^j\right) W_j + V^j \left(\partial^i W_j\right). \quad (I.78)$$

Let us now consider the right hand side. The component i of the first term is

$$[(\mathbf{W} \cdot \nabla) \mathbf{V}]^i = W^j \partial_j V^i. \quad (I.79)$$

The second term is similar to this with $\mathbf{V}$ and $\mathbf{W}$ exchanged. The component i of the third term is

$$[\mathbf{W} \times (\nabla \times \mathbf{V})]^i = \varepsilon^{ijk} W_j (\nabla \times \mathbf{V})_k = \varepsilon^{ijk} W_j \varepsilon_{klm} \partial^l V^m = \varepsilon^{kij} \varepsilon_{klm} W_j \partial^l V^m$$
$$= \left(\delta^i_l \delta^j_m - \delta^i_m \delta^j_l \right) W_j \partial^l V^m = W_j \partial^i V^j - W_j \partial^j V^i . \quad (I.80)$$

The fourth term is similar to the third term with $\mathbf{V}$ and $\mathbf{W}$ exchanged. If we combine Eqs. (I.79) and (I.80), we have

$$[(\mathbf{W} \cdot \nabla) \mathbf{V}]^i + [\mathbf{W} \times (\nabla \times \mathbf{V})]^i = W_j \left(\partial^i V^j \right) . \quad (I.81)$$

Similarly, we have

$$[(\mathbf{V} \cdot \nabla) \mathbf{W}]^i + [\mathbf{V} \times (\nabla \times \mathbf{W})]^i = V_j \left(\partial^i W^j \right) , \quad (I.82)$$

and the sum of Eqs. (I.81) and (I.82) gives the expression in Eq. (I.78).

Problem 4.3: The Faraday tensor in the first reference frame is

$$||F_{\mu\nu}|| = \begin{pmatrix} 0 & -E & 0 & 0 \\ E & 0 & 0 & 0 \\ 0 & 0 & 0 & 0 \\ 0 & 0 & 0 & 0 \end{pmatrix} , \quad (I.83)$$

The Faraday tensor in the second reference frame can be calculated from

$$F'_{\mu\nu} = \Lambda^\alpha_\mu \Lambda^\beta_\nu F_{\alpha\beta} = \Lambda^t_\mu \Lambda^x_\nu F_{tx} + \Lambda^x_\mu \Lambda^t_\nu F_{xt} , \quad (I.84)$$

where

$$||\Lambda^\mu_\nu|| = \begin{pmatrix} \gamma & -\gamma\beta & 0 & 0 \\ -\gamma\beta & \gamma & 0 & 0 \\ 0 & 0 & 1 & 0 \\ 0 & 0 & 0 & 1 \end{pmatrix} , \quad (I.85)$$

$\beta = v/c$ and $\gamma = 1/\sqrt{1 - \beta^2}$. The result is that $F'_{\mu\nu} = F_{\mu\nu}$.

Problem 4.4: The faraday tensor is given in Eq. (I.83), so the only non-vanishing components are $F_{xt} = -F_{tx} = E$ and $F^{tx} = -F^{xt} = E$. We have $F^{\rho\sigma} F_{\rho\sigma} = -2E^2$. The energy momentum tensor is

$$||T^{\mu\nu}|| = \frac{1}{4\pi} ||F^{\mu\rho} F^\nu_\rho|| - \frac{1}{16\pi} ||\eta^{\mu\nu}|| F^{\rho\sigma} F_{\rho\sigma} = \frac{E^2}{8\pi} \begin{pmatrix} 1 & 0 & 0 & 0 \\ 0 & -1 & 0 & 0 \\ 0 & 0 & 1 & 0 \\ 0 & 0 & 0 & 1 \end{pmatrix} , \quad (I.86)$$

and its trace vanishes, $T^\mu_\mu = E^2/8\pi \left(-1 - 1 + 1 + 1\right) = 0$.

I.5 Chapter 5

Problem 5.1: From Eq. (5.50) we have

$$
\begin{aligned}
\nabla_\mu A_{\alpha\beta} &= \frac{\partial A_{\alpha\beta}}{\partial x^\mu} - \Gamma^\sigma_{\mu\alpha} A_{\sigma\beta} - \Gamma^\sigma_{\mu\beta} A_{\alpha\sigma} \, , \\
\nabla_\mu A^{\alpha\beta} &= \frac{\partial A^{\alpha\beta}}{\partial x^\mu} + \Gamma^\alpha_{\mu\sigma} A^{\sigma\beta} + \Gamma^\beta_{\mu\sigma} A^{\alpha\sigma} \, , \\
\nabla_\mu A^\alpha{}_\beta &= \frac{\partial A^\alpha{}_\beta}{\partial x^\mu} + \Gamma^\alpha_{\mu\sigma} A^\sigma{}_\beta - \Gamma^\sigma_{\mu\beta} A^\alpha{}_\sigma \, , \\
\nabla_\mu A_\alpha{}^\beta &= \frac{\partial A_\alpha{}^\beta}{\partial x^\mu} - \Gamma^\sigma_{\mu\alpha} A_\sigma{}^\beta + \Gamma^\beta_{\mu\sigma} A_\alpha{}^\sigma \, .
\end{aligned}
\tag{I.87}
$$

Problem 5.2: It is straightforward to compute these quantities with Cartesian coordinates, where the metric tensor is $\eta_{\mu\nu} = \mathrm{diag}(-1, 1, 1, 1)$. In Cartesian coordinates, all the components of the Riemann tensor vanish, and therefore all the components of the Ricci tensor and the scalar curvature vanish as well. From the transformation rules for tensors and scalars, we see that all the components of the Riemann tensor, the Ricci tensor, and the scalar curvature are identically zero in the Minkowski spacetime.

Problem 5.3: We write $R_{\alpha\mu\beta\nu}$ with the help of Eq. (5.76)

$$
\begin{aligned}
R_{\alpha\mu\beta\nu} &= \frac{1}{2} \left(\frac{\partial^2 g_{\alpha\nu}}{\partial x^\mu \partial x^\beta} + \frac{\partial^2 g_{\mu\beta}}{\partial x^\alpha \partial x^\nu} - \frac{\partial^2 g_{\alpha\beta}}{\partial x^\mu \partial x^\nu} - \frac{\partial^2 g_{\mu\nu}}{\partial x^\alpha \partial x^\beta} \right) \\
&\quad + g_{\kappa\lambda} \left(\Gamma^\lambda_{\mu\beta} \Gamma^\kappa_{\alpha\nu} - \Gamma^\lambda_{\mu\nu} \Gamma^\kappa_{\alpha\beta} \right) \, .
\end{aligned}
\tag{I.88}
$$

The Ricci tensor is

$$
\begin{aligned}
R_{\mu\nu} &= g^{\alpha\beta} R_{\alpha\mu\beta\nu} \\
&= \frac{1}{2} g^{\alpha\beta} \left(\frac{\partial^2 g_{\alpha\nu}}{\partial x^\mu \partial x^\beta} + \frac{\partial^2 g_{\mu\beta}}{\partial x^\alpha \partial x^\nu} \right) - \frac{1}{2} g^{\alpha\beta} \frac{\partial^2 g_{\alpha\beta}}{\partial x^\mu \partial x^\nu} - \frac{1}{2} g^{\alpha\beta} \frac{\partial^2 g_{\mu\nu}}{\partial x^\alpha \partial x^\beta} \\
&\quad + \frac{1}{2} g^{\alpha\beta} g_{\kappa\lambda} \Gamma^\lambda_{\mu\beta} \Gamma^\kappa_{\alpha\nu} - \frac{1}{2} g^{\alpha\beta} \Gamma^\lambda_{\mu\nu} \Gamma^\kappa_{\alpha\beta} \, ,
\end{aligned}
\tag{I.89}
$$

and is explicitly symmetric in the indices μ and ν.

I.6   Chapter 6

Problem 6.1: Since the metric matrix is diagonal, it is straightforward to find the vierbeins. $E_{(\alpha)}^{\mu}$ will be zero for $\alpha \neq \mu$ and $1/\sqrt{|g_{\mu\mu}|}$ for $\alpha = \mu$. We thus have

$$
E_{(t)} = \left(\frac{1}{\sqrt{f}}, 0, 0, 0 \right) , \quad E_{(r)} = \left(0, \sqrt{f}, 0, 0 \right) ,
$$

$$
E_{(\theta)} = \left(0, 0, \frac{1}{r}, 0 \right) , \quad E_{(\phi)} = \left(0, 0, 0, \frac{1}{r \sin \theta} \right) . \tag{I.90}
$$

Problem 6.2: If the observer has constant spatial coordinates, in the line element $dx^i = 0$ and therefore

$$
d\tau^2 = -g_{tt}dt^2 = \left(1 - \frac{r_{\text{Sch}}r}{r^2 + a^2 \cos^2 \theta} \right) dt^2 . \tag{I.91}
$$

Note that this requires $r > r_{\text{sl}} = r_{\text{Sch}}/2 + \sqrt{r_{\text{Sch}}^2/4 - a^2 \cos^2 \theta}$. For $r < r_{\text{sl}}$ we have $g_{tt} > 0$ and therefore an observer with constant spatial coordinates would follow a space-like trajectory, which is not allowed. Observers with $r < r_{\text{sl}}$ are allowed but they have to move. We will discuss this point in Sect. 10.3.3.

Problem 6.3: For a general reference frame, we simply have to replace the partial derivative ∂_μ with the covariant derivative ∇_μ. The result is that the equation now reads $\nabla_\mu J^\mu = 0$. Note that $\partial_\mu J^\mu = 0$ holds in an inertial reference frame in flat spacetime in Cartesian coordinates. $\nabla_\mu J^\mu = 0$ holds for any other case, including when we do not have a Cartesian coordinate system, or when the spacetime is flat but the reference frame is not inertial, or in a curved spacetime.

In a general reference frame, we have

$$
\nabla_\mu J^\mu = \frac{\partial J^\mu}{\partial x^\mu} + \Gamma_{\mu\nu}^{\mu} J^\nu = 0 . \tag{I.92}
$$

To write this expression in an inertial reference frame in flat spacetime in spherical coordinates, it is necessary to evaluate the Christoffel symbols and then plug them into Eq. (I.92). Alternatively, it is possible to employ the formula (5.64) and write

$$
\begin{aligned}
\nabla_\mu J^\mu &= \frac{1}{r^2 \sin \theta} \frac{\partial}{\partial t} \left(J^t r^2 \sin \theta \right) + \frac{1}{r^2 \sin \theta} \frac{\partial}{\partial r} \left(J^r r^2 \sin \theta \right) \\
&\quad + \frac{1}{r^2 \sin \theta} \frac{\partial}{\partial \theta} \left(J^\theta r^2 \sin \theta \right) + \frac{1}{r^2 \sin \theta} \frac{\partial}{\partial \phi} \left(J^\phi r^2 \sin \theta \right) \\
&= \frac{\partial J^t}{\partial t} + \frac{2 J^r}{r} + \frac{\partial J^r}{\partial r} + \frac{\partial J^\theta}{\partial \theta} + \cot \theta J^\theta + \frac{\partial J^\phi}{\partial \phi} .
\end{aligned} \tag{I.93}
$$

Problem 6.4: In the case of minimal coupling, we can just apply our standard recipe and replace partial derivatives with covariant derivatives. The result is

$$\left(\nabla_\mu \partial^\mu - \frac{m^2 c^2}{\hbar^2}\right)\phi = 0 \,. \tag{I.94}$$

Note that we have $\nabla_\mu \partial^\mu$ instead of $\nabla_\mu \nabla^\mu$ because ϕ is a scalar and therefore $\nabla^\mu \phi = \partial^\mu \phi$. Since $\partial^\mu \phi$s are the components of a vector field, we need the covariant derivative and therefore we write $\nabla_\mu \partial^\mu \phi$. If in flat spacetime in Cartesian coordinates we had the expression

$$\partial_\mu \partial^\mu A^\nu \,, \tag{I.95}$$

where A^ν is some vector field, then the generalization to curved spacetime would be

$$\nabla_\mu \nabla^\mu A^\nu \,. \tag{I.96}$$

In the case of non-minimal coupling, we have an extra term coming from $\partial \mathscr{L}/\partial \phi$ and the field equation becomes

$$\left(\nabla_\mu \partial^\mu - \frac{m^2 c^2}{\hbar^2} + \frac{2\xi R}{\hbar}\right)\phi = 0 \,. \tag{I.97}$$

Problem 6.5: We have to change sign in front of the term with $g^{\mu\nu}$. The Lagrangian density reads

$$\mathscr{L} = \frac{\hbar}{2} g^{\mu\nu} \left(\partial_\mu \phi\right) \left(\partial_\nu \phi\right) - \frac{1}{2}\frac{m^2 c^2}{\hbar}\phi^2 + \xi R \phi^2 \,. \tag{I.98}$$

The Klein–Gordon equation becomes

$$\left(\nabla_\mu \partial^\mu + \frac{m^2 c^2}{\hbar^2} - \frac{2\xi R}{\hbar}\right)\phi = 0 \,. \tag{I.99}$$

Problem 6.6: The Minkowski metric $\eta^{\mu\nu}$ has to be replaced by the general expression for the metric tensor $g^{\mu\nu}$. The energy-momentum tensor of a perfect fluid reads

$$T^{\mu\nu} = (\varepsilon + P)\frac{U^\mu U^\nu}{c^2} + P g^{\mu\nu} \,. \tag{I.100}$$

I.7   Chapter 7

Problem 7.1: We already know from Sect. 7.4 that the action

$$S = -\frac{\hbar}{2c} \int \left[g^{\mu\nu} (\partial_\mu \phi) (\partial_\nu \phi) + \frac{m^2 c^2}{\hbar^2} \phi^2 \right] \sqrt{-g} \, d^4 x \,, \tag{I.101}$$

leads to the energy-momentum tensor

$$T^{\mu\nu} = \hbar (\partial^\mu \phi) (\partial^\nu \phi) - \frac{\hbar}{2} g^{\mu\nu} \left[g^{\rho\sigma} (\partial_\rho \phi) (\partial_\sigma \phi) + \frac{m^2 c^2}{\hbar^2} \phi^2 \right] . \tag{I.102}$$

Now we have to evaluate the contribution from

$$S_{R\phi^2} = \frac{1}{c} \int \xi R \phi^2 \sqrt{-g} \, d^4 x \,. \tag{I.103}$$

Instead of Eq. (7.38), now we have

$$\delta S_{R\phi^2} = \frac{1}{c} \int \xi \phi^2 \left(\frac{1}{2} g^{\mu\nu} R - R^{\mu\nu} \right) \sqrt{-g} \, (\delta g_{\mu\nu}) \, d^4 x$$
$$+ \frac{1}{c} \int \xi \phi^2 \nabla_\rho H^\rho \sqrt{-g} \, d^4 x \,. \tag{I.104}$$

The first term on the right hand side leads to the left hand side of the Einstein equations with an effective Einstein constant $\kappa_{\text{eff}} = 1/(2\xi\phi^2)$. The second term on the right hand side cannot be ignored now, as we did in Sect. 7.3, and contributes to the energy-momentum tensor of the scalar field. In Eq. (7.37), H^ρ is written in terms of $\delta \Gamma^\kappa_{\mu\nu}$ and now we have to write it extracting $\delta g_{\mu\nu}$. After some tedious but straightforward calculation, we can recast the second term on the right hand side in Eq. (I.104) in the form (7.42) with the energy-momentum tensor

$$T^{\mu\nu}_{R\phi^2} = -2\xi \, (g^{\mu\nu} \Box - \nabla^\mu \partial^\nu) \, \phi^2 \,, \tag{I.105}$$

where $\Box = \nabla_\sigma \partial^\sigma$. The final energy-momentum tensor of the scalar field ϕ is $T^{\mu\nu}_\phi = T^{\mu\nu} + T^{\mu\nu}_{R\phi^2}$, where $T^{\mu\nu}$ is given in Eq. (I.102).

Problem 7.2: The equations of motion for the gravity sector are the Einstein equations and can be obtained by considering the variation $g_{\mu\nu} \to g'_{\mu\nu} = g_{\mu\nu} + \delta g_{\mu\nu}$. The result is

$$2\xi\phi^2 \left(R_{\mu\nu} - \frac{1}{2} g_{\mu\nu} R \right) = T^\phi_{\mu\nu} \,, \tag{I.106}$$

where $T^\phi_{\mu\nu}$ is the energy-momentum tensor of the scalar field found in Problem 7.1.

The equations of motion for the matter sector can be obtained by considering the variations in ϕ and $\partial_\mu \phi$. The result is

$$\left(\Box - \frac{m^2 c^2}{\hbar^2} + \frac{2\xi R}{\hbar} \right) \phi = 0 \,, \tag{I.107}$$

where $\Box = \nabla_\mu \partial^\mu$.

Problem 7.3: Without a cosmological constant, the action is

$$S = \frac{1}{2\kappa c} \int R\sqrt{-g}\, d^4x + S_m . \tag{I.108}$$

When we consider the variation $g_{\mu\nu} \to g'_{\mu\nu} = g_{\mu\nu} + \delta g_{\mu\nu}$, we find

$$\delta S = \frac{1}{2\kappa c} \int \left(\frac{1}{2} g^{\mu\nu} R - R^{\mu\nu} + \kappa T^{\mu\nu} \right) \sqrt{-g}\, (\delta g_{\mu\nu})\, d^4x , \tag{I.109}$$

which leads the the Einstein equations without a cosmological constant.

Now we want to include the cosmological constant Λ. The variation $g_{\mu\nu} \to g'_{\mu\nu} = g_{\mu\nu} + \delta g_{\mu\nu}$ should lead to

$$\delta S = \frac{1}{2\kappa c} \int \left(\frac{1}{2} g^{\mu\nu} R - R^{\mu\nu} - \Lambda g^{\mu\nu} + \kappa T^{\mu\nu} \right) \sqrt{-g}\, (\delta g_{\mu\nu})\, d^4x . \tag{I.110}$$

From Eq. (7.29), we see that the Einstein–Hilbert action should be

$$S'_{EH} = \frac{1}{2\kappa c} \int (R - 2\Lambda)\, \sqrt{-g}\, d^4x . \tag{I.111}$$

I.8   Chapter 8

Problem 8.1: We write the geodesic equations as

$$\frac{d}{d\tau} \left(g_{\mu\nu} \dot{x}^\nu \right) = \frac{1}{2} \frac{\partial g_{\nu\rho}}{\partial x^\mu} \dot{x}^\nu \dot{x}^\rho , \tag{I.112}$$

where the dot stands for the derivative with respect to the proper time of the particle, τ. For simplicity and without loss of generality, we consider orbits in the equatorial plane, so $\theta = \pi/2$ and $\dot{\theta} = 0$. In the case of circular orbits, we have $\dot{r} = \ddot{r} = 0$ and for $\mu = r$ Eq. (I.112) reduces to

$$\frac{\partial g_{tt}}{\partial r} \dot{t}^2 + \frac{\partial g_{\phi\phi}}{\partial r} \dot{\phi}^2 = 0 \tag{I.113}$$

when we consider the Schwarzschild metric, because only the diagonal metric coefficients are non-vanishing. The angular velocity of the particle is

$$\Omega(r) = \frac{\dot{\phi}}{\dot{t}} = \sqrt{-\frac{\partial_r g_{tt}}{\partial_r g_{\phi\phi}}} = \sqrt{\frac{r_S}{2r^3}} . \tag{I.114}$$

Fig. I.1 Penrose diagram
for the Minkowski
spacetime, trajectory of the
massive particle (red
arrows), and trajectory of the
electromagnetic pulse (blue
arrow)

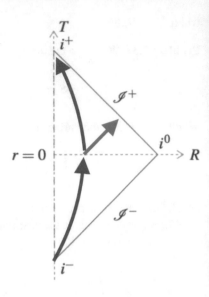

From $g_{\mu\nu}\dot{x}^\mu\dot{x}^\nu = -1$ with $\dot{r} = \dot{\theta} = 0$, we can write

$$g_{tt}\dot{t}^2 + g_{\phi\phi}\dot{\phi}^2 = \dot{t}^2\left(g_{tt} + \Omega^2 g_{\phi\phi}\right) = -1,$$

$$\Rightarrow \dot{t} = \frac{1}{\sqrt{-g_{tt} - \Omega^2 g_{\phi\phi}}} = \sqrt{\frac{2r}{2r - 3r_S}}. \qquad (I.115)$$

Since $\dot{t} = dt/d\tau$, the relation between the particle proper time τ and the coordinate time t is

$$\Delta t = \Delta \tau \sqrt{\frac{2r}{2r - 3r_S}}. \qquad (I.116)$$

Note that for $r \to 3r_S/2$ we have $\Delta \tau \to 0$. As we will see in Sect. 10.3.1, $r = 3r_S/2$ is the photon orbit of the Schwarzschild spacetime. Massive particles can orbit at the photon orbit in the limit $v \to c$ and therefore have Lorentz factor $\gamma \to \infty$.

Problem 8.2: The possible trajectory of the massive particle is shown in red in Fig. I.1, while the trajectory of the electromagnetic pulse is in blue. Note that the trajectory of the massive particle starts from past time-like infinity, ends at future time-like infinity, and it is always inside the light-cone of the particle (the particle velocity is lower than the speed of light). The trajectory of the electromagnetic pulse is a straight line at 45°, starts at $t = 0$, and reaches future null infinity.

Problem 8.3: The future-light cones of an event in region I, of an event inside the black hole (region II), and of an event inside the white hole (region IV) are shown in, respectively, the top, central, and bottom panels in Fig. I.2. Time-like and null trajectories in region I can either fall to the singularity of the black hole at $r = 0$ or reach future time-like infinity (time-like trajectories) and future null infinity (null trajectories). All time-like and null trajectories in region II must end at the singularity of the black hole at $r = 0$. Time-like and null trajectories in region IV may go to region I, region II, or region III.

I.9 Chapter 10

Problem 10.1: From the line element in Eq. (10.4) we see that the metric tensor of the Reissner–Nordström spacetime is

$$||g_{\mu\nu}|| = \begin{pmatrix} -f & 0 & 0 & 0 \\ 0 & \frac{1}{f} & 0 & 0 \\ 0 & 0 & r^2 & 0 \\ 0 & 0 & 0 & r^2 \sin^2\theta \end{pmatrix}, \tag{I.117}$$

where (in units in which $G_N = c = 4\pi\varepsilon_0 = 1$)

$$f = 1 - \frac{2M}{r} + \frac{Q^2}{r^2}. \tag{I.118}$$

By definition of inverse metric $g_{\mu\nu}g^{\nu\rho} = \delta^\rho_\mu$. Since the metric matrix in Eq. (I.117) is diagonal, the inverse is

$$||g^{\mu\nu}|| = \begin{pmatrix} -\frac{1}{f} & 0 & 0 & 0 \\ 0 & f & 0 & 0 \\ 0 & 0 & \frac{1}{r^2} & 0 \\ 0 & 0 & 0 & \frac{1}{r^2 \sin^2\theta} \end{pmatrix}. \tag{I.119}$$

Problem 10.2: From the line element in Eq. (10.6) we see that the metric tensor of the Kerr spacetime is

$$||g_{\mu\nu}|| = \begin{pmatrix} -\left(1 - \frac{2Mr}{\Sigma}\right) & 0 & 0 & -\frac{2aMr\sin^2\theta}{\Sigma} \\ 0 & \frac{\Sigma}{\Delta} & 0 & 0 \\ 0 & 0 & \Sigma & 0 \\ -\frac{2aMr\sin^2\theta}{\Sigma} & 0 & 0 & \left(r^2 + a^2 + \frac{2a^2Mr\sin^2\theta}{\Sigma}\right)\sin^2\theta \end{pmatrix}. \tag{I.120}$$

We have to find the inverse matrix. For the metric coefficients involving at least one index r or θ, it is straightforward because we can still treat the matrix as diagonal. For the metric coefficients g_{tt}, $g_{t\phi}$, $g_{\phi t}$, and $g_{\phi\phi}$ the problem reduces to find the inverse matrix of a symmetric matrix 2×2. For a general symmetric matrix 2×2

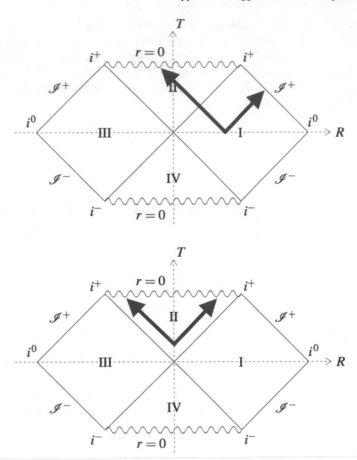

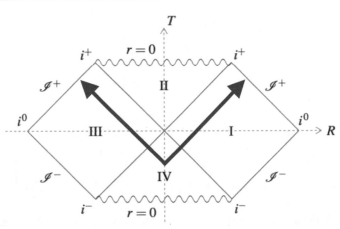

Fig. I.2 Schwarzschild spacetime. The top diagram shows the future light-cone of an event in region I, the central diagram shows the future light-cone of an event inside the black hole (region II), and the bottom diagram shows the future light-cone of an event inside the white hole (region IV)

$$A = \begin{pmatrix} a & b \\ b & c \end{pmatrix}, \tag{I.121}$$

the inverse matrix is

$$A^{-1} = \frac{1}{ac - b^2} \begin{pmatrix} c & -b \\ -b & a \end{pmatrix} = \frac{1}{\det|A|} \begin{pmatrix} c & -b \\ -b & a \end{pmatrix}. \tag{I.122}$$

For the metric of the Kerr spacetime in Boyer–Lindquist coordinates, we find

$$||g^{\mu\nu}|| = \begin{pmatrix} -\frac{(r^2+a^2)^2-a^2\Delta\sin^2\theta}{\Sigma\Delta} & 0 & 0 & -\frac{2aMr}{\Sigma\Delta} \\ 0 & \frac{\Delta}{\Sigma} & 0 & 0 \\ 0 & 0 & \frac{1}{\Sigma} & 0 \\ -\frac{2aMr}{\Sigma\Delta} & 0 & 0 & \frac{\Delta-a^2\sin^2\theta}{\Sigma\Delta\sin^2\theta} \end{pmatrix}. \tag{I.123}$$

Problem 10.3: The left hand side of Eq. (10.21) can be written as

$$\frac{d}{d\tau}\left(g_{\mu\nu}\dot{x}^\nu\right) = \frac{\partial g_{\mu\nu}}{\partial x^\rho}\dot{x}^\nu\dot{x}^\rho + g_{\mu\nu}\ddot{x}^\nu$$

$$= \frac{1}{2}\frac{\partial g_{\mu\nu}}{\partial x^\rho}\dot{x}^\nu\dot{x}^\rho + \frac{1}{2}\frac{\partial g_{\mu\rho}}{\partial x^\nu}\dot{x}^\nu\dot{x}^\rho + g_{\mu\nu}\ddot{x}^\nu. \tag{I.124}$$

Equation (10.21) thus becomes

$$g_{\mu\nu}\ddot{x}^\nu + \frac{1}{2}\frac{\partial g_{\mu\nu}}{\partial x^\rho}\dot{x}^\nu\dot{x}^\rho + \frac{1}{2}\frac{\partial g_{\mu\rho}}{\partial x^\nu}\dot{x}^\nu\dot{x}^\rho = \frac{1}{2}\frac{\partial g_{\nu\rho}}{\partial x^\mu}\dot{x}^\nu\dot{x}^\rho \tag{I.125}$$

We move the term on the right hand side to the left hand side and we multiply everything by $g^{\sigma\mu}$ summing over the repeated index μ. We find

$$\ddot{x}^\sigma + \frac{1}{2}g^{\sigma\mu}\left(\frac{1}{2}\frac{\partial g_{\mu\nu}}{\partial x^\rho} + \frac{1}{2}\frac{\partial g_{\rho\mu}}{\partial x^\nu} - \frac{\partial g_{\rho\nu}}{\partial x^\mu}\right)\dot{x}^\nu\dot{x}^\rho = \ddot{x}^\sigma + \Gamma^\sigma_{\nu\rho}\dot{x}^\nu\dot{x}^\rho = 0, \tag{I.126}$$

which are the geodesic equations in their standard form.

Problem 10.4: Let us model Earth as a uniform sphere. Its moment of inertia is

$$I = \frac{2}{5}MR^2, \tag{I.127}$$

where $M = 6.0 \cdot 10^{24}$ kg is Earth's mass and $R = 6.4 \cdot 10^6$ m is Earth's radius. Earth's spin angular momentum is $J = I\omega$, where $\omega = 7.3 \cdot 10^{-5}$ rad/s is Earth's angular velocity. The spin parameter of Earth is

$$a_* = \frac{cJ}{G_N M^2} = 897 \gg 1. \tag{I.128}$$

Note that $a_* \gg 1$ does not imply that the effect of frame dragging is strong, because Earth's physical radius is much larger than Earth's gravitational radius.

I.10   Chapter 11

Problem 11.1: If we want to do the calculations by hand, we can start deriving the geodesic equations in the Friedmann–Robertson–Walker metric in order to get the Christoffel symbols, then we calculate the Ricci tensor and the scalar curvature, and eventually we write the *tt* component of the Einstein equations in which matter is described by a perfect fluid in its rest-frame. We find the first Friedmann equation.

Alternatively, we can use the RGTC package presented in Appendix E. We initialize the code

```
<< EDCRGTCcode.m
```

We define the coordinates

```
Coord = {t,r,θ,φ};
```

We define the non-vanishing metric coefficients

```
gtt = - 1;
grr = a[t]^2/(1 - k r^2);
gpp = a[t]^2 r^2;
gvv = a[t]^2  r^2 Sin[θ]^2;
```

and then the metric

```
g = {{gtt, 0, 0, 0}, {0, grr, 0, 0}, {0, 0, gpp, 0},
     { 0, 0, 0, gvv }};
```

We launch the code with the command

```
RGtensors[g, Coord, {1, 1, 1}]
```

At this point, we ask the code to provide us the *tt* component of the Einstein tensor G_t^t (note: an upper index and a lower index because this is what the code calculates as default)

```
EUd[[1,1]]
```

The output is

$$- 3 \frac{\dot{a}^2 + k}{a^2} . \tag{I.129}$$

Since we have used units in which $c = 1$, we reintroduce the speed of light

$$-3\frac{\dot{a}^2 + kc^2}{a^2c^2}.$$ (I.130)

The tt component of the Einstein equations (with an upper index and a lower index) is

$$-3\frac{\dot{a}^2 + kc^2}{a^2c^2} = G_t^t = \frac{8\pi G_N}{c^4}T_t^t = -\frac{8\pi G_N}{c^4}\rho,$$ (I.131)

and we find the first Friedmann equation.

Problem 11.2: As in the previous exercise, we can calculate these quantities either by hand or with a program like the RGTC package. In the latter case, we proceed with the same commands as in the previous exercise and then we ask the code to provide us the Kretschmann scalar

```
Kretschmann = Simplify[ Sum[
   Rdddd[[i, j, k, l]]*RUddd[[i, m, n, o]]*gUU[[j, m]]
      *gUU[[k, n]]*gUU[[l, o]],
   {i, 1, 4}, {j, 1, 4}, {k, 1, 4}, {l, 1, 4},
      {m, 1, 4}, {n, 1, 4}, {o, 1, 4} ] ]
```

and the scalar curvature

```
ScalarCurvature = Simplify[ Sum[ Rdd[[i, j]]*gUU[[i, j]],
                  {i, 1, 4}, {j, 1, 4} ] ]
```

Reintroducing the speed of light c, we obtain the expressions in Eqs. (11.3) and (11.4).

Problem 11.3: Let us consider a small change in either the value of the matter energy density ρ, the cosmological constant Λ, or the scale factor a. The result is that the universe either starts expanding forever ($a \to \infty$) or recollapses to a singular solution ($a = 0$).

Problem 11.4: If we include a radiation component, we have to add the following energy density to the list in Eq. (11.62)

$$\rho_r = \rho_r^0 (1 + z)^4,$$ (I.132)

and Eq. (11.63) becomes

$$H^2 = H_0^2 \left[\Omega_r^0 (1 + z)^4 + \Omega_m^0 (1 + z)^3 + \Omega_\Lambda^0 + \Omega_k^0 (1 + z)^2 \right].$$ (I.133)

Now $\Omega_k^0 = 1 - \Omega_r^0 - \Omega_m^0 - \Omega_\Lambda^0$ and Eq. (11.67) becomes

$$\tau = \frac{1}{H_0} \int_0^\infty \frac{d\tilde{z}}{1 + \tilde{z}} \frac{1}{\sqrt{\tilde{z}(2 + \tilde{z})(1 + \tilde{z})^2 \Omega_r^0 + (1 + \Omega_m^0 \tilde{z})(1 + \tilde{z})^2 - \tilde{z}(2 + \tilde{z})\Omega_\Lambda^0}}.$$ (I.134)

Let us consider the situation in our Universe. If we ignore the contribution from radiation, we have $\Omega_m^0 = 0.31$ and $\Omega_\Lambda^0 = 0.69$, and the integral gives the numerical factor 0.9553. The contribution of radiation today is $\Omega_r^0 = 5 \cdot 10^{-5}$. If we take this contribution into account, the integral in Eq. (I.134) gives the numerical factor 0.9551. Note that here we are ignoring the possibility that some matter is relativistic at some early time and becomes non-relativistic at a later time.

I.11   Chapter 12

Problem 12.1: For $M = 10^6 \, M_\odot$, the maximum frequency is $\nu_{max} \sim 10$ mHz. For $M = 10^9 \, M_\odot$, we have $\nu_{max} \sim 10$ nHz. This is consistent with the expected signal from these objects in Fig. 12.4.

I.12   Chapter 13

Problem 13.1: Let us use units in which $c = \hbar = 1$ for simplicity. The area of the event horizon is $A_H \sim r_g^2$, where $r_g = G_N M$ is the gravitational radius of the black hole. Since the black hole temperature is $T_{BH} \sim 1/r_g$, the black hole luminosity is

$$L_{BH} \sim A_H T_{BH}^4 \sim \frac{1}{r_g^2} = \frac{1}{G_N^2 M^2} \, . \tag{I.135}$$

We write $L_{BH} = dM/dt$ and Eq. (I.135) gives

$$G_N^2 M^2 \, dM \sim dt \, . \tag{I.136}$$

We integrate both sides and we get a rough estimate of the evaporation time

$$\tau_{evap} = \int dt \sim \int G_N^2 M^2 \, dM = \frac{1}{3} G_N^2 M_0^3 \, , \tag{I.137}$$

where M_0 is the initial mass of the black hole. Since $G_N^2 = T_{Pl}/M_{Pl}^3$, we have

$$\tau_{evap} \sim \left(\frac{M_0}{M_{Pl}} \right)^3 T_{Pl} \sim 10^{-44} \left(\frac{M_0}{10^{-5} \, g} \right)^3 s \, , \tag{I.138}$$

For $M_0 = M_\odot \sim 10^{33}$ g, we find $\tau_{evap} \sim 10^{70}$ s, which is about 10^{63} years and is much longer than the age of the Universe (about 10^{10} years). A more accurate calculation would lead to $\tau_{evap} \sim 10^{74}$ s.

Subject Index

© Springer Nature Singapore Pte Ltd. 2018, corrected publication 2020
C. Bambi, *Introduction to General Relativity*, Undergraduate Lecture Notes
in Physics, https://doi.org/10.1007/978-981-13-1090-4

Printed in the United States
by Baker & Taylor Publisher Services